Differential Geometry of Curves and Surfaces

Differential Geometry of Curves and Surfaces

Masaaki Umehara • Kotaro Yamada

Tokyo Institute of Technology, Japan

Translated by

Wayne Rossman

Kobe University, Japan

World Scientific

NEW JERSEY • LONDON • SINGAPORE • BEIJING • SHANGHAI • HONG KONG • TAIPEI • CHENNAI • TOKYO

Published by

World Scientific Publishing Co. Pte. Ltd.
5 Toh Tuck Link, Singapore 596224
USA office: 27 Warren Street, Suite 401-402, Hackensack, NJ 07601
UK office: 57 Shelton Street, Covent Garden, London WC2H 9HE

Library of Congress Cataloging-in-Publication Data
Names: Umehara, Masaaki. | Yamada, Kotaro, 1961– | Rossman, Wayne, 1965– translator.
Title: Differential geometry of curves and surfaces / by Masaaki Umehara (Tokyo Institute of Technology, Japan), Kotaro Yamada (Tokyo Institute of Technology, Japan) ; translated by Wayne Rossman (Kobe University, Japan).
Other titles: Kyokusen to kyokumen. English
Description: Kaiteiban = Revised edition [1st English edition]. | New Jersey : World Scientific, 2017. | Includes bibliographical references and index.
Identifiers: LCCN 2016059832| ISBN 9789814740234 (hardcover : alk. paper) | ISBN 9789814740241 (pbk. : alk. paper)
Subjects: LCSH: Geometry, Differential--Textbooks. | Curves on surfaces--Textbooks. | Curves--Textbooks. | Surfaces--Textbooks.
Classification: LCC QA641 .U4413 2017 | DDC 516.3/6--dc23
LC record available at https://lccn.loc.gov/2016059832

British Library Cataloguing-in-Publication Data
A catalogue record for this book is available from the British Library.

Printed in Singapore

Preface

This book is an English translation of the second Japanese edition of our text on differential geometry of curves and surfaces published by Shokabo Co. Ltd. Although there are already quite a few textbooks on this subject, we hope that students and researchers interested in geometry or topology will find this book to be an accessible and informative introduction and reference text for the subject.

The contents from Sections 1 to 11 (Chapters I and II) are intended for a half-year course on curves and surfaces, which starts by introducing the geometry of planar curves and continue up to the Gauss-Bonnet theorem for surfaces in 3-dimensional Euclidean space. In fact, these first eleven sections are closely connected to each other, and readers will be able to understand them as a single contiguous story in a natural way. The sequential relationships between sections in Chapters I and II are shown in the diagram given at the end of the table of contents.

On the other hand, Sections 12–19 (Chapter III) have been prepared for more advanced readers who are interested in or are already familiar with manifolds, and these sections can be viewed as roughly independent of each other.

A teacher using this textbook for a course can omit the sections or chapters which are marked with asterisks, and then will be able to give the lectures without excessive time pressure. The exercises and remarks that are not intended for beginners are also marked with asterisks, and these can also be omitted on a first reading. In fact, by leaving out these advanced topics, the reader can continue his reading without stress.

There are two appendices: Appendix A presents fundamental knowledge from linear algebra and calculus that is helpful for reading the first eleven sections of this book, and Appendix B introduces more advanced and important topics within curve and surface theory. The book is, of course, primarily written for those who are just beginning to learn geometry and topology, but it also contains the following topics, which are well-known but usually not discussed so thoroughly in other textbooks:

(1) The classification of closed regular curves by regular homotopy (Section 3),
(2) The four-vertex theorem for simple closed curves (Section 4),
(3) Cycloids as brachistochrones and cycloidal pendulums, and their evolutes (Sections B.1 and 18),
(4) Several properties of principal curvature flows (Section 15),
(5) Proofs of existence of curvature line coordinates, asymptotic line coordinates and isothermal coordinates (Sections B.5 and 16),
(6) Criteria of several types of singularities in curves and surfaces (Section B.9).

The authors hope that each of these topics will be helpful introductions for the readers that will prove useful for their further studies. Moreover, there are three sections that are newly added to the contents, which are not found in the Japanese edition: "developability of flat surfaces" in Section B.6 in Appendix B, which was in the first Japanese edition, but was removed from the second edition; "a proof of the isoperimetric inequality" in Section B.3; "the classification of surfaces of constant principal curvatures" in Proposition B.7.4 of Section B.7. These three topics are well-known but frequently omitted in courses on this subject. In this sense, this English edition might be considered as a complete version of our Japanese textbook.

Finally, the authors are grateful to Wayne Rossman, who not only worked on the translation, but also gave us several valuable comments to make the English edition more comprehensive, and to Joseph Cho for very carefully checking this textbook before publication.

Masaaki Umehara and Kotaro Yamada

March, 2017

Preface to the Japanese first edition

Following the discovery of calculus by Newton and Leibniz in the 17th century, the geometry of curves and surfaces became a classical and traditional subject in mathematics. In modern times, this subject is of interest not only to students of mathematical sciences, but also to many who are concerned with computer graphics. However, this classical subject is often not included as part of the standard curriculum in mathematical courses, since so much time is required to explain manifold theory, one of the foundational subjects in mathematics. In spite of this situation, the authors believe it is educationally beneficial, for students aiming to learn about manifolds, to first learn the geometry of curves and surfaces and to have experience with the Gauss-Bonnet formula.

Several books on this subject have already been published in Japan. One example is the book entitled "Differential Geometry of Curves and Surfaces" by Shoshichi Kobayashi, which is an excellent book in this subject, and was used by the authors when they were students. In spite of the existence of such books, the authors decided to write this book for two reasons: one is that there are no books for half-year courses on this subject, and the other is the authors' desire to produce a new introductory textbook with which the readers can learn not only fundamental material but also enthusiastically learn a number of deeper results in differential geometry and topology.

This book is not only designed for a half-year course on this subject, but also for seminars for undergraduate students. The readers of this book should have knowledge of calculus and linear algebra at the undergraduate level. The authors believe that such readers can understand this textbook and solve the exercises without guidance, and can understand properties of the curvature functions of planar curves, the geometry of space curves, the local theory of surfaces and the Gauss-Bonnet theorem, without any great difficulty. Moreover, the readers will also be able to learn about several global properties of planar curves, as well as important pieces of knowledge about surfaces. At the end of each section, exercises are included which are helpful for understanding the contents of those sections.

A teacher using this textbook for a course can omit the sections or chapters which are marked with asterisks, and then will be able to give the lectures without excessive time pressure. The exercises and remarks that are not intended for beginners are also marked with asterisks, and these can also be omitted on a first reading. In fact, by leaving out these advanced topics, the reader can continue his reading without stress.

The authors give suitable comments and remarks in the places where further information is required, and they hope that this book will be of help not only for students of mathematics but also for students in other fields.

The first intended purpose of this book is contained in Sections 1–11. Chapter III is intended for readers aiming to learn manifold theory, and the authors hope this chapter will assist in understanding the relationship between surface theory and manifold theory. Even if a reader finds Chapter III difficult to penetrate in a first reading, he should not be discouraged, and hold off on reading this chapter until he has acquired more understanding of manifolds. In this book, there are two appendices, one for the required knowledge from calculus and linear algebra, and the other for advanced topics. The readers can refer to them in accordance with their needs and interests.

The authors recommend the readers also refer to other related textbooks on the same subject. The book of Shoshichi Kobayashi mentioned before is, of course, one such reference, but there are several other excellent books which would be useful for assisting in understanding the contents of this book from different points of view. For example, "Geometry of Surfaces" written by Tadashi Nagano is useful for learning about several interesting connections between surface theory and manifold theory. From the book "An Overview on Differential Geometry" written by Chuji Adachi, readers can study classical topics in the theory of curves and surfaces not mentioned in this book. Recently, Katsuei Kenmotsu published a book "Surfaces with Constant Mean Curvature" including advanced topics on constant mean curvature surfaces. Seiki Nishikawa recently published "Geometry" which covers interesting topics on planar curves.

The authors thank Professor Hajime Urakawa for helpful advice on the first draft of this book, and who encouraged the authors to write this book. The authors thank Wayne Rossman for pointing out several typographical errors, and Takashi Sato, who was a student at Hiroshima University, for his comments. The authors also would like to thank the many students in classes and seminars during the preparation of this book. Finally, the authors deeply thank the publisher Shuji Hosoki for handling the book production procedures.

Masaaki Umehara and Kotaro Yamada

April, 2002

Preface to the Japanese second edition

In the past twelve years since writing the first edition, the authors have received many valuable comments from readers, and have corrected typographical errors with each new printing.

The authors reluctantly removed "the fundamental theorem of surfaces", "properties of geodesic curvature" and "indices of umbilics" from the text due to page limitations. Also, while using the material in the text for classes and seminars, a number of places were found where it would be suitable to add new material and further explanations, and so the authors were delighted to have the opportunity to write a second edition.

In this revision, the authors made changes in definitions, proofs and exercises where necessary, while keeping the framework of the first edition intact. Only the section on the "developability of zero Gaussian curvature surfaces" was deleted, and the following materials were reinserted or added:

- the fundamental theorem of surface theory and its proof,
- properties of geodesic curvature,
- examples of minimal surfaces,
- a classification of rotationally symmetric surfaces of constant negative Gaussian curvature,
- indices of umbilics as an application of the Poincaré-Hopf index formula.

If one puts a glass on a table by a window, then one might find a figure on the table made by the sunlight passing through the glass, and one might recognize singularities in the figure (see the book of Izumiya and Takeuchi). The authors have recently investigated singularities of curves and surfaces and decided to introduce the criteria of typical singularities appearing on curves and surfaces in one section of this book. Although the proofs of the criteria are not given, these criteria would be useful for readers when they encounter singularities and would like to specify the types of singularities. A typical example having singularities is a cycloid, which is a curve generated by a circle rolling on a straight line. Cycloids were applied by Christiaan Huygens to make cycloidal pendulums to improve chronometers, and were also discovered by Johann Bernoulli as solutions of the brachistochrone problem. These properties of cycloids are encountered in several places in this book, and readers who wish to can learn the important properties of cycloids.

In spite of these revisions, the way of using the book is essentially the same as the previous edition, and readers who enjoyed the previous edition will find this new edition comfortable.

During the preparation of this revision, Yuri Akiyama and Toshifumi Fujiyama, who were students at the Tokyo Institute of Technology, gave us valuable comments. The authors are also grateful to the publisher Shuji Hosoki as well as the publishers Tatuya Ono and Taro Kume.

Masaaki Umehara and Kotaro Yamada

January, 2015

Contents

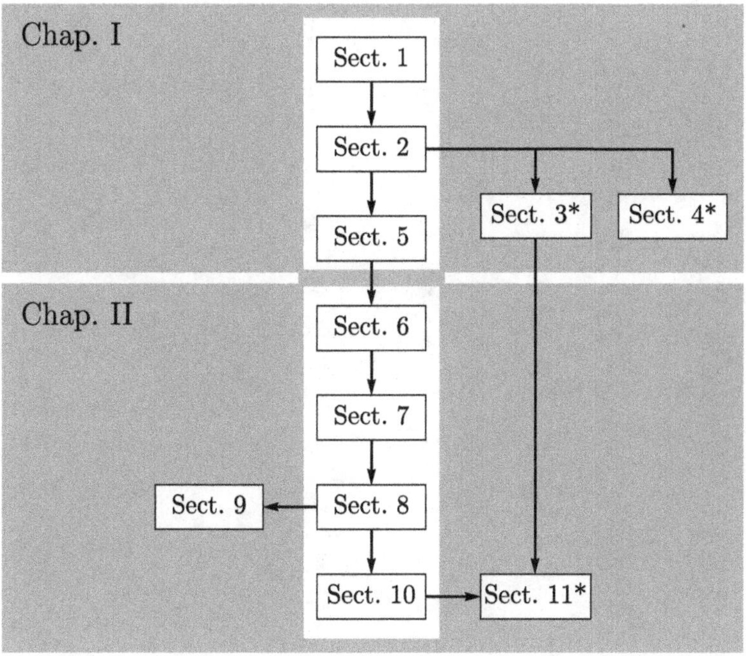

Chap. I
Sect. 1
Sect. 2
Sect. 3*
Sect. 4*
Sect. 5
Chap. II
Sect. 6
Sect. 7
Sect. 8
Sect. 9
Sect. 10
Sect. 11*

Chapter I

Curves

Students at the college and university level are familiar with graphs of functions and parametrized curves. In this chapter, we define the curvature of plane curves and space curves, which measures the "amount of bending" of the curve, and introduce its geometric meanings. As applications, the rotation index for closed plane curves and properties of spirals are introduced. While space curves are closely related to surface theory, which is treated in Chapter II, the rotation index for plane curves is also used in the proof of the Gauss-Bonnet theorem for 2-dimensional manifolds in Chapter III.

1. What exactly is a "curve"?

With pencil and paper, one can draw a variety of curves.

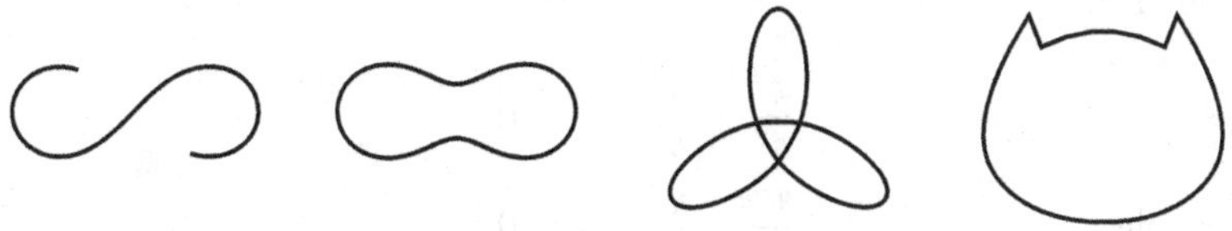

(i) (ii) (iii) (iv)

Fig. 1.1 Various curves.

The first picture (i) in Fig. 1.1 is an example of a *smooth curve*. The second picture (ii) is also of a smooth curve, but now without endpoints, since traveling along the curve will bring you back to where you started. This type of curve is called a *closed curve*. Like in the picture (iii), we allow crossing points, which we can also call *self-intersections*, and we still regard it as a smooth closed curve. The picture (iv) is a closed curve, but as it has sharp angles at particular points, it is not smooth at those points. This type of curve is called a *piecewise smooth curve* (cf. page 225 in Appendix B.3). In this case, we are dealing with a curve that is still smooth almost everywhere. We now introduce mathematical methods for representing curves in the plane $\boldsymbol{R}^2$.

Graphs of functions. In this text, we use the word *smooth* to refer to maps that are of class C^∞ (see Appendix A.1 for details). The graph of a smooth function $y = f(x)$ in the coordinate plane $\boldsymbol{R}^2$ is an example of a smooth curve. However, general curves are not necessarily representable as graphs of functions. For example, the graph of $y = \sqrt{1 - x^2}$ is the upper half of a circle of radius 1 with center at the origin, but the full circle cannot be expressed as the graph of any single function, because $y = \pm\sqrt{1 - x^2}$ cannot be called a single function.

The implicit function representation for curves. The points in the plane satisfying the equation $x^2 + y^2 = 1$ are those lying on the curve that is the circle of radius 1 with center $(0, 0)$. In general, if we take a function $F(x, y)$ of two variables and collect together all points in the plane satisfying $F(x, y) = 0$ to produce a curve, we call this the *implicit function representation* for curves by $F(x, y)$. As a second example, taking a function $f(x)$ and defining $F(x, y) = y - f(x)$, the resulting curve will be the graph of $y = f(x)$.

Example 1.1. Here we give examples of curves, represented implicitly by the solution sets of the following four equations (a and b are positive reals), see Fig. 1.2:

Ellipse: $$\frac{x^2}{a^2} + \frac{y^2}{b^2} - 1 = 0,$$

Hyperbola: $$\frac{x^2}{a^2} - \frac{y^2}{b^2} - 1 = 0,$$

Lemniscate[1]: $$(x^2 + y^2)^2 - a^2(x^2 - y^2) = 0,$$

Astroid[2]: $$(a^2 - x^2 - y^2)^3 - 27a^2x^2y^2 = 0.$$

[1]Each point on the lemniscate satisfies that the product of its distances to the two points $(\pm a/\sqrt{2}, 0)$ is $a^2/2$. This curve was studied by Jacob Bernoulli (1654–1705).

[2]The *astroid* is a locus of one point of a circle with radius $a/4$ rotating around the inside of a circle of radius a. See Problem **1** of Appendix B.1.

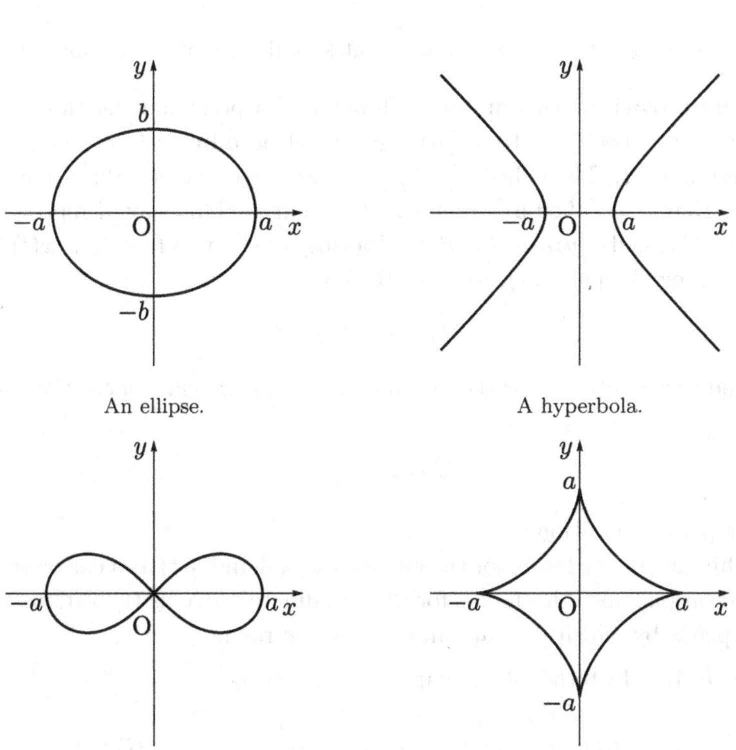

An ellipse. A hyperbola.

The lemniscate. The astroid.

Fig. 1.2

For a smooth function $F(x, y)$ of two variables, we consider the figure produced by the solutions of $F(x, y) = 0$. For a point (x_0, y_0) in this figure (that is, $F(x_0, y_0) = 0$ is satisfied), if

$$F_y(x_0, y_0) = \frac{\partial F}{\partial y}(x_0, y_0) \neq 0$$

holds, points in the figure near (x_0, y_0) can be given as the graph $y = f(x)$ of a function $f(x)$, and in particular the figure has no self-intersections near (x_0, y_0) (cf. Theorem A.1.6 (the implicit function theorem) in Appendix A.1). Similarly, if $F_x(x_0, y_0) \neq 0$, then, switching the roles of x and y, we find that the figure is the graph of some function $x = g(y)$ near (x_0, y_0).

On the other hand, if $F_x(x_0, y_0) = F_y(x_0, y_0) = 0$, then the point is called a *singular point*. For example, four singular points appear in the astroid in Example 1.1, where the curve is cusped, or pointed. Another singularity is seen at $(0, 0)$ in the lemniscate, where the curve crosses itself.

There are many varieties of shapes that singular points can appear as.

Parametrizations of curves. Thinking of a point moving through the plane with respect to a time variable t, we denote by $(x(t), y(t))$ the point depending on t. Then the trajectory of these points can be imagined as a curve. Here, t is called a *parameter* of the curve. This way of representing a curve is called a *parametrization.* Pairing up the two functions $x(t)$, $y(t)$ and naming the pair $\gamma(t)$, we have the curve

$$\gamma(t) = (x(t), y(t)). \tag{1.1}$$

For the curve obtained as the graph of $y = f(x)$, we can choose the special form

$$\gamma(t) = (t, f(t))$$

for its parametrization.

Thinking of the figure for the curve as a road and $\gamma(t)$ as a car traversing it, the various possible choices for the coordinate t give us the various times and speeds by which the car can traverse the road.

Example 1.2. In terms of the parameter t, setting[3]

$$x(t) := \frac{1-t^2}{1+t^2}, \qquad y(t) := \frac{2t}{1+t^2} \qquad (t \in \boldsymbol{R}) \tag{1.2}$$

gives us a curve that is the radius 1 circle with center at the origin and with the point $(-1, 0)$ removed. When both coordinates of a point in the plane are rational numbers, we say it is a *rational point.* A point on the circle of radius 1 centered at the origin different from $(-1, 0)$ is a rational point if and only if it is represented as in (1.2) with a rational number t.

In terms of another parameter s, we set

$$x(s) := \cos s, \qquad y(s) := \sin s \qquad (-\pi \leq s \leq \pi). \tag{1.3}$$

If we restrict s to the open interval $(-\pi, \pi)$, we again get the same circle as in (1.2) with one point removed. When $(x(t), y(t))$ and $(x(s), y(s))$ are the same point, the parameters s and t are related by $t = \tan(s/2)$.

In general, a curve is assumed to be represented as a smooth parametrization, with parameter t, by

$$\gamma(t) = (x(t), y(t)) \qquad (a \leq t \leq b).$$

[3] The symbol ":=" means "the object on the left-hand side is defined to be the object on the right-hand side".

Here, we say that a function is *smooth over a closed interval* $[a, b]$ if it extends to a function defined and smooth on an open interval containing $[a, b]$.

If we have a strictly increasing smooth function $t \colon [c, d] \to [a, b]$ such that $t(c) = a$, $t(d) = b$ and $dt/du > 0$ on the interval $[c, d]$, then the curve defined by $\tilde{\gamma}(u) = \gamma(t(u))$ $(c \leq u \leq d)$ will give the same set of points as the curve $\gamma(t)$. In this sense the two curves are the same, and we say that $\tilde{\gamma}$ is obtained from γ by a *change of parametrization*. When there is no risk of misinterpretation, we may write $\tilde{\gamma}(u)$ simply as $\gamma(u)$.

Example 1.3. The curves in (1.1) can be parametrized as follows:

$$
\begin{array}{llll}
\text{Ellipse:} & x(t) := a\cos t, & y(t) := b\sin t & (0 \leq t \leq 2\pi), \\
\text{Hyperbola:} & x(t) := a\cosh t, & y(t) := b\sinh t & (t \in \boldsymbol{R}), \\
\text{Lemniscate:} & x(t) := \dfrac{a\cos t}{1+\sin^2 t}, & y(t) := \dfrac{a\sin t\cos t}{1+\sin^2 t} & (0 \leq t \leq 2\pi), \\
\text{Astroid:} & x(t) := a\cos^3 t, & y(t) := a\sin^3 t & (-\pi \leq t \leq \pi),
\end{array}
$$

where $\cosh t$, $\sinh t$ are hyperbolic functions (see page 195 in Appendix A.1).

Given a parametrization $\gamma(t) = (x(t), y(t))$ for a curve, and viewing the curve as the trajectory of the point $\gamma(t)$ that is moving with respect to t, and in turn regarding t as representing time, we then have the *velocity vector*

$$
\dot{\gamma}(t) := (\dot{x}(t), \dot{y}(t)) \qquad \left(\dot{x} = \frac{dx}{dt}, \dot{y} = \frac{dy}{dt}\right),
$$

where an overhead dot "˙" denotes differentiation with respect to t. If the velocity vector $\dot{\gamma}(c)$ vanishes, the value $t = c$ of the parameter or the image $\gamma(c)$ is called a *singular point* (cf. Fig. 1.3). The singular point appearing in Fig. 1.3 is called a *cusp*. See Appendix B.9 for a criterion for cusps.

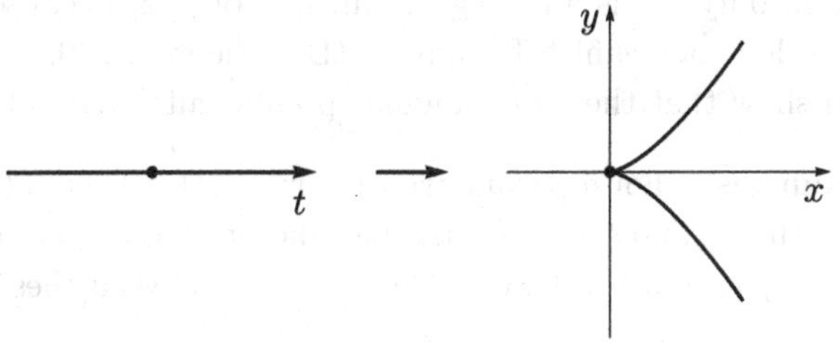

Fig. 1.3 The singular point of the map $t \mapsto (t^2, t^3)$ and its image.

For a given curve, the word "singular point" may have different meanings, depending on whether it is represented by an implicit function or as a parametrized curve. For example, the origin $(0, 0)$ is a singular point in the implicit function representation of the lemniscate in Example 1.1, which corresponds to the points $t = \pm\pi/2$ of the parametrization of the same lemniscate in Example 1.3, which are not singular points of the parametrization.

On the other hand, a point where the velocity vector $\dot{\gamma}(t)$ is not the zero vector is called a *regular point.* A curve $\gamma(t)$ defined on an interval $[a, b]$ is called a *regular curve* if all points are regular points. If $\gamma(t)$ is a regular curve, $\dot{\gamma}(t)$ points in the direction tangent to the curve at the point $\gamma(t)$. In particular, at a value $t = t_0$ where $\dot{x}(t_0) \neq 0$, the inverse function theorem (Theorem A.1.5 in Appendix A.1) tells us that for some neighborhood of $x(t_0)$, the function $x = x(t)$ has an inverse function $t = g(x)$. In that case, writing $f(x) = y(g(x))$, some portion of the curve $\gamma(t)$ sufficiently close to the point $(x(t_0), y(t_0))$ can be described as the graph of a smooth function $y = f(x)$. In the same way, by switching the roles of x and y, if $\dot{y}(t_0) \neq 0$, there exists a function $h(y)$ depending on y so that a part of the curve is the same as the graph of $x = h(y)$.

Recall that a point where the velocity vector (i.e. the tangent vector) of the curve $\gamma(t) = (x(t), y(t))$ vanishes, that is, $\dot{x}(t) = \dot{y}(t) = 0$, is a singular point. It is possible that the curve looks like it is not smooth near that point, for example, it can be sharply pointed there, even though it actually is a smooth mapping. For example, differentiating the parametrization of the astroid in Example 1.3

$$\gamma(t) := a(\cos^3 t, \sin^3 t) \qquad (t \in \boldsymbol{R},\ a > 0), \tag{1.4}$$

the velocity vector

$$\dot{\gamma}(t) = 3a \cos t \sin t(-\cos t, \sin t)$$

vanishes if and only if t is an integral multiple of $\pi/2$, because the vector $(-\cos t, \sin t)$ does not vanish for any t. (By Theorem B.9.1 in Appendix B.9, one can show that these are singular points called *cusps.*)

Length of curves. For a parametrized curve $\gamma(t) = (x(t), y(t))$ $(a \leq t \leq b)$, take two values t and $t + \Delta t$ for the parameter, with Δt close to zero, to get two points $\gamma(t)$ and $\gamma(t + \Delta t)$. The distance between these two points is

$$|\gamma(t + \Delta t) - \gamma(t)|.$$

Writing the change in the two coordinate functions $x(t)$, $y(t)$ of these two points as

$$\Delta x := x(t+\Delta t) - x(t), \qquad \Delta y := y(t+\Delta t) - y(t),$$

we have

$$|\gamma(t+\Delta t) - \gamma(t)| = \sqrt{(\Delta x)^2 + (\Delta y)^2} = \sqrt{\left(\frac{\Delta x}{\Delta t}\right)^2 + \left(\frac{\Delta y}{\Delta t}\right)^2}\ \Delta t.$$

Adding up many such short distances and taking the limit of such sums as those distances approach zero, we arrive at the *length*

$$\mathcal{L}(\gamma) = \int_a^b \sqrt{\left(\frac{dx}{dt}\right)^2 + \left(\frac{dy}{dt}\right)^2}\ dt = \int_a^b \sqrt{\dot{x}^2 + \dot{y}^2}\ dt \tag{1.5}$$

of the curve.[4] This length does not depend on the choice of the parametrization for the regular curve (Problem **2** at the end of this section asks for a proof of this).

The integrand in (1.5) is the norm $|\dot{\gamma}| := \sqrt{\dot{x}(t)^2 + \dot{y}(t)^2}$ of the velocity vector $\dot{\gamma}$, so we can write

$$\mathcal{L}(\gamma) = \int_a^b |\dot{\gamma}(t)|\ dt. \tag{1.6}$$

In particular, recalling that the graph of a function $y = f(x)$ $(a \leq x \leq b)$ can be regarded as a parametrized curve $\gamma(t) = (t, f(t))$, the length in (1.6) becomes

$$\mathcal{L}(\gamma) = \int_a^b \sqrt{1 + \left(\frac{dy}{dx}\right)^2}\ dx.$$

We could also consider curves coming from polar graphs $r = r(\theta)$ of functions r depending on θ, given in terms of the polar coordinates (r, θ) of the plane. Such a curve can be parametrized as $\gamma(\theta) = (r(\theta)\cos\theta, r(\theta)\sin\theta)$, $(a \leq \theta \leq b)$. The length of this curve is then

$$\mathcal{L}(\gamma) = \int_a^b \sqrt{r^2 + \left(\frac{dr}{d\theta}\right)^2}\ d\theta. \tag{1.7}$$

In this way, we can compute the length of curves using integrals, but those integrals will not necessarily be expressible in terms of elementary functions even when the curve can be parametrized by elementary functions. Here, *elementary functions* are those that can be described, using a finite

[4] We do not give a rigorous proof here. See, for example, [37, Problems in Chap. 13].

number of arithmetic operations and compositions, in terms of polynomials, rational functions, n'th root functions, exponential functions, logarithmic functions, trigonometric functions and inverse trigonometric functions.

For example, even for the graph $y = 1/x$, it is known that the length cannot be represented in terms of elementary functions. But by knowing the equation above, we can approximate the length by approximating this integral (using a computer, for example). Next we give an example where we can compute the length explicitly:

Example 1.4. The graph of the function ($c > 0$)

$$y = \frac{1}{a}\cosh ax \qquad (-c \le x \le c) \tag{1.8}$$

gives a regular curve called a *catenary*, and it is exactly the curve that one would get by hanging a string of uniform density between two fixed points (see Fig. 1.4). Here a is a fixed positive real number, and $x = \pm c$

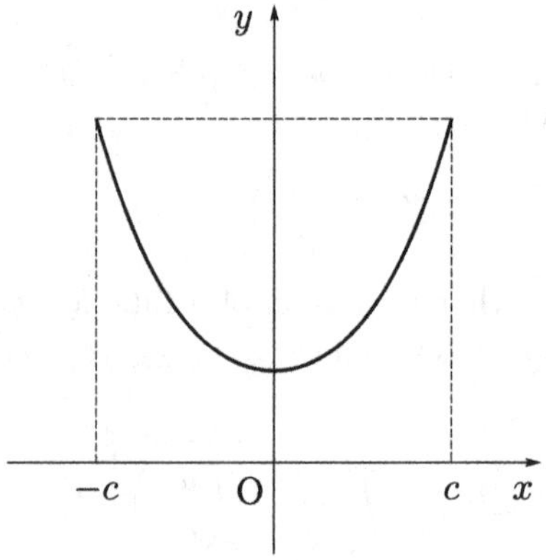

Fig. 1.4 The catenary.

give the fixed endpoints of the hanging string. By equation (1.8), we have $dy/dx = \sinh ax$, and so the length l of the curve is

$$l = \int_{-c}^{c} \sqrt{1 + \sinh^2 ax}\, dx = 2\int_0^c \cosh ax\, dx = \frac{2}{a}\sinh ac. \tag{1.9}$$

Using this, one can observe that for a given length l (> 0) of the string and a distance $2c$ ($2c < l$) between the two endpoints, there is only one positive constant a that satisfies (1.9).

A closed curve without self-intersection points is called a *simple closed curve*, and if it is regular, it is called a *simple closed regular curve*. It is known as the *Jordan closed curve theorem*[5] that a simple closed curve on the plane separates the plane into two connected open subsets, one bounded and the other not.

Definition 1.5. For a given simple closed curve in $\boldsymbol{R}^2$, the bounded (resp. unbounded) region[6] whose boundary is a given curve is called the *interior* (resp. *exterior*) of the curve.

The Jordan closed curve theorem can be proven in a rather simple way for polygons (cf. [4]). A proof for general continuous curves can be found in the textbook [8, Sec. 18] on algebraic topology.

For a simple closed curve with length l, the area A of the interior of the curve satisfies the inequality

$$4\pi A \leq l^2, \tag{1.10}$$

which is called the *isoperimetric inequality*, for which equality holds if and only if the curve is a circle. See Appendix B.3 for the proof, noting that this appendix is dependent on material in the next section.

Exercises 1

1 Find the length of the portion of the parabola $y = x^2$ where $|x| \leq 1$.

2 Show that the length of a regular curve as given in (1.5) does not depend on the choice of parametrization.

3 (1) For $0 < b < a$, show that the length of the ellipse $(a\cos t, b\sin t)$, $0 \leq t \leq 2\pi)$, is given by

$$a\int_0^{2\pi} \sqrt{1 - \varepsilon^2\cos^2 t}\, dt \qquad \left(\varepsilon := \sqrt{\frac{a^2 - b^2}{a^2}}\right).$$

Here ε $(0 \leq \varepsilon < 1)$ is the *eccentricity* of the ellipse. When $\varepsilon \neq 0$, this integral cannot be computed in terms of elementary functions. When $\varepsilon = 0$ we have a circle, and when $\varepsilon = 1$ we have a line segment.

(2) A longitude line of the earth is the intersection of the earth and a plane containing the axis of the earth (the line joining the north pole and the south pole), which is an almost circular ellipse whose minor-axis is contained in the axis of the earth and whose major-axis is contained in the

---

[5] Jordan, Camille (1838–1922).

[6] A connected open subset of $\boldsymbol{R}^2$ or $\boldsymbol{R}^3$ is called a *region* or *domain* (cf. page 194 in Appendix A.1).

equatorial plane. Using the approximation[7]

$$\sqrt{1-x} \approx 1 - \frac{x}{2}$$

for sufficiently small x (which is obtained by Taylor's formula, cf. equation (A.1.1) in Appendix A.1), approximate the length of this ellipse, and estimate the error using the remainder term of Taylor's formula. Here, the semi-major-axis a and the semi-minor-axis b of this ellipse are $a = 6377.397\,\text{km}$ and $b = 6356.079\,\text{km}$, respectively.

4 Taking a circle of radius a (> 0) and rolling it without slippage along the horizontal x-axis, any given point on the circle traces out an image curve called a *cycloid* (see the next figure). We choose that given point so that it starts at the origin $(0, 0)$ in the x-axis, and we let t denote the oriented angle between the spoke from the center of the circle to the given point and the vertical direction. We choose t so that it is zero when the given point is at $(0, 0)$ and is continuously increasing as the circle rolls to the right. Then the resulting cycloid can be parametrized as

$$\gamma(t) = a(t - \sin t, 1 - \cos t).$$

As the circle rotates completely around once, t increases from 0 to 2π, giving us one period of the cycloid. Show that the length of one period of the cycloid is eight times the radius of the circle.

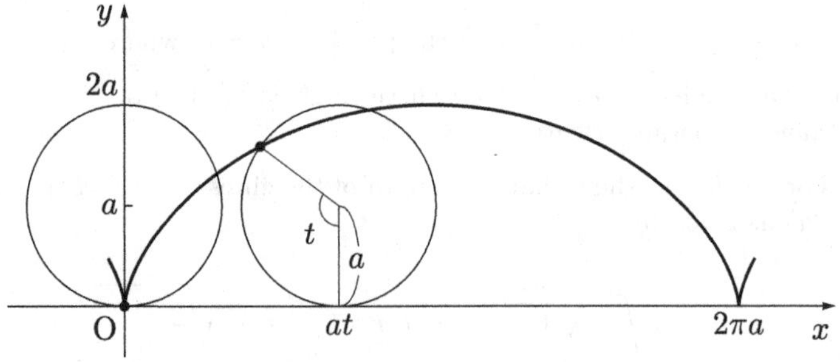

2. Curvature and the Frenet formula

Here we introduce the notion of curvature, which is a measure of the "amount of bending", of a curve. From now on, we consider only parametrized regular curves, that is, the velocity vector of the curve never vanishes. First we define arc-length parameters, the most standard way of giving parameters.

[7] The symbol "$\approx$" denotes "approximately equal".

Arc-length parameters. Looking at Example 1.2, we can see that there is more than one way to parametrize a regular curve. This causes us to wonder what the most natural way to choose a parametrization would be. Using the parametrization given in the previous section, we could take a starting point on the regular curve, and take the distance from that point along the regular curve as one natural choice of parametrization.

For a regular curve $\gamma(t)$ $(a \leq t \leq b)$,

$$s(t) = \int_a^t |\dot{\gamma}(u)|\, du \tag{2.1}$$

is the length of the image under $\gamma(t)$ of the interval $[a, t]$. Taking the derivative of this equation and noting that $\dot{\gamma} \neq \mathbf{0}$, we have

$$\frac{ds}{dt} = |\dot{\gamma}(t)| > 0.$$

Hence, if l is the length of the regular curve $\gamma(t)$ over the closed interval $[a, b]$, then $s(t)$ is a strictly increasing function from the interval $[a, b]$ to the interval $[0, l]$, and thus the inverse correspondence $[0, l] \ni s \mapsto t(s) \in [a, b]$ exists. By the inverse function theorem (Theorem A.1.5 in Appendix A.1), $t(s)$ is also smooth. Then we can parametrize γ by s as well:

$$\gamma(s) := \gamma(t(s)) \qquad (0 \leq s \leq l).$$

We call this s an *arc-length parameter* of the regular curve.

Up to now we have been using the dot " $\dot{}$ " to represent the derivative with respect to t, but when we have an arc-length parameter s, we will denote the derivative with respect to s by using a " $'$ ". The purpose of this change of notation is to help clarify when we are using an arc-length parameter. By the chain rule for the derivative of a composition of functions, we have

$$\gamma'(s) = \frac{d\gamma}{ds} = \frac{d\gamma}{dt}\frac{dt}{ds} = \frac{\dot{\gamma}(t)}{|\dot{\gamma}(t)|}. \tag{2.2}$$

Therefore we have $|\gamma'(s)| = 1$. In other words, a curve parametrized by arc-length always has unit speed.

Conversely, suppose we have a curve $\gamma(t)$ whose speed is always 1, that is, $|\dot{\gamma}| = 1$ holds. Then, by (2.1), the distance along the curve from some starting point differs from t only by an additive constant. Thus, an *arc-length parameter* means a parameter of unit speed, that is, the norm of the velocity vector is 1.

Thinking of the curve as a road, with a car moving along it with respect to time t, the car is moving at constant speed 1 when t is an arc-length parameter, and a general parametrization corresponds to travel where speed can vary.

For a curve $\gamma(s) = (x(s), y(s))$ parametrized by the arc-length s,

$$e(s) := \gamma'(s) = (x'(s), y'(s)) \tag{2.3}$$

is a *unit tangent vector* of $\gamma(s)$. Furthermore,

$$n(s) := (-y'(s), x'(s)) \tag{2.4}$$

is a unit vector perpendicular to $e(s)$, on the left side of $e(s)$ (see Fig. 2.1). This is called the leftward *unit normal vector* of $\gamma(s)$, at each point of $\gamma(s)$, with respect to the direction of travel given by s.

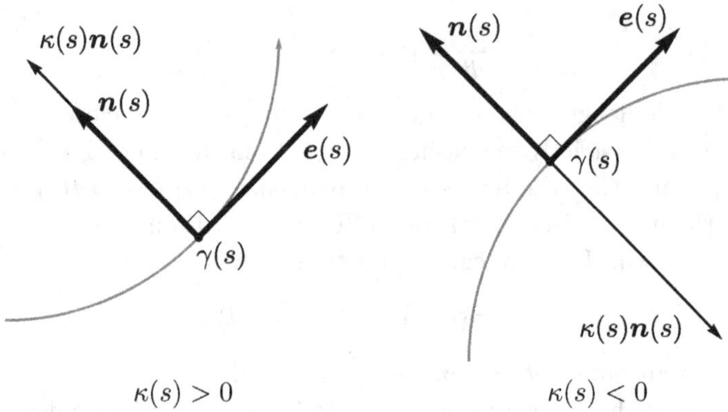

Fig. 2.1 Sign of the curvature and the shape of the curve.

Definition of curvature. Now, parametrizing the curve $\gamma(s)$ by an arc-length parameter s, the velocity vector will have length 1. In other words,

$$\gamma'(s) \cdot \gamma'(s) = 1$$

holds. Here the " · " means the *inner product* (also called the *scalar product*) for vectors in the plane. Differentiating both sides of this equation with respect to s using Corollary A.3.10 in Appendix A.3, we have

$$2\gamma''(s) \cdot \gamma'(s) = 0,$$

so the acceleration vector $\gamma''(s)$ is perpendicular to $\gamma'(s)$. This means that $\gamma''(s)$ is proportional to $n(s)$. The ratio $\kappa(s)$ of $\gamma''(s)$ to $n(s)$ is called the *curvature* of the curve at $\gamma(s)$. In other words, we have

$$\gamma''(s) = \kappa(s)n(s). \tag{2.5}$$

For each given choice of s, the curvature becomes a real number, and so the curvature $\kappa(s)$ is a function of s; hence, we can call it the *curvature function*. When $\kappa(s)$ is positive (negative), the curve is turning to the left (right) near $\gamma(s)$. For example, while driving a car, if the steering is not turned at all, the car is moving straight forward on a path with curvature 0, and when the steering wheel is turned toward the left or right, the curvature of the path is positive or negative, respectively (see Fig. 2.1).

Considering $\boldsymbol{e}$ and $\boldsymbol{n}$ as column vectors, together they form a 2×2 orthogonal matrix $(\boldsymbol{e}, \boldsymbol{n})$. Since this matrix has determinant 1, we have that (see Problem **2** at the end of this section)

$$\kappa(s) = \det(\gamma', \gamma''), \tag{2.6}$$

where "det" denotes the determinant of the matrix.

Example 2.1. For a line parallel to a unit vector $\boldsymbol{v}$, we can take a point on the line with position vector $\boldsymbol{c}$. Then any other point on the line can be given as $\gamma(s) = \boldsymbol{c} + s\boldsymbol{v}$, and s becomes an arc-length parameter for the line. In particular, $\gamma''(s) = \boldsymbol{0}$, so the curvature is identically zero.

Let us compute the curvature of a circle. For a positive constant a,

$$\gamma(s) = \left(a \cos \frac{s}{a}, a \sin \frac{s}{a}\right)$$

is a leftward-turning parametrization of a circle of radius a, and its velocity vector

$$\gamma'(s) = \left(-\sin \frac{s}{a}, \cos \frac{s}{a}\right)$$

has norm 1, so s is an arc-length parameter. By (2.4), the unit normal vector is

$$\boldsymbol{n}(s) = \left(-\cos \frac{s}{a}, -\sin \frac{s}{a}\right)$$

and we have

$$\gamma''(s) = \frac{1}{a}\left(-\cos \frac{s}{a}, -\sin \frac{s}{a}\right) = \frac{1}{a}\boldsymbol{n}(s).$$

Thus a leftward-turning circle of radius a (> 0) has constant curvature $\kappa(s) = 1/a$. When the circle is rightward-turning and of radius a (> 0), the curvature would be $-1/a$.

When the parameter is not an arc-length parameter, we can find the curvature function by changing to an arc-length parameter. However, an arc-length parameter is not written in terms of elementary functions generally, so we will also need a formula for the curvature that applies when the parameter is not necessarily arc-length. If t is any parameter for a regular curve $\gamma(t) = (x(t), y(t))$, the curvature function $\kappa(t)$ is

$$\kappa(t) = \frac{\dot{x}\ddot{y} - \dot{y}\ddot{x}}{(\dot{x}^2 + \dot{y}^2)^{3/2}} = \frac{\det(\dot{\gamma}, \ddot{\gamma})}{|\dot{\gamma}|^3}. \tag{2.7}$$

Proof of equation (2.7). Using (2.2) and the chain rule for differentiating compositions of functions, we have

$$\gamma' = \frac{\dot{\gamma}}{|\dot{\gamma}|}, \qquad \gamma'' = \frac{\ddot{\gamma}}{|\dot{\gamma}|^2} + \left(\frac{1}{|\dot{\gamma}|}\right)' \dot{\gamma}.$$

Substituting these into (2.6) and using the properties of determinants, we arrive at the desired equation. □

In particular, the regular curve given by the graph of a function $y = f(x)$ has curvature

$$\kappa(x) = \frac{\ddot{y}}{(1 + \dot{y}^2)^{3/2}} \qquad \left(\dot{y} = \frac{dy}{dx},\ \ddot{y} = \frac{d^2y}{dx^2}\right). \tag{2.8}$$

Example 2.2. By (2.7), the ellipse $\gamma(t) = (a\cos t, b\sin t)$ $(a, b > 0)$ has curvature

$$\kappa(t) = \frac{ab}{(a^2\sin^2 t + b^2\cos^2 t)^{3/2}}. \tag{2.9}$$

When $a > b$, the curvature has a maximum value a/b^2 at the points $(\pm a, 0)$ where the ellipse intersects the x-axis, and a minimum value b/a^2 at the points $(0, \pm b)$ where the ellipse intersects the y-axis.

The osculating circle. At a point on a regular curve, the line that comes closest to "approximating" the curve near that point is the tangent line at that point. Now, instead of a line, let us consider what circle would best "approximate" the curve near a given point on the curve.

Let us take a point $\gamma(s)$ on the curve where the curvature $\kappa(s)$ is not zero. We can then place a circle of radius $1/|\kappa(s)|$ so that it is tangent to the curve at $\gamma(s)$, and we can give the circle a parametrization so that its parameter is increasing in the same direction as the s along $\gamma(s)$. When $\kappa(s) > 0$, we place the circle on the left-hand side of the curve, and when

$\kappa(s) < 0$, we place it on the right-hand side. We call this circle the *osculating circle* of the curve γ at the point $\gamma(s)$; see Fig. 2.2. The osculating circle has radius $1/|\kappa(s)|$ and center

$$\gamma(s) + \frac{1}{\kappa(s)}\boldsymbol{n}(s).$$

The locus of the center of osculating circles is called the *evolute* (or *focal curve* or *caustic*) of $\gamma(s)$. See Appendix B.1 for the construction of the evolutes and relationships with the original curves.

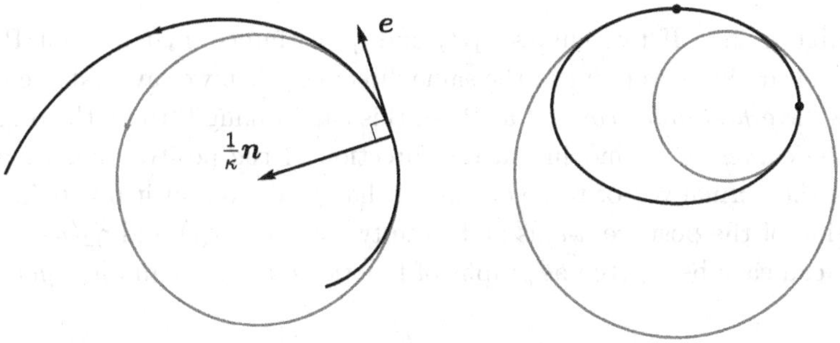

Fig. 2.2 The osculating circles (lighter shaded) of curves (darker shaded).

At a point of $\gamma(s)$ where the curvature is 0, we say that the tangent line is the osculating circle (a circle with infinite radius).

We give the osculating circle a parametrization in the same direction as the original curve. Then by Example 2.1, we see that *the original curve and the osculating circle will both have the same curvature at the point where they touch tangentially.* If one is driving a car, and fixes the steering wheel in the position it is in at one moment, the car will start driving along a circle. That circle is the osculating circle of the path of the car at the moment the steering wheel was held fixed.

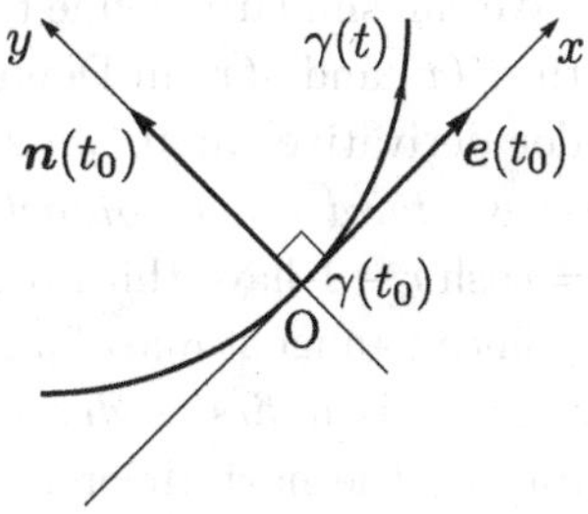

Fig. 2.3

Theorem 2.4 on page 16 will show us that the osculating circle really is the circle that best approximates the original curve at each point. But first we must clearly explain what we mean by "the best approximating circle". As we are considering regular curves, $\gamma(t) = (x(t), y(t))$ has the non-zero tangent vector $\dot{\gamma}(t_0)$ when t takes the value t_0 (cf. page 6). Let us further suppose that $\gamma(t_0)$ is the origin and $\dot{\gamma}(t_0)$ points in the direction of the x-axis, Then the normal vector at $t = t_0$ on the left-hand side of the curve will point in the direction of the y-axis. We can thus write the tangent vector as $\dot{\gamma}(t_0) = (a, 0)$ $(a := \dot{x}(t_0) > 0)$, see Fig. 2.3. Therefore, by what we stated in Section 1 (on page 6), for values of t close to t_0 (i.e. for a portion of the curve near the origin), the curve can be written as a graph $y = f(x)$. We use these facts in the next definition.

Definition 2.3. If two curves $\gamma_1(t)$ and $\gamma_2(u)$ intersect at a point P so that their derivatives point in the same direction, then we say that the two curves have *first order contact* at P. In this case, taking P to be the origin, and the curves to be moving in the direction of the positive x-axis, and taking the normal vector to be on the left-hand side so that it points in the direction of the positive y-axis at P, the two curves $\gamma_1(t)$ and $\gamma_2(u)$ given above can each be written as graphs of functions $y = f(x)$ and $y = g(x)$ so that

$$f(0) = g(0) = 0, \qquad \frac{df}{dx}(0) = \frac{dg}{dx}(0) = 0.$$

Then, if we further have

$$\frac{d^2 f}{dx^2}(0) = \frac{d^2 g}{dx^2}(0),$$

we say that the two curves have *second order contact* (see the left-hand side of Fig. 2.4).

We can similarly define the notion of higher order contact. For example, if the $f(x)$ and $g(x)$ in Definition 2.3 have the same first, second and third order derivatives at 0, we say that the two corresponding curves γ_1 and γ_2 have *third order contact.* The parabola $y = x^2/2$ and the catenary $y = \cosh x - 1$ have third order contact at the origin. (See Fig. 2.4.)

Since a tangent line of a curve makes first order contact with that curve, we say it is a *first order approximation* to the curve. Using the same phrasing, the next theorem tells us that the osculating circle is a *second order approximation* to the curve.

Theorem 2.4. *The osculating circle at a point $\gamma(t_0)$ of a regular curve $\gamma(t)$ makes second order contact with the curve at that point. Conversely, any circle that makes second order contact with $\gamma(t)$ at $\gamma(t_0)$ must be the osculating circle at that point.*

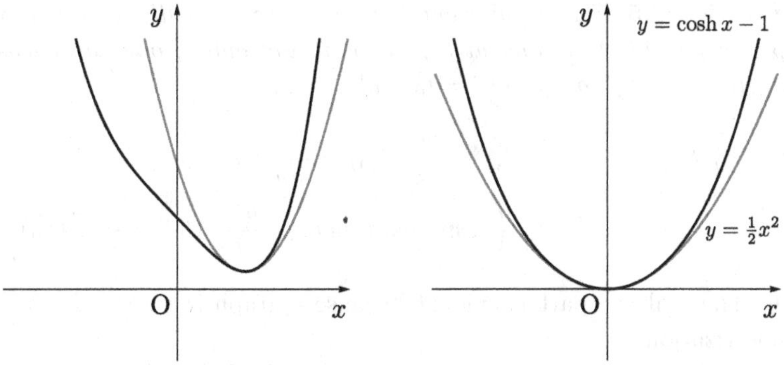

Second order contact of two curves.

Third order contact of a parabola and the catenary.

Fig. 2.4

Proof. Take a coordinate system (x, y) on the plane as in Definition 2.3 at $\gamma(t_0)$, and then the curve can be represented as a graph $y = f(x)$, where $f(0) = \frac{df}{dx}(0) = 0$. So by (2.8), the Taylor expansion of $f(x)$ at $x = 0$ is

$$f(x) = \frac{\kappa(t_0)}{2}x^2 + o(x^2). \tag{2.10}$$

Here, $o(\cdot)$ is Landau's symbol (cf. Appendix A.1).

The graph $y = f(x)$ has second order contact with the x-axis (i.e., the graph of $y = 0$) at the origin if and only if $\kappa(t_0) = 0$. So, we consider only the case $\kappa(t_0) \neq 0$ from now on. By reflecting the curve across the x-axis if necessary, we may assume $\kappa(t_0) > 0$ without loss of generality. Here, the circle of radius a turning to the left and tangent to the x-axis at the origin can be represented as a graph $y = a - \sqrt{a^2 - x^2}$, which can be expanded as

$$y = \frac{1}{2a}x^2 + o(x^2) \tag{2.11}$$

by (2.10) and the fact that the curvature of the circle is $1/a$. Comparing (2.10) and (2.11), we can conclude that the curve and the circle have second order contact at the origin if and only if $a = 1/\kappa(t_0)$. □

The following two propositions give important properties, which will be used in Section 4.

To test for the contact order of regular curves from the definition, one has to express the curves as graphs. To avoid this procedure, the following criteria is useful:

Proposition 2.5. *Two regular curves $\gamma_1(t)$ and $\gamma_2(u)$ will make first order (resp. second order) contact at a point* P *if and only if one can choose a parameter u for γ_2 so that at $t = t_0$ and $u = u_0$,*

$$\text{(2.12)} \quad \gamma_1(t_0) = \gamma_2(u_0) = \mathrm{P}, \qquad \frac{d\gamma_1}{dt}(t_0) = \frac{d\gamma_2}{du}(u_0)$$
$$\left(\textit{resp. additionally} \quad \frac{d^2\gamma_1}{dt^2}(t_0) = \frac{d^2\gamma_2}{du^2}(u_0)\right).$$

Proof. The 'only if' part is evident because a graph is a special case of a parametrization.

We shall now show the 'if' part: We denote by "$\dot{}$" differentiation with respect to the parameter t, and "$'$" with respect to u (though the parameter u is not necessarily arc-length, we use this notation to distinguish the two variables t and u). For the first order contact, the conclusion is evident by Definition 2.3. We show when the two curves have second order contact.

Set $\gamma_1(t) = (x(t), y(t))$, $\gamma_2(u) = (\tilde{x}(u), \tilde{y}(u))$ and express the two curves as graphs $y = f(x)$, $\tilde{y} = g(\tilde{x})$, respectively, as in Definition 2.3. Then $y(t) = f(x(t))$ and $\tilde{y}(u) = g(\tilde{x}(u))$ hold, and thus we have

$$\frac{df}{dx} = \frac{\dot{y}}{\dot{x}}, \qquad \frac{dg}{dx} = \frac{\tilde{y}'}{\tilde{x}'}, \qquad \frac{d^2 f}{dx^2} = \frac{\ddot{y}\dot{x} - \dot{y}\ddot{x}}{\dot{x}^3}, \qquad \frac{d^2 g}{dx^2} = \frac{\tilde{y}''\tilde{x}' - \tilde{y}'\tilde{x}''}{\tilde{x}'^3},$$

by the chain rule. Hence

$$\frac{d^2 f}{dx^2}(0) = \frac{d^2 g}{dx^2}(0)$$

if $\dot{\gamma}_1(t_0) = \gamma_2'(u_0)$ and $\ddot{\gamma}_1(t_0) = \gamma_2''(u_0)$, that is, the two curves have second order contact. □

To test for higher contact order, it is sufficient to find certain parametrizations and compare higher order derivatives (cf. (1) in Problem **9**). The notion of contact order is invariant under (general) change of coordinates, seen as follows:

For a planar region $D \subset \boldsymbol{R}^2$, we consider a smooth map $\varphi\colon D \to \boldsymbol{R}^2$, written as $\varphi(x, y) = (\xi(x, y), \eta(x, y))$. We call this map a *diffeomorphism*, or a *coordinate change*, if φ is a one-to-one map whose inverse map $\varphi^{-1}\colon \varphi(D) \to D$ is also smooth. When the Jacobian, that is, the determinant of the Jacobi matrix $\left(\begin{smallmatrix} \xi_x & \xi_y \\ \eta_x & \eta_y \end{smallmatrix}\right)$ is positive, φ is said to be *orientation preserving*, or a *positive coordinate change*. In this case, the left-side region of the curve on the xy-plane is mapped to the left-side region in the $\xi\eta$-plane of the corresponding curve. On the other hand, if the Jacobian is negative, φ is called *orientation reversing*, or a *negative coordinate change.*

For example, if we set $x = r\cos\theta$, $y = r\sin\theta$ for $r > 0$ and θ, then (x, y) is a point on the plane with *polar coordinate* (r, θ). This correspondence is regarded as a map from the $r\theta$-plane to the xy-plane as

$$\varphi\colon (r,\theta) \longmapsto (x,y) = (r\cos\theta, r\sin\theta).$$

If one takes a region $D := \{(r,\theta) \mid r > 0,\ -\pi < \theta < \pi\}$ of this mapping, then $\varphi(D)$ is the xy-plane excluding the negative part of the x-axis, and φ gives a bijection of D to $\varphi(D)$. Hence there is an inverse map $\psi\colon \varphi(D) \to D$. Moreover, since the Jacobian of φ is

$$\det\begin{pmatrix} x_r & x_\theta \\ y_r & y_\theta \end{pmatrix} = \det\begin{pmatrix} \cos\theta & -r\sin\theta \\ \sin\theta & r\cos\theta \end{pmatrix} = r > 0,$$

ψ is smooth because of the inverse function theorem (Theorem A.1.5 in Appendix A.1). Hence $\varphi\colon (r,\theta) \mapsto (x,y)$ is an orientation preserving diffeomorphism.

Given a diffeomorphism φ from one region D to another region $\varphi(D) \subset \boldsymbol{R}^2$, a curve $\gamma(t)$ in D becomes the curve $\tilde\gamma(t) = \varphi \circ \gamma(t)$ in $\varphi(D)$, where $\varphi \circ \gamma$ denotes the composition of φ and γ, that is,

$$\varphi \circ \gamma(t) = \varphi(\gamma(t)).$$

We show in the next proposition that the property of tangency of regular curves is preserved under such a diffeomorphism φ. (See Fig. 2.5.)

Proposition 2.6. *Let $\varphi\colon D \to \varphi(D)$ be a diffeomorphism defined on a planar region D. Then if two regular curves $\gamma_1(t)$ and $\gamma_2(u)$ in the region D have first order (resp. second order) contact at a point* P *in D, the image curves $\tilde\gamma_1(t) = \varphi\circ\gamma_1(t)$ and $\tilde\gamma_2(u) = \varphi\circ\gamma_2(u)$ will also have first order (resp. second order) contact, now at the point φ*(P).

Proof. Writing φ as $\varphi(x,y) = (\xi(x,y), \eta(x,y))$, for the curve $\gamma(t) = (x(t), y(t))$ in the region D we have the corresponding curve

$$\tilde\gamma(t) = \varphi\circ\gamma(t) = (\xi(x(t),y(t)), \eta(x(t),y(t))).$$

Applying the chain rule,

$$\begin{aligned} &\frac{d\xi}{dt} = \xi_x\dot x + \xi_y\dot y, \qquad \frac{d\eta}{dt} = \eta_x\dot x + \eta_y\dot y, \\ &\frac{d^2\xi}{dt^2} = \xi_x\ddot x + \xi_y\ddot y + \xi_{xx}\dot x^2 + 2\xi_{xy}\dot x\dot y + \xi_{yy}\dot y^2, \\ &\frac{d^2\eta}{dt^2} = \eta_x\ddot x + \eta_y\ddot y + \eta_{xx}\dot x^2 + 2\eta_{xy}\dot x\dot y + \eta_{yy}\dot y^2. \end{aligned} \tag{2.13}$$

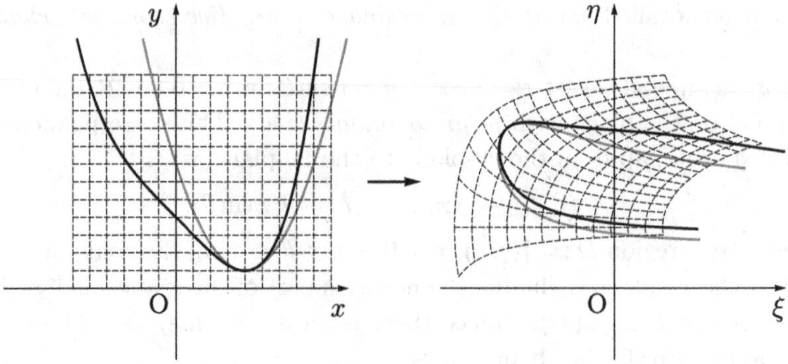

Fig. 2.5 Diffeomorphism and contact of curves: Two regular curves in the xy-plane having second order contact are mapped to curves in the $\xi\eta$-plane having second order contact, under the diffeomorphism $(x, y) \mapsto \varphi(x, y) = (\xi, \eta)$. Here, the dotted lines in the right-hand figure are the images under φ of dotted lines parallel to coordinate axes in the left-hand figure.

Assume $\gamma_1(t) := (x_1(t), y_1(t))$ and $\gamma_2(u) := (x_2(u), y_2(u))$ have first order contact at $\mathrm{P} = \gamma_1(t_0) = \gamma_2(u_0)$ so that the parameters t and u satisfy (2.12) in Proposition 2.5. We write $\tilde{\gamma}_1(t) = (\xi_1(t), \eta_1(t))$ and $\tilde{\gamma}_2(u) = (\xi_2(u), \eta_2(u))$, that is,

$$\xi_1(t) = \xi(x_1(t), y_1(t)), \qquad \eta_1(t) = \eta(x_1(t), y_1(t)),$$
$$\xi_2(u) = \xi(x_2(u), y_2(u)), \qquad \eta_2(u) = \eta(x_2(u), y_2(u)).$$

Then we have

$$\xi_1(t_0) = \xi(x_1(t_0), y_1(t_0)) = \xi(x_2(u_0), y_2(u_0)) = \xi_2(u_0),$$
$$\eta_1(t_0) = \eta(x_1(t_0), y_1(t_0)) = \eta(x_2(u_0), y_2(u_0)) = \eta_2(u_0),$$

by the first equality of (2.12), and hence $\tilde{\gamma}_1(t_0) = \tilde{\gamma}_2(u_0)$ $\big(= \varphi(\mathrm{P})\big)$ holds. Moreover, using (2.13) and (2.12),

$$\begin{aligned}\frac{d\xi_1}{dt}(t_0) &= \xi_x(x_1(t_0), y_1(t_0))\frac{dx_1}{dt}(t_0) + \xi_y(x_1(t_0), y_1(t_0))\frac{dy_1}{dt}(t_0)\\ &= \xi_x(x_2(u_0), y_2(u_0))\frac{dx_2}{du}(u_0) + \xi_y(x_2(u_0), y_2(u_0))\frac{dy_2}{du}(u_0)\\ &= \frac{d\xi_2}{du}(u_0), \qquad \text{and} \qquad \frac{d\eta_1}{dt}(t_0) = \frac{d\eta_2}{du}(u_0)\end{aligned}$$

hold. Thus, we have $\frac{d\tilde{\gamma}_1}{dt}(t_0) = \frac{d\tilde{\gamma}_2}{du}(u_0)$. Hence by Proposition 2.5, $\tilde{\gamma}_1$ and $\tilde{\gamma}_2$ have first order contact at $\varphi(\mathrm{P})$. The case of second order contact can be obtained in a similar way. □

A similar result can be stated for third or higher order contact as well (see (2) in Problem **9** at the end of this section).

Corollary 2.7. *The curvature of a regular curve is invariant under rotations and translations.*

In addition to rotations and translations, a reflection about a line is a congruence of the plane (cf. Appendix A.3). By a reflection, the absolute value of the curvature is invariant, but the sign is reversed. See Problem **4**.

Proof. By a rotation or a translation, a circle turning right (resp. left) is mapped to a circle turning right (resp. left). Since rotations and translations are diffeomorphisms of $\boldsymbol{R}^2$, the order of contact of a curve and a circle is preserved by these transformations, because of Proposition 2.6. Hence by Theorem 2.4, applying a rotation and a translation to a curve, the osculating circle is likewise mapped to the osculating circle of the image curve. Since the radius of a circle is preserved by rotations and translations, the conclusion follows. □

Corollary 2.7 can be also proved by direct calculations (cf. Problem **6** of this section).

The Frenet formula. Given a regular curve $\gamma(s) = (x(s), y(s))$ parametrized by an arc-length parameter s, equation (2.5) can be rewritten as:

$$x''(s) = -\kappa(s)\, y'(s), \qquad y''(s) = \kappa(s)\, x'(s). \tag{2.14}$$

From this and (2.3) and (2.4), we know that $\boldsymbol{n}'(s) = -\kappa(s)\boldsymbol{e}(s)$, so, together with (2.5), we have that

$$\boldsymbol{e}'(s) = \kappa(s)\boldsymbol{n}(s), \qquad \boldsymbol{n}'(s) = -\kappa(s)\boldsymbol{e}(s). \tag{2.15}$$

These two equations are called the *Frenet formula.*[1]

It is natural to ask, for a given function, whether there always exists a regular curve whose curvature function is equal to the given function, and whether it is unique if it exists. The following theorem is the answer to this question.

Theorem 2.8 (The fundamental theorem for plane curves). *For a smooth function $\kappa(s)$ defined for s in the interval $0 \le s \le l$, there exists a regular curve $\gamma(s)$ with arc-length parameter $s \in [0, l]$ and curvature $\kappa(s)$. Furthermore, this curve is uniquely determined up to rotations and translations of the plane.*

[1] Frenet, Jean Frédéric (1816–1900).

Proof. Existence of the curve seems natural when one imagines driving a car, steering in accordance with the curvature function, while driving at unit speed.

We can give a precise proof as follows: Let $\gamma(s)$ be a curve parametrized by an arc-length parameter s, and $\boldsymbol{e}(s)$, $\boldsymbol{n}(s)$ the unit tangent vector and the leftward unit normal vector, respectively. Then one can take a 2×2-matrix-valued function $\mathcal{F}(s) := (\boldsymbol{e}(s), \boldsymbol{n}(s))$, where we consider $\boldsymbol{e}(s)$ and $\boldsymbol{n}(s)$ to be column vectors. Then the Frenet formula (2.15) is expressed as

$$\mathcal{F}' = \mathcal{F}\Omega, \qquad \Omega := \begin{pmatrix} 0 & -\kappa \\ \kappa & 0 \end{pmatrix}. \tag{2.16}$$

This equation can be regarded as a linear ordinary differential equation of the unknown matrix-valued function $\mathcal{F}(s)$, where $\kappa(s)$ is a given data (cf. Appendix A.2). Then there exists a unique solution $\mathcal{F}(s)$ of (2.16) satisfying the initial condition

$$\mathcal{F}(0) := (\boldsymbol{e}(0), \boldsymbol{n}(0)) = I, \qquad I := \begin{pmatrix} 1 & 0 \\ 0 & 1 \end{pmatrix},$$

because of Theorem A.2.2 in Appendix A.2. For each s, $\mathcal{F}(s)$ is an orthogonal matrix with determinant 1. In fact, since Ω is a skew-symmetric matrix, Proposition A.3.9 in Appendix A.3 yields that

$$(\mathcal{F}\mathcal{F}^T)' = \mathcal{F}'\mathcal{F}^T + \mathcal{F}\mathcal{F}'^T = \mathcal{F}\Omega\mathcal{F}^T + \mathcal{F}\Omega^T\mathcal{F}^T = O, \tag{2.17}$$

where $\mathcal{F}^T$ is the transposition of $\mathcal{F}$, and O denotes the 2×2 zero matrix. Hence $\mathcal{F}(s)\mathcal{F}(s)^T$ does not depend on s, i.e., is a constant matrix:

$$\mathcal{F}(s)\mathcal{F}(s)^T = \mathcal{F}(0)\mathcal{F}(0)^T = I.$$

This implies that $\mathcal{F}(s)$ is an orthogonal matrix for each s. Then the determinant of $\mathcal{F}(s)$ is ± 1 (cf. (A.3.11) in Appendix A.3), and by the fact $\mathcal{F}(0) = I$, we can conclude that the determinant of $\mathcal{F}(s)$ is 1 by continuity of $\det \mathcal{F}(s)$. Thus $\mathcal{F}(s)$ is an orthogonal matrix of determinant 1, and hence $\boldsymbol{n}(s)$ points leftward with respect to $\boldsymbol{e}(s)$ (cf. (3) of Problem **2**).

Let

$$\gamma(s) := \int_0^s \boldsymbol{e}(t)\,dt.$$

Then $\gamma(0) = \mathbf{0}$ holds, and $\gamma(s)$ is the desired curve. In fact, $\boldsymbol{e}(s)$ is the unit tangent vector of $\gamma(s)$ since $\gamma'(s) = \boldsymbol{e}(s)$. Moreover, since $\mathcal{F}(s)$ is an orthogonal matrix of determinant 1, $\boldsymbol{n}(s)$ is the left-hand unit normal vector. Then equation (2.16) for $\mathcal{F}(s) = (\boldsymbol{e}(s), \boldsymbol{n}(s))$ is equivalent to the Frenet formula for $\gamma(s)$, and then $\kappa(s)$ is a curvature function.

Next we prove uniqueness: Let $\Gamma(s)$ $(0 \leq s \leq l)$ be a curve parametrized by arc-length whose curvature function coincides with the curvature $\kappa(s)$ of $\gamma(s)$. By Corollary 2.7, we may assume that

$$\gamma(0) = \Gamma(0) = \begin{pmatrix} 0 \\ 0 \end{pmatrix}, \ \boldsymbol{e}(0) = \boldsymbol{E}(0) = \begin{pmatrix} 1 \\ 0 \end{pmatrix}, \ \boldsymbol{n}(0) = \boldsymbol{N}(0) = \begin{pmatrix} 0 \\ 1 \end{pmatrix} \tag{2.18}$$

by applying rotations and translations, without loss of generality. Here, $\boldsymbol{E}(s)$ and $\boldsymbol{N}(s)$ are the unit tangent vector and the left-hand unit normal vector of $\Gamma(s)$, respectively. Then $\mathcal{G} := (\boldsymbol{E}, \boldsymbol{N})$ satisfies (2.16) as well as $\mathcal{F}$, and the initial values of $\mathcal{F}$ and $\mathcal{G}$ coincide. Thus by the uniqueness of solutions of ordinary differential equations, $\mathcal{G}(s) = \mathcal{F}(s)$ holds for each s. Moreover,

$$\gamma(s) = \int_0^s \boldsymbol{e}(t)\,dt = \int_0^s \boldsymbol{E}(t)dt = \Gamma(s),$$

that is, the two curves are the same. □

Remark 2.9. For a given curvature function $\kappa(s)$, the desired regular curve $\gamma(s)$ can be obtained explicitly as follows:

$$\gamma(s) = \int_0^s \left(\cos\left(\int_0^t \kappa(u)\,du\right),\ \sin\left(\int_0^t \kappa(u)\,du\right)\right) dt. \tag{2.19}$$

In fact, $|d\gamma/ds| = 1$ holds, that is, s is an arc-length parameter, and computing $d^2\gamma/ds^2$, we have that $\kappa(s)$ is the curvature of $\gamma(s)$ (cf. Problem **7**).

By Theorem 2.8 and Example 2.1 (page 13), we have the following corollary.

Corollary 2.10. *The curvature function of a regular curve is constant if and only if the curve is a circle or a line.*

The four-vertex theorem for convex curves. At the end of this section we give an example of an application of the Frenet formula.

In Section 1, a curve was said to be closed when traveling along the curve would bring you back to where you started. More precisely, closed curves are defined as follows: If a curve $\gamma(t)$ defined on the entire real line $\boldsymbol{R}$ has a period l (> 0), that is,

$$\gamma(t + l) = \gamma(t) \qquad (t \in \boldsymbol{R}), \tag{2.20}$$

then the restriction of γ to the closed interval $[0, l]$ is said to be a *closed curve*. In other words, a curve $\gamma : [a, b] \to \boldsymbol{R}^2$ is said to be a closed curve

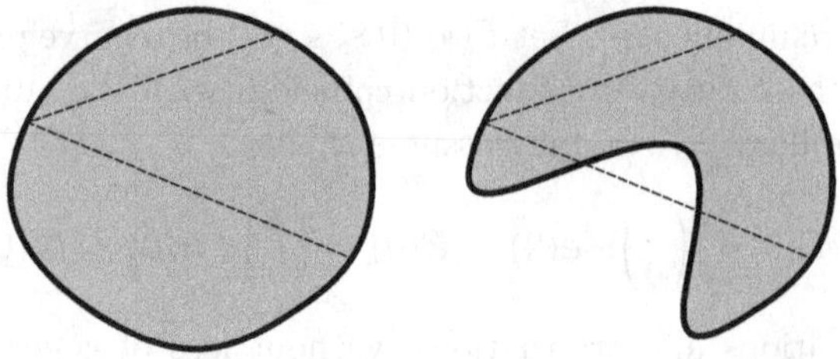

A convex curve. A non-convex simple closed curve.

Fig. 2.6 Convex and non-convex closed curves.

if γ can be extended to a periodic map with period $l := b - a$ satisfying (2.20). In this case, smooth periodicity

$$\gamma^{(n)}(0) = \lim_{t \to 0+} \gamma^{(n)}(t) = \lim_{t \to l^-} \gamma^{(n)}(t) = \gamma^{(n)}(l)$$

for all non-negative integers n, holds, and the curvature function $\kappa(t)$ is also a periodic function on $\boldsymbol{R}$ with period l.

As mentioned in the last part of Section 1, a closed regular curve that does not intersect itself is called a *simple closed regular curve.*

We say that a simple closed regular curve is *convex* (resp. *strictly convex*) if, for any two distinct points chosen on the curve, the line segment between them is contained in the closure of the interior region (resp. the interior region itself) (cf. Definition 1.5 in Section 1) of the curve (Fig. 2.6). For example, circles and ellipses are strictly convex. Also, it is known that any simple closed regular curve for which the curvature function does not change sign (resp. does not have any zeros) is convex (resp. strictly convex). See Theorem B.2.1 and Corollary B.2.2 in Appendix B.2.

A *vertex* is a point on a curve where the curvature $\kappa(s)$ attains a local maximum or minimum.[2] On a closed regular curve which is not a circle, the curvature function can be considered as a periodic function defined on the entire real line, and it must have at least one point of maximum value and one point of minimum value within one period. Furthermore, if, within one period, there are a finite number of points where the maximum or minimum is obtained, then the number of maximum points will equal the number of minimum points. So if the closed regular curve has only a finite number of vertices, then it will have an even number of vertices (see Fig. 2.7).

We have the following theorem about the vertices of convex curves.

[2]In this text, we say that a function $\kappa(s)$ takes a local maximum (resp. local minimum) at $s = s_0$ if there exists an interval $[a, b]$ such that $s_0 \in (a, b)$, and $\kappa(s_0)$ is the maximum (resp. minimum) of $\kappa(s)$ on $[a, b]$, and $\kappa(a)$, $\kappa(b) < \kappa(s_0)$ (resp. $\kappa(a)$, $\kappa(b) > \kappa(s_0)$).

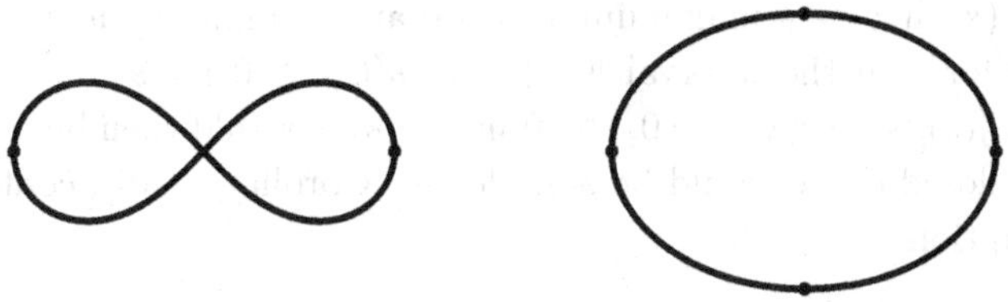

Fig. 2.7 Vertices of closed curves.

Theorem 2.11. *There exist at least four vertices on any strictly convex curve that is not a round circle.*

This theorem is called the *four-vertex theorem*, and was discovered by Syamadas Mukhopadhyaya[3] [27] in 1909. Later, in 1912, Adolph Kneser[4] [18] showed that the theorem still holds even when the simple closed regular curve is not assumed to be convex, and in Section 4 (Theorem 4.4) we introduce the proof of that result.

Using the example of driving a car, the four-vertex theorem says this: Driving once around a non-circular circuit with no intersections, one must change the direction that one is turning the steering wheel at least four times. As seen from Example 2.2, the ellipse has exactly four vertices, so we know that the value 4 in Theorem 2.11 is the best value that can be obtained.

On the other hand, there are closed regular curves (with self-intersections) whose number of vertices is 2. For example, the lemniscate (Example 1.3 of Section 1) has two vertices. See the left side of Fig. 2.7 and (3) in Problem **5** in Section 4.

Proof of Theorem 2.11. We argue by way of contradiction. Suppose that the number of vertices is less than 4. Since the number of vertices is even, the number is less than or equal to 2. Here, since there are maximum and minimum values of the curvature function, the number of vertices must be 2. We represent the closed curve by $\gamma(s)$, $0 \leq s \leq l$, for an arc-length parameter s. Letting $\gamma(0)$ and $\gamma(v)$, $0 < v < l$, denote the two vertex points, and we may assume without loss of generality that the curvature

[3]Mukhopadhyaya, Syamadas (1866–1937).
[4]Kneser, Adolf (1862–1930).

function $\kappa(s)$ attains a maximum at $s = 0$ and a minimum at $s = v$. Then $\kappa'(s) \le 0$ for s in the interval $[0, v]$, and $\kappa'(s) \ge 0$ for $s \in [v, l]$. In the case that $\kappa'(s)$ is everywhere 0, the function $\kappa(s)$ would then be a constant. Then the closed curve would be a circle (see Corollary 2.10), contradicting our assumption.

Let

$$ax + by + c = 0 \qquad (a,\ b,\ c \text{ are constants}) \tag{2.21}$$

be the equation for the straight line from $\gamma(0)$ to $\gamma(v)$. As the curve is convex, the line segment between these two points must lie in the interior region of the curve (cf. Definition 1.5), and then the curve $\gamma(s) = (x(s), y(s))$ is split into two parts by the line segment, one part where $0 \le s \le v$ lying to one side, and the other part where $v \le s \le l$ lying to the other side. The function $ax(s) + by(s) + c$ is non-negative on one of the intervals $[0, v]$ and $[v, l]$, and is non-positive on the other. Hence $\kappa'(s)(ax(s) + by(s) + c)$ does not change sign on $0 \le s \le l$, and is not identically zero. Then the integral

$$J := \int_0^l \kappa'(s)\,(ax(s) + by(s) + c)\,ds$$

will not be zero. On the other hand, integrating by parts and using (2.14) and (2.20), we have

$$\begin{aligned} J &= -\int_0^l \kappa(s)\,(ax'(s) + by'(s))\,ds \\ &= \int_0^l (-ay''(s) + bx''(s))ds = [-ay'(s) + bx'(s)]_{s=0}^{s=l} = 0. \end{aligned}$$

The last equality above holds because the curve is closed. This gives us a contradiction, proving the result. □

Exercises 2

1 An $n \times n$ matrix A is called an *orthogonal matrix* if A^T is the inverse matrix of A, where A^T denotes the transposition of A (cf. Appendix A.3). When setting $n = 2$, show that a 2×2 orthogonal matrix can be written in the form

$$\begin{pmatrix} \cos\alpha & -\sin\alpha \\ \sin\alpha & \cos\alpha \end{pmatrix} \qquad \text{(when the determinant is } +1\text{)},$$

$$\begin{pmatrix} \cos\alpha & \sin\alpha \\ \sin\alpha & -\cos\alpha \end{pmatrix} \qquad \text{(when the determinant is } -1\text{)},$$

where α is a real number.

2 (1) For a unit vector $\boldsymbol{e}$ in the plane, let $\boldsymbol{n}$ be the unit vector perpendicular to $\boldsymbol{e}$. Show that the matrix $(\boldsymbol{e}, \boldsymbol{n})$ (the 2×2 matrix whose first column is $\boldsymbol{e}$ and second is $\boldsymbol{n}$) is an orthogonal matrix (cf. Appendix A.3).
(2) Let $A = (\boldsymbol{e}, \boldsymbol{n})$ be an orthogonal matrix, where $\boldsymbol{e}$ and $\boldsymbol{n}$ are column vectors. Show that $\{\boldsymbol{e}, \boldsymbol{n}\}$ is a pair of unit vectors that are perpendicular, i.e., form an *orthonormal basis.*
(3) Continuing from part (2), show that the determinant of A is $+1$ if and only if $\boldsymbol{n}$ is obtained by counterclockwise 90 degree rotation of $\boldsymbol{e}$.

3 Consider a car whose axles are pictured on the right, with axle length ε, and with distance Δ between axle midpoints, and with front axle turned by an angle θ from the forward direction. Compute the curvature of the path that the tires will trace out as the car moves.

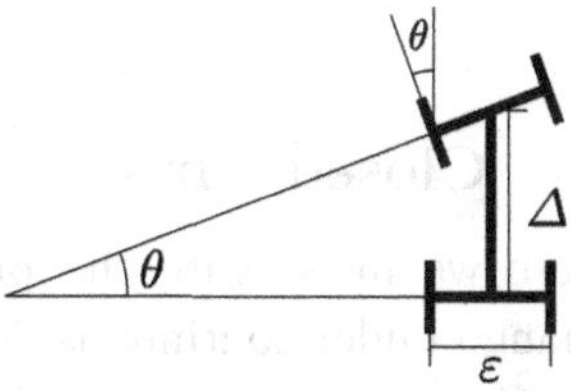

4 Show that reflecting a regular curve across a straight line results in changing the sign of its curvature.

5 For the cycloid $\gamma(t) = a(t - \sin t, 1 - \cos t)$ $(0 \leq t \leq 2\pi)$ as in Problem **4** of Section 1, do the following:

(1) Compute the curvature function using the formula (2.7).
(2) For the interval $[0, t]$, find an arc-length parameter $s(t)$ and its inverse function $t(s)$.
(3) Writing $\gamma(t)$ as $\gamma(s)$, in terms of its arc-length parameter, compute the curvature. And verify that it coincides with the result of (1).

6 Show by direct computation that the curvature of a regular curve is not changed by translations and rotations.

7 For the curve $\gamma(s)$ given in (2.19) in the proof of Theorem 2.8, confirm that s is an arc-length parameter and that $\kappa(s)$ is the curvature.

8 Let $\boldsymbol{e}(s)$ be the unit tangent vector to a curve $\gamma(s)$ with arc-length parameter s. Let $\boldsymbol{n}(s)$ be the unit normal vector to the left-hand side of $\boldsymbol{e}$. Notating the curvature by $\kappa(s)$, show that for s near 0, we have

$$\begin{aligned}\gamma(s) =& \gamma(0) + s\boldsymbol{e}(0) + \frac{s^2}{2}\kappa(0)\boldsymbol{n}(0) \\ &+ \frac{s^3}{6}(-\kappa(0)^2\boldsymbol{e}(0) + \kappa'(0)\boldsymbol{n}(0)) + o(s^3).\end{aligned}$$

Here o is Landau's symbol as in Appendix A.1.

9* (1) Show that two regular curves $\gamma_1(t)$, $\gamma_2(u)$ have third order contact at P if and only if one can choose a parametrization of γ_2 such that $\gamma_1(t_0) = \gamma_2(u_0) = \mathrm{P}$ and

$$\frac{d\gamma_1}{dt}(t_0) = \frac{d\gamma_2}{du}(u_0), \ \frac{d^2\gamma_1}{dt^2}(t_0) = \frac{d^2\gamma_2}{du^2}(u_0), \ \frac{d^3\gamma_1}{dt^3}(t_0) = \frac{d^3\gamma_2}{du^3}(u_0)$$

hold.

(2) Let $\varphi\colon D \to \varphi(D)$ be a diffeomorphism from a region D to $\varphi(D)$ in the plane. Show that for two curves γ_1 and γ_2 in D having third order contact at $\mathrm{P} \in D$, $\varphi \circ \gamma_1$ and $\varphi \circ \gamma_2$ have third order contact at $\varphi(\mathrm{P})$.

(3) Show that a curve $\gamma(t)$ has a third order contact at $\gamma(t_0)$ with its osculating circle at t_0 if and only if the derivative $\dot{\kappa}(t_0)$ of the curvature vanishes.

3. Closed curves*

Here we investigate the properties of closed regular curves that do not change under continuous deformations.

Given two closed regular curves, if there exists a smooth deformation from one to the other so that, at each moment of the deformation, the curve at that moment has a nowhere vanishing velocity vector, then we say that the two regular curves are *equivalent* or *regular homotopic* to each other. More precisely, we define two closed regular curves to be regular homotopic as follows: Changing parameter if necessary, two closed regular curves $\gamma_1(s)$ and $\gamma_2(s)$ can be parametrized by the same parameter s $(0 \le s \le l)$. Then, γ_1 and γ_2 are regular homotopic if and only if there exists a family of closed curves $\sigma_t : [0, l] \to \boldsymbol{R}^2$ $(0 \le t \le 1)$ so that:

(i) For each t, $\sigma_t(s)$ $(0 \le s \le l)$ is a smooth closed curve whose velocity vector $\frac{d\sigma_t}{ds}(s)$ never vanishes.

(ii) $\sigma_0(s) = \gamma_1(s)$ and $\sigma_1(s) = \gamma_2(s)$. In other words, when t moves from 0 to 1, the curves σ_t move from γ_1 to γ_2.

(iii) $\sigma_t(s)$ is continuous with respect to both t and s.

Here we do not assume that s is an arc-length parameter.

This deformation gives an equivalence relation for closed regular curves, and every closed regular curve lies in one of the classes shown in Fig. 3.1. In order to check that this is so, we need to introduce the notion of rotation index for closed regular curves.

The rotation index. For an arc-length parametrized closed curve $\gamma(s)$ $(0 \le s \le l)$ with curvature function $\kappa(s)$, we call

$$T_\gamma := \int_0^l \kappa(s)\, ds, \qquad i_\gamma := \frac{T_\gamma}{2\pi}$$

the *total curvature* and the *rotation index* of the curve γ, respectively.

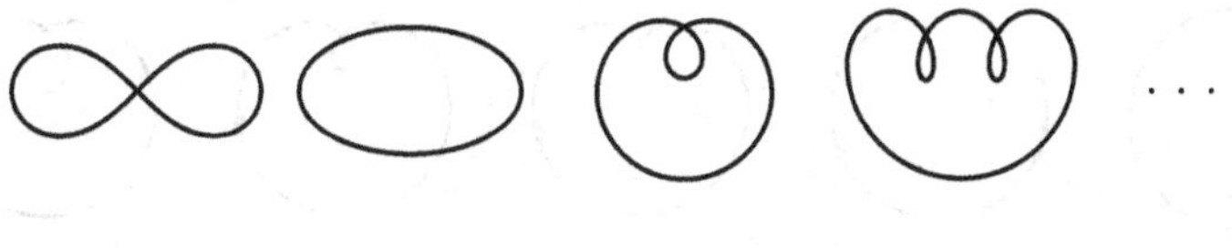

Fig. 3.1 Representatives of regular homotopy classes of closed regular curves.

Proposition 3.1. *The rotation index i_γ of a closed regular curve γ is always an integer.*

Proof. Parametrizing the closed regular curve as $\gamma(s)$, $(0 \le s \le l)$, it can be represented as in (2.19) in Remark 2.9:

$$\gamma(s) = \int_0^s (\cos\theta(t), \sin\theta(t))\, dt,$$

where

$$\theta(s) := \int_0^s \kappa(u)\, du. \tag{3.1}$$

This gives $\gamma'(s) = (\cos\theta(s), \sin\theta(s))$, and so $\theta(s)$ gives us the angle between $\gamma'(s)$ and $\gamma'(0) = (1, 0)$. Here, because we have the condition (2.20) for closed curves, which implies $\gamma'(l) = \gamma'(0)$, it follows that $\theta(l) - \theta(0)$ is an integer multiple of 2π. Therefore i_γ is an integer. $\square$

Because the rotation index is an integer, it cannot change under continuous deformations of a closed regular curve described in the beginning of this section. So if two closed regular curves are regular homotopic, then they have the same rotation index. However, if we allow deformations under which the velocity vector can become zero at some point (that is, a singular point) of some curve during the deformation, the curvature cannot be defined at that point, and it is then possible for the winding number to change.

For example, Fig. 3.2 shows a deformation, where the velocity vector becomes zero at one point in the curve (iii), called the *cardioid*. The initial curve (i) of this deformation is a counterclockwise wrapping circle with rotation index 1, but the final curve (v), which is called the *limaçon*, has rotation index 2 and so has a different rotation index than the initial curve.

Denote by S^1 the unit circle in the plane, that is, the circle centered at the origin with radius 1. For a given regular curve $\gamma(s)$ parametrized by

Fig. 3.2 A deformation of closed curves admitting a singular point: $\sigma_t(s) := (1 - 2t\sin s)(\cos s, \sin s)$ $(0 \leq s \leq 2\pi,\ 0 \leq t \leq 1)$.

arc-length, the velocity $\gamma'(s)$ is a unit vector. Therefore, if we translate the velocity vector so that its tail is at the origin, then we have $\gamma'(s) \in S^1$. In this way, we have the mapping

$$[0, l] \ni s \longmapsto \gamma'(s) \in S^1, \tag{3.2}$$

which we call the *Gauss map*[1] of the curve, where S^1 denotes the unit circle centered at the origin of $\boldsymbol{R}^2$. By the proof of Lemma 3.1, when we travel once around a closed regular curve, the number of times that the Gauss map will travel around the unit circle is equal to the rotation index.

Imagine that the closed regular curve $\gamma(t)$ is a closed circuit, and a car is driving along the curve such that the position of the car at the time t is $\gamma(t)$. Then the Gauss map is a meter which measures the direction in the xy-plane at time t. The rotation index is the number of times this meter loops around when the car traverses the entire $\gamma(t)$ once. If the number i_γ is positive (resp. negative), the meter turns i_γ times (resp. $|i_\gamma|$ times) counterclockwise (clockwise).

In particular, for the case of simple closed curves, we have the following theorem.

Theorem 3.2. *Every simple closed regular curve has rotation index either* 1 *or* -1.

Proof. Although this appears to be a rather obvious assertion, the proof is not so simple. The following proof is due to [11]. Translate a line parallel to the x-axis from far below in the plane in the upward direction, and let d be such a line which touches the curve for the first time, and let P be one of the tangent points in the curve (see Fig. 3.3).

Without loss of generality, we can take an arc-length parameter s $(0 \leq s \leq l)$ for $\gamma(s)$ so that $\gamma(0) = \gamma(l) = \mathrm{P}$. Changing the direction of the curve if necessary, we may assume that $\gamma'(0) = (1, 0)$. Defining the map $\boldsymbol{w}$ on a closed domain $\overline{D} := \{(s,t) \,|\, 0 \leq s \leq t \leq l\}$ in the st-plane by

[1] Gauss, Carl Friedrich (1777–1855).

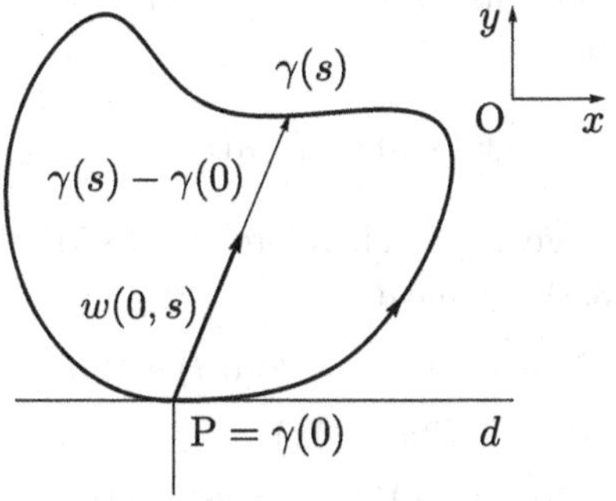

Fig. 3.3

$$\boldsymbol{w}(s,t) := \begin{cases} \gamma'(s) & \text{(when } s = t\text{)}, \\ -\gamma'(0) & \text{(when } s = 0 \text{ and } t = l\text{)}, \\ \dfrac{\gamma(t)-\gamma(s)}{|\gamma(t)-\gamma(s)|} & \text{(otherwise)}, \end{cases}$$

this $\boldsymbol{w}$ is a continuous map from $\overline{D}$ to the unit circle S^1. Then there exists a unique continuous function Θ on $\overline{D}$ such that

$$\boldsymbol{w}(s,t) = (\cos\Theta(s,t), \sin\Theta(s,t)), \qquad \Theta(0,0) = 0. \tag{3.3}$$

The function $\Theta(s,t)$ represents the angle of the vector $\boldsymbol{w}(s,t)$ with $(1,0)$. Since the angle has the ambiguity of integer multiples of 2π, it cannot be determined uniquely, in general. But if we allow the angle $\Theta(s,t)$ to take arbitrary real number values beyond the interval $[0,2\pi]$, and choose values $\Theta(s,t)$ which vary continuously in (s,t), then the function $\Theta(s,t)$ satisfies (3.3). (The existence of this Θ is usually shown using the notion of covering spaces, and the fact that $\overline{D}$ is simply connected (cf. [8, 35]). However, noticing that $\boldsymbol{w}(s,t)$ is of class C^1, one can show the existence of the required function Θ without using the notion of covering spaces. See Problem 4.)

Since γ is a closed curve, $\gamma(l) = \gamma(0)$ holds. So we have $\boldsymbol{w}(s,l) = -\boldsymbol{w}(0,s)$, and we have that $\Theta(s,l) - \Theta(0,s) = \pi \times$ (an odd integer). Since the function $\Theta(s,l) - \Theta(0,s)$ of s is continuous, there exists an integer n independent of s so that we can write

$$\Theta(s,l) - \Theta(0,s) = (2n+1)\pi \qquad (0 \le s \le l). \tag{3.4}$$

On the other hand, because $\boldsymbol{w}(s,s) = \gamma'(s)$, and because $\Theta(0,0) = 0$ (cf. (3.3)), we have that

$$\Theta(s,s) = \int_0^s \kappa(u)\,du \tag{3.5}$$

holds (see (3.1) on page 29). Therefore, by (3.5), the definition of the rotation index and (3.4), we have

$$\begin{aligned} 2\pi i_\gamma &= \Theta(l,l) - \Theta(0,0) = (\Theta(0,l) + (2n+1)\pi) - \Theta(0,0) \\ &= (\Theta(0,l) + (2n+1)\pi) \\ &\quad + (\Theta(0,l) - \Theta(0,0) - (2n+1)\pi) - \Theta(0,0) \\ &= 2(\Theta(0,l) - \Theta(0,0)). \end{aligned} \tag{3.6}$$

Now, because of the way we have chosen P, we can take a half-line extending out from P in the direction of the negative y-axis, and the closed curve will never intersect that half-line except at P itself. Hence $\boldsymbol{w}(0,s)$ will never become $(0,-1)$ (see Fig. 3.3). Hence we have

$$\Theta(0,s) + \frac{\pi}{2} \neq 2\pi m \qquad (m = 0, \pm 1, \pm 2, \dots).$$

Then, noticing that $\Theta(0,0) = 0$, the function $\Theta(0,s)$ in s satisfies

$$-\frac{\pi}{2} < \Theta(0,s) < \frac{3\pi}{2} \qquad (0 \le s \le l),$$

because it is continuous. Moreover, by (3.4), $\Theta(0,l) - \Theta(0,0) = \pi$ holds. Then by (3.6), we have $i_\gamma = 1$. □

Theorem 3.3 (Whitney's Theorem[2] [42]). *A necessary and sufficient condition for two closed regular curves $\gamma_1(s)$ and $\gamma_2(s)$ to be regular homotopic is that they have the same rotation index.*

Proof. The necessity of the equality of the rotation indices is clear. Conversely, suppose that the two closed regular curves γ_1 and γ_2 have the same rotation index:

$$i_{\gamma_1} = i_{\gamma_2} = m.$$

We will construct a family $\sigma_t(s)$ of curves satisfying the conditions (i)–(iii) for regular homotopy. Because homotheties do not change the rotation index, we may assume without loss of generality that both curves have length 1 and have arc-length parametrizations $\gamma_1(s)$, $\gamma_2(s)$ $(0 \le s \le 1)$. Letting $\kappa_1(s)$ and $\kappa_2(s)$ be the respective curvatures of γ_1 and γ_2, from

[2] Whitney, Hassler (1907–1989).

equation (2.19) in the proof of the fundamental theorem for plane curves (Theorem 2.8), we see that we may assume $\gamma_j(0) = \mathbf{0}$ and

$$\gamma_j(s) = \int_0^s (\cos\theta_j(u), \sin\theta_j(u))\,du, \quad \theta_j(s) := \int_0^s \kappa_j(u)\,du \qquad (j = 1, 2).$$

Since κ_1 and κ_2 are periodic functions with period 1, from the definition of the rotation index, we have

$$\theta_j(s+1) = \int_0^1 \kappa_j(u)\,du + \int_1^{s+1} \kappa_j(u)\,du = 2\pi m + \theta_j(s) \quad (j = 1, 2). \tag{3.7}$$

Here, we set

$$\varphi_t(s) := (1-t)\theta_1(s) + t\theta_2(s) \qquad (0 \le t \le 1). \tag{3.8}$$

By (3.7), it holds that $\varphi_t(s+1) = \varphi_t(s) + 2\pi m$. Then defining

$$\boldsymbol{v}_t(s) := (\cos\varphi_t(s), \sin\varphi_t(s)),$$

$$\tilde{\boldsymbol{v}}_t(s) := \boldsymbol{v}_t(s) - \int_0^1 \boldsymbol{v}_t(u)\,du,$$

$$\sigma_t(s) := \int_0^s \tilde{\boldsymbol{v}}_t(u)\,du,$$

one can show that, for each t, $\boldsymbol{v}_t(s)$, $\tilde{\boldsymbol{v}}_t(s)$ and $\sigma_t(s)$ are periodic in s with period 1, and then σ_t is a closed curve. Because $\gamma_j(0) = \mathbf{0}$ $(j = 1, 2)$, we have $\tilde{\boldsymbol{v}}_t = \boldsymbol{v}_t$ for $t = 0, 1$, and it follows that $\sigma_0 = \gamma_1$, $\sigma_1 = \gamma_2$. That is, σ_t is a deformation from γ_1 to γ_2. In other words, $\sigma_t(s)$ satisfies the first part of the condition (i) and the conditions (ii) and (iii) of the definition of regular homotopy. So it is sufficient to show that $\sigma_t(s)$ satisfies the second part of the condition (i), that is, $\tilde{\boldsymbol{v}}_t = d\sigma_t/ds$ does not vanish anywhere in the deformation. By definition, we have

$$\frac{d\sigma_t(s)}{ds} = \tilde{\boldsymbol{v}}_t(s) = \boldsymbol{v}_t(s) - \int_0^1 \boldsymbol{v}_t(u)\,du,$$

and the triangle inequality for integrals (cf. Theorem A.1.4 in Appendix A.1) gives

$$\left|\int_0^1 \boldsymbol{v}_t(u)\,du\right| \le \int_0^1 |\boldsymbol{v}_t(u)|\,du = 1. \tag{3.9}$$

In order for equality to hold, $\boldsymbol{v}_t(s)$ $(0 \le s \le 1)$ must be a constant vector. (Here, we used the fact that $|\boldsymbol{v}_t| = 1$.)

When the rotation index m is not 0, (3.8) implies that φ_t moves from 0 to $2\pi m$ when s moves 0 to 1, and then $\boldsymbol{v}_t$ cannot be a constant vector. Consequently, equality does not hold in (3.9), implying

$$\left|\frac{d\sigma_t}{ds}\right| \ge |\boldsymbol{v}_t| - \left|\int_0^1 \boldsymbol{v}_t(u)\,du\right| > |\boldsymbol{v}_t| - 1 = 0,$$

and so we know $d\sigma_t/ds$ does not vanish for all s.

Next, we consider the case that the rotation index is $m = 0$. If $\boldsymbol{v}_t$ is a constant vector, then the fact $\varphi_t(0) = 0$ implies that $\varphi_t(s) = (1-t)\theta_1(s) + t\theta_2(s)$ is zero for all s. In particular, two functions θ_1 and θ_2 differ by only constant multiplication. If the curvatures of γ_1, γ_2 are identically zero, they cannot be closed curves. Hence θ_1 and θ_2 are not constant functions. If θ_1 is proportional to θ_2, translating the parameter s of γ_2 if necessary, one can change θ_2 with an additional constant, and then cause θ_1 and θ_2 not to be proportional. Thus, the vector $\boldsymbol{v}_t$ is not constant in t. Hence in a similar way to the case above, one can show $d\sigma_t/ds \neq \mathbf{0}$. □

By Theorem 3.3, we can find the rotation index of a given curve without computation. In particular, for the curves in Fig. 3.1, the leftmost curve is regular homotopic to the lemniscate and so has rotation index 0, by Problem **1**. Also, the other curves in that figure are regular homotopic to curves that wrap finitely many times about a circle, so their rotation indices can be found as well. Also, for closed regular curves $\gamma(s)$ $(0 \leq s \leq l)$ that are C^1, we have existence of a continuous function $\theta\colon [0, l] \to \boldsymbol{R}$ so that $\gamma'(s) = (\cos\theta(s), \sin\theta(s))$ holds, and so we can define the rotation index by $2\pi i_\gamma := \theta(l) - \theta(0)$. Then, by adjusting the proofs of Theorems 3.2 and 3.3 a bit, we can reprove those results for the case of C^1-regular curves.

Corollary 3.4. *A simple closed curve is regular homotopic to the circle.*

Proof. A circle turning right (resp. left) has the rotation index 1 (resp. -1). On the other hand, a given simple closed curve γ has the rotation index 1 or -1 by Theorem 3.2. Hence by Theorem 3.3, it is regular homotopic to a circle. □

Advanced Topic: Whitney's formula. Though it still looks difficult to find the rotation index for a given closed curve, the following Whitney's formula (Theorem 3.5) provides a way to calculate the rotation index by counting self-intersections of the curve.

A point where a closed curve crosses itself is called a *self-intersection*. A closed regular curve is said to be *generic* if it has a finite number of self-intersections, and for each self-intersection, two pieces of the curve are crossing transversely (are not tangent each other) (Fig. 3.4). An arbitrary closed regular curve can be deformed to a generic curve by a small deformation.

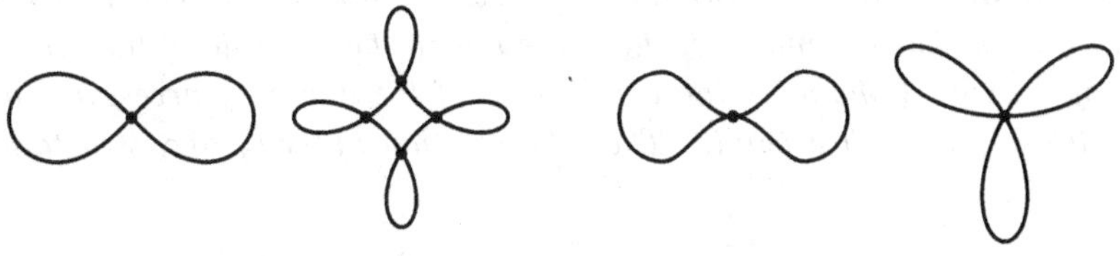

Generic curves. Non-generic curves.

Fig. 3.4

Take a point P on an oriented generic curve. When the curve passes through a self-intersection Q for the first time after starting at P, if another piece of the curve crosses from left to right (resp. right to left) with respect to the direction of travel, the self-intersection Q is called a *positive self-intersection* (resp. *negative self-intersection*) with respect to the base point P (see Fig. 3.5).

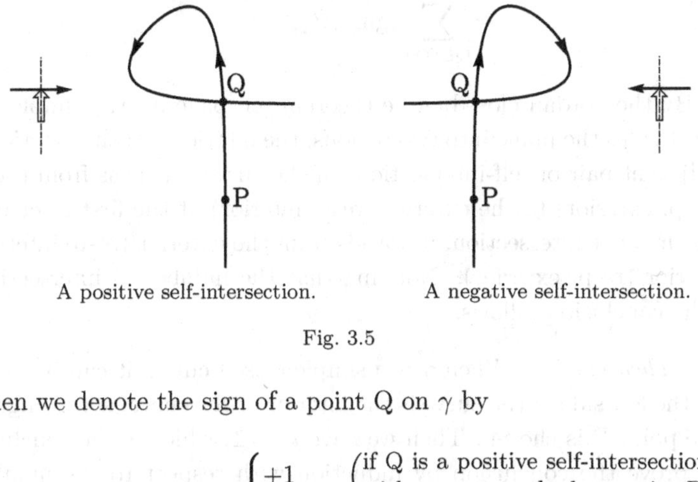

A positive self-intersection. A negative self-intersection.

Fig. 3.5

Then we denote the sign of a point Q on γ by

$$\operatorname{sgn}_\gamma(Q) = \operatorname{sgn}_{P,\gamma}(Q) := \begin{cases} +1 & \begin{pmatrix}\text{if Q is a positive self-intersection} \\ \text{with respect to the base point P}\end{pmatrix}, \\ -1 & \begin{pmatrix}\text{if Q is a negative self-intersection} \\ \text{with respect to the base point P}\end{pmatrix}, \\ 0 & (\text{if Q is not a self-intersection}). \end{cases}$$

The sign of a self-intersection depends on the choice of base point P, and the sign is reversed when the orientation of the curve is reversed.

The following is a counting formula for the rotation index:

Theorem 3.5 (Whitney's formula [42]). *Take a point* P *on a generic closed curve* γ *such that the image of the curve lies to the left side of the tangent line of* γ *at* P. *(Changing the orientation of the curve, if necessary, one can always find such a point.) Then the rotation index* i_γ *of* γ *is obtained by*

$$i_\gamma = 1 + \sum_{Q \in \gamma} \operatorname{sgn}_{P,\gamma}(Q).$$

To prove the theorem, we prepare the following lemma:

Lemma 3.6. *Assume an oriented generic closed curve* γ *and an oriented simple closed regular curve* σ *cross transversely (are not tangent) at a finite number of points, and all self-intersections of* γ *do not lie on* σ. *At an intersection* Q *of these two curves, if* γ *crosses* σ *from left to right (resp. right to left) with respect to the direction of* σ, *we set* $\operatorname{sgn}_{\gamma,\sigma}(Q) = +1$ *(resp.* $\operatorname{sgn}_{\gamma,\sigma}(Q) = -1$*). Then*

$$\sum_{Q \in \gamma \cap \sigma} \operatorname{sgn}_{\gamma,\sigma}(Q) = 0.$$

Proof. By the Jordan closed curve theorem (cf. page 9), the simple closed curve σ divides the plane into two regions, the interior and the exterior. For each adjacent pair of self-intersections, if the curve γ travels from the interior (resp. exterior) to the exterior (resp. interior) at the first intersection, then at the next intersection, it travels from the exterior (resp. interior) to the interior (resp. exterior). Noticing that the number of intersections is even, the conclusion follows. □

Proof of Theorem 3.5. When γ is a simple closed curve, it can be oriented so that the left side of the curve is the interior of the curve, according to the way the point P is chosen. Then we have $i_\gamma = 1$, which is the conclusion.

We prove the conclusion by induction with respect to the number of self-intersections. Assume the conclusion holds for curves with $n-1$ self-intersections, and take a closed curve γ having n self-intersections. We can find a self-intersection Q_1 on the curve γ such that the part of the curve starting at Q_1 and returning again to Q_1 is without self-intersection. By rounding the angle smoothly (see Proposition B.5.5 of Appendix B.5 for the way to round the angle) at Q_1, this closed subarc of γ turns out to be a smooth simple closed regular curve, which is denoted by γ_1. We take the first Q_1 among such self-intersections starting from the base point P (Fig. 3.6). Then γ_1 does not pass through self-intersections between P and

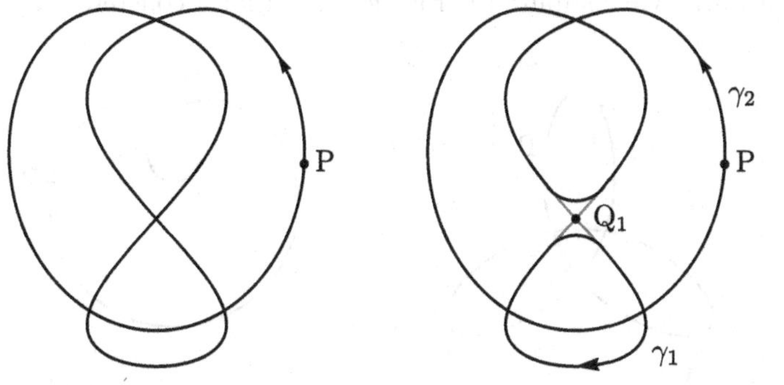

Fig. 3.6

Q_1. In fact, if there exists such a self-intersection, the point has the same property as Q_1, which contradicts the way we chose Q_1. Moreover, γ_1 does not pass through the base point P, because the curve passes through each self-intersection twice before returning to P.

On the other hand, let γ_2 be a closed regular curve which is on the opposite side of γ_1 where we rounded the vertical angle at Q_1. The integrals of the curvature on the rounded angles are equal to the increment of the angles at the rounded parts (cf. (3.1)). Since these angles at the rounded parts are equal, these increments on γ_1 and γ_2 cancel, which implies $i_\gamma = i_{\gamma_1} + i_{\gamma_2}$. Since the rotation index of the simple closed regular curve γ_1 is the sign of the self-intersection Q_1, by the inductive assumption, we have

$$\begin{aligned} i_\gamma = i_{\gamma_1} + i_{\gamma_2} &= \mathrm{sgn}_{\mathrm{P},\gamma}(\mathrm{Q}_1) + i_{\gamma_2} \\ &= \mathrm{sgn}_{\mathrm{P},\gamma}(\mathrm{Q}_1) + 1 + \sum_{\mathrm{Q}\in\gamma_2} \mathrm{sgn}_{\mathrm{P},\gamma_2}(\mathrm{Q}). \end{aligned}$$

By Lemma 3.6, the sum of the signs of intersections of γ_2 with γ_1 vanishes. So we have

$$\sum_{\mathrm{R}\in\gamma_1\cap\gamma_2} \mathrm{sgn}_{\gamma_2,\gamma_1}(\mathrm{R}) = 0.$$

Here, γ_1 does not pass through self-intersections between P and Q_1. Thus for each intersection R of γ_2 with γ_1, it holds that $\mathrm{sgn}_{\mathrm{P},\gamma}(\mathrm{Q}) = \mathrm{sgn}_{\gamma_2,\gamma_1}(\mathrm{R})$. Hence

$$\begin{aligned} \sum_{\mathrm{Q}\in\gamma} \mathrm{sgn}_{\mathrm{P},\gamma}(\mathrm{Q}) &= \mathrm{sgn}_{\mathrm{P},\gamma}(\mathrm{Q}_1) + \sum_{\mathrm{R}\in\gamma_1\cap\gamma_2} \mathrm{sgn}_{\gamma_2,\gamma_1}(\mathrm{R}) + \sum_{\mathrm{S}\in\gamma_2} \mathrm{sgn}_{\mathrm{P},\gamma_2}(\mathrm{S}) \\ &= \mathrm{sgn}_{\mathrm{P},\gamma}(\mathrm{Q}_1) + \sum_{\mathrm{S}\in\gamma_2} \mathrm{sgn}_{\mathrm{P},\gamma_2}(\mathrm{S}) = i_\gamma - 1, \end{aligned}$$

which proves the theorem. □

Example 3.7. We compute the rotation index for a concrete example.

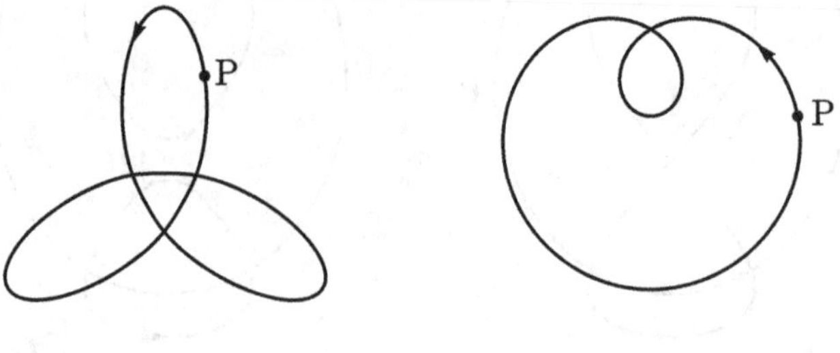

Fig. 3.7

The left side of Fig. 3.7 has 3 self-intersections. Taking P as a base point, the signs of these self-intersections are $+$, $-$, $+$, in order of appearance. Thus by Theorem 3.5, the rotation index is $1 + (1 - 1 + 1) = 2$. On the other hand, the rotation index of the right side figure is 2, because the sign of the self-intersection is $+$. Therefore these two curves are regular homotopic, by Theorem 3.3. One can verify this fact by drawing pictures of a smooth deformation.

Exercises 3

1 Compute the rotation indices of the ellipse and the lemniscate from the definition of the rotation index.

2* Assume a C^∞-function $\varphi(t)$ defined on $\boldsymbol{R}$ satisfies $\varphi(0) = 0$. If we set

$$\psi(t) := \begin{cases} \varphi(t)/t & \text{(when } t \neq 0\text{)}, \\ \dot{\varphi}(0) & \text{(when } t = 0\text{)}, \end{cases}$$

then show that $\psi(t)$ is a C^∞-function. Here, $\dot{\varphi} := d\varphi/dt$.

3* (1) For a C^∞-function $f(t)$ defined on $\boldsymbol{R}$, we set

$$F(s,t) := \begin{cases} \dfrac{f(t) - f(s)}{t - s} & \text{(when } s \neq t\text{)}, \\ \dot{f}(s) & \text{(when } s = t\text{)}. \end{cases}$$

Show that F is a C^∞-function on $\boldsymbol{R}^2$ using Problem **2**.

(2) Let $\gamma(s)$ $(0 \leq s \leq l)$ be a simple closed curve and consider it as an

l-periodic curve defined on $\boldsymbol{R}$. If we set

$$\tilde{\boldsymbol{w}}(s,t) = \begin{cases} \dfrac{\gamma(t)-\gamma(s)}{t-s}\dfrac{|t-s|}{|\gamma(t)-\gamma(s)|} & (s-l<t<s+l, s\neq t), \\ \gamma'(s)\left(=\dfrac{\gamma'(s)}{|\gamma'(s)|}\right) & (s=t), \\ -\dfrac{\gamma(t)-\gamma(s)}{t-(s+l)}\dfrac{|t-(s+l)|}{|\gamma(t)-\gamma(s)|} & (s+l<t<s+2l, s\neq t), \\ -\gamma'(s) & (s+l=t), \end{cases}$$

show that $\tilde{\boldsymbol{w}}$ is C^∞ on the domain $\{(s,t)\,|\,s-l<t<s+2l\}$, and the restriction of $\tilde{\boldsymbol{w}}$ to $\overline{D}=\{(s,t)\,|\,0\le s\le t\le l\}$ coincides with the $\boldsymbol{w}$ in the proof of Theorem 3.2.

4* For a positive constant l, let $\boldsymbol{w}\colon \overline{D}\to\boldsymbol{R}^2$ be a C^∞-map $\boldsymbol{w}\colon\overline{D}\to\boldsymbol{R}^2$ defined on a closed domain $\overline{D}=\{(s,t)\mid 0\le s\le t\le l\}$ of the st-plane satisfying $|\boldsymbol{w}(s,t)|=1$ and $\boldsymbol{w}(0,0)=(1,0)$. Show, in the following way, that there exists a continuous function Θ on $\overline{D}$ satisfying $\boldsymbol{w}(s,t)=(\cos\Theta(s,t),\sin\Theta(s,t))$ and $\Theta(0,0)=0$.

(1) Let $\boldsymbol{e}\colon I\ni s\longmapsto \boldsymbol{e}(s)\in\boldsymbol{R}^2$ be a C^∞-map defined on an interval I containing the origin and satisfying $|\boldsymbol{e}(s)|=1$ for all $s\in I$. If $\boldsymbol{e}(0)=(\cos\alpha,\sin\alpha)$ for a constant $\alpha\in\boldsymbol{R}$, show that the function $\theta(s)$ defined by

$$\theta(s) := \int_0^s \det(\boldsymbol{e}(u),\boldsymbol{e}'(u))\,du+\alpha$$

satisfies $\boldsymbol{e}(s)=(\cos\theta(s),\sin\theta(s))$.

(2) Show that the function

$$\theta(t) := \int_0^t \det(\boldsymbol{w}(0,v),\boldsymbol{w}_2(0,v))\,dv \qquad \left(\boldsymbol{w}_2 := \frac{\partial\boldsymbol{w}}{\partial t}\right)$$

defined on the interval $0\le t\le l$ satisfies

$$\boldsymbol{w}(0,t)=(\cos\theta(t),\sin\theta(t)).$$

(3) Let

$$\Theta(s,t)=\int_0^s \det(\boldsymbol{w}(u,t),\boldsymbol{w}_1(u,t))\,du+\int_0^t\det(\boldsymbol{w}(0,v),\boldsymbol{w}_2(0,v))\,dv$$
$$\left(\boldsymbol{w}_1 := \frac{\partial\boldsymbol{w}}{\partial s},\quad \boldsymbol{w}_2 := \frac{\partial\boldsymbol{w}}{\partial t}\right)$$

for $(s,t)\in\overline{D}$. Show that $\Theta(s,t)$ is the desired function.

4. Geometry of spirals*

We first introduce two representative examples of spiral curves, the *Archimedean spiral* and the *logarithmic spiral.*

The Archimedean spiral. Taking polar coordinates (r, θ) for the plane, the curve satisfying

$$r = a\theta \qquad (a > 0 \text{ is a constant, and } \theta > 0)$$

is called the *Archimedean spiral* (see Fig. 4.1, left side). If one were to imagine a horizontal plane rotating about one fixed point at a constant rate, and if one were to take a nail and press it gently onto that plane by starting at the fixed point and moving at constant speed out in one unchanging direction, one would scratch out such a spiral. In ancient Greece, Archimedes[1] investigated the properties of this spiral curve (see Problem **1** at the end of this section).

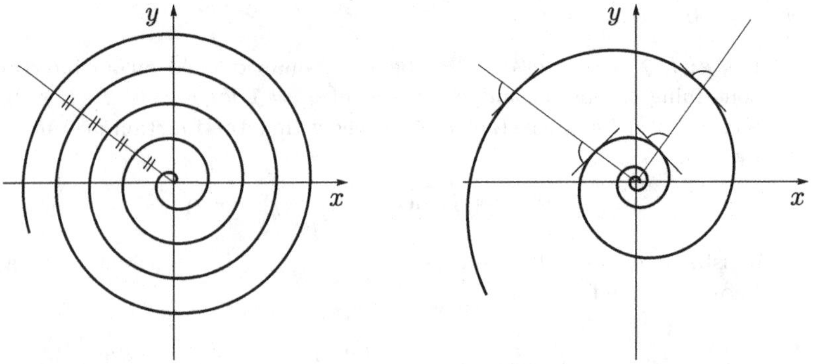

Fig. 4.1

The logarithmic spiral. In polar coordinates (r, θ), the curve given by

$$r = a^{\theta} \qquad (a > 1 \text{ is a constant, } \ \theta \in \boldsymbol{R})$$

is called the *logarithmic spiral.* When drawing a half-line out from the origin, the half-line will always intersect the spiral at the same angle, regardless of choice of direction of the half-line (see the right-hand side of Fig. 4.1). One can find the logarithmic spiral in nature, for example, in spiral shells or in the shape of galaxies. Similarity transformation of the

[1]Archimedes (B.C. 287–212).

logarithmic spiral is congruent to the original spiral (this property is called *self-similarity*; cf. Problem **2** at the end of this section).

In this text, a regular curve is called a *spiral* if the derivative of the curvature function never vanishes. A spiral is called a *positive spiral* (resp. *negative spiral*) if the curvature function is increasing (resp. decreasing), see Fig. 4.2.

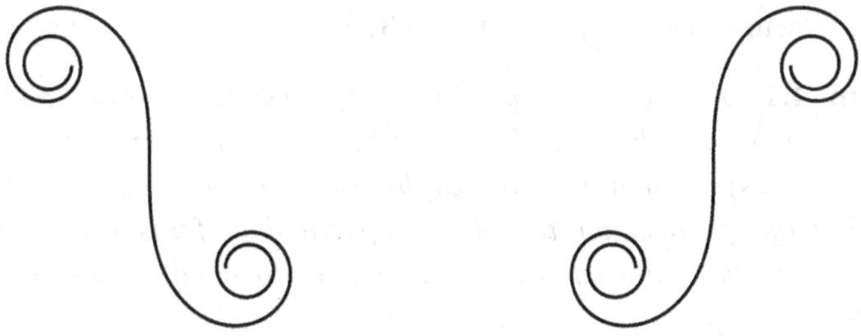

A positive spiral. A negative spiral.

Fig. 4.2

The mirror image of a positive spiral is a negative spiral. On the other hand, a positive spiral cannot be superimposed on a negative spiral by a congruent transformation of $\boldsymbol{R}^2$ unless reflection is allowed, because rotations and translations preserve the curvature function (Corollary 2.7 in Section 2).

Imagine driving on a spiral road. One can say that

a positive spiral: keep turning the wheel to the left,
a negative spiral: keep turning the wheel to the right,

as a demonstration of what the spirals are. In fact, when one stops turning the wheel, the car moves in a straight line or a circle, and to drive along a spiral, one must keep turning the wheel. For example, look at the positive spiral on the left side of Fig. 4.2: Starting from the state where the wheel is turned to the right side, we could turn the wheel to the left little by little. Then the curvature is negative at first, and it increases to 0, the inflection point. When we continue to turn the wheel to the left, the curvature becomes positive.

It should be remarked that the positivity and negativity of spirals does not depend on the direction of the curves (Fig. 4.2). In fact, let $\gamma(s)$ $(0 \le s \le l)$ be a positive spiral parametrized by the arc-length s. Then the curvature function $\kappa(s)$ is increasing. Here, consider the curve $\tilde{\gamma}(s) := \gamma(l-s)$ $(0 \le s \le l)$ obtained by changing the direction of $\gamma(s)$. Then by (2.6) on page 13, the curvature function $\tilde{\kappa}(s)$ of $\tilde{\gamma}(s)$ is

$$\tilde{\kappa}(s) = \det(\tilde{\gamma}'(s), \tilde{\gamma}''(s)) = -\det(\gamma'(l-s), \gamma''(l-s)) = -\kappa(l-s),$$

which is an increasing function, because $\kappa(l-s)$ is decreasing. That is, the curve obtained by changing direction of a positive spiral is also a positive spiral (cf. Problem **3**).

The following theorem holds (Fig. 4.3).

Theorem 4.1. *Let $\gamma(s)$ be a positive (resp. negative) spiral, and let the osculating circle at each point of the spiral have the same orientation as the spiral. Then $\gamma(s)$ tangentially crosses the osculating circle from the circle's right side to its left (resp. left to right). In particular, if γ is positively curved at s, then γ moves from outside (resp. inside) to inside (resp. outside) of the circle.*

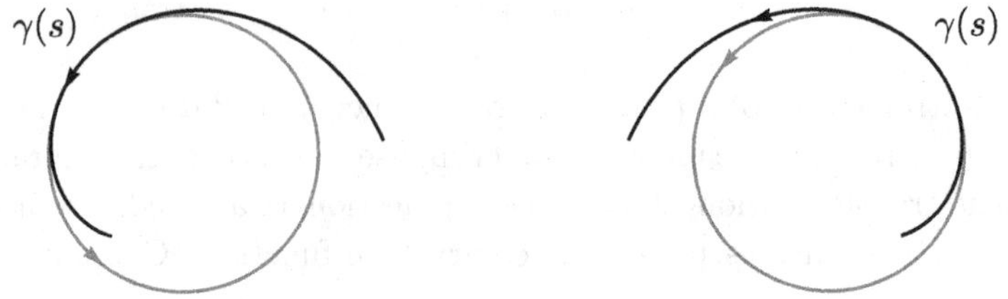

A positive spiral. A negative spiral.

Fig. 4.3 Spirals and osculating circles.

Proof. A driver traveling along a positive (resp. negative) spiral keeps turning the wheel to the left (resp. right). If he would stop turning the wheel, then the car follows the osculating circle at the point. Then one can see intuitively that the trajectory of the car crosses the circle from right to left (resp. left to right).

We give a rigorous proof: For a given positive spiral $\gamma(s)$ and a point $\gamma(s_0)$ of the spiral, choose an xy-coordinate system such that the point $\gamma(s_0)$ is the origin, the unit tangent vector $\boldsymbol{e}(s_0) := \gamma'(s_0)$ is the direction of the x-axis, and the direction of the unit normal vector $\boldsymbol{n}(s_0)$ is the direction of the y-axis. Then the curve can be expressed as a graph $y = f(x)$ (cf. page 19). In particular, since $f(0) = 0$ and $\dot{f}(0) := \frac{df}{dx}(0) = 0$ hold, (2.8) implies that the curvature and the derivative of the curvature at the origin are expressed as

$$\kappa(0) = \frac{\ddot{f}(0)}{(1+\dot{f}(0)^2)^{3/2}} = \ddot{f}(0), \qquad \dot{\kappa}(0) = \dddot{f}(0),$$

where the dot means the derivative with respect to x. Then by Taylor's formula (Theorem A.1.1 in Appendix A.1),

$$f(x) = \frac{1}{2}\kappa(0)x^2 + \frac{1}{6}\dot{\kappa}(0)x^3 + o(x^3) \tag{4.1}$$

holds, where $o(x^3)$ is Landau's symbol. On the other hand, the osculating circle of the curve can be expressed as a graph $y = g(x)$ near the origin, and we have

$$g(x) = \frac{1}{2}\kappa(0)x^2 + o(x^3)$$

by replacing f in (4.1) with g, since the curvature of the circle $y = g(x)$ is constant. Hence we have

$$f(x) - g(x) = \frac{1}{6}\dot{\kappa}(0)x^3 + o(x^3).$$

Here, $\dot{\kappa}(0) > 0$, because the graph $f(x)$ is a positive spiral. Then, as x increases, $f(x) - g(x)$ changes sign from negative to positive at the origin. This implies that $\gamma(s)$ tangentially crosses the osculating circle from right to left. □

Transformations mapping spirals to spirals. We identify the complex plane $\boldsymbol{C}$ with the plane $\boldsymbol{R}^2$ by the correspondence

$$\boldsymbol{R}^2 \ni (x,y) \longleftrightarrow x + iy \in \boldsymbol{C} \qquad \left(i := \sqrt{-1}\right).$$

For complex numbers a, b, c, d satisfying $ad - bc \neq 0$, a transformation T which maps $z \in \boldsymbol{C}$ to $w = T(z)$ via

$$z \longmapsto w = T(z) = \frac{az+b}{cz+d} \tag{4.2}$$

is called a *Möbius*[2] *transformation*, or a *linear fractional transformation*. When $c \neq 0$, though the Möbius transformation (4.2) cannot be defined at $z = -d/c$, it gives a diffeomorphism of the domain in $\boldsymbol{C}$ excluding the point $-d/c$ into $\boldsymbol{C}$. Both the composition of Möbius transformations, and the inverse of a Möbius transformation, are also Möbius transformations.

[2] Möbius, August Ferdinand (1790–1868).

Any Möbius transformation can be expressed as a composition of the following four basic transformations (cf. Problem **6** at the end of this section):

$$(4.3)\quad \begin{cases} \text{A translation} & z \mapsto z + d \ (d \in \boldsymbol{C}), \\ \text{A rotation} & z \mapsto e^{i\theta} z \quad (\theta \in \boldsymbol{R}), \\ \text{A similarity transformation} & z \mapsto rz \quad (r \in \boldsymbol{R} \setminus \{0\}), \\ \text{Conjugation composed with inversion} & z \mapsto 1/z. \end{cases}$$

Here we explain inversion: Let C be a circle of radius ρ in the plane centered at the origin O. Take a map which maps each point P on the plane different from O to the point Q on the half-line starting at O passing through P such that $\mathrm{OP} \cdot \mathrm{OQ} = \rho^2$. Such a map is called *inversion* with respect to the circle C. The inversion on the complex plane $\boldsymbol{C}$ with respect to the unit circle centered at the origin is the map $z \mapsto 1/\bar{z}$, where $\bar{z}$ is the complex conjugation of z.

Theorem 4.2. *A Möbius transformation maps each positive spiral to a positive spiral, and maps each negative spiral to a negative spiral. Furthermore, the osculating circles of a regular curve are mapped to the osculating circles at the corresponding points of the image curve.*

For example, by the transformation $z \mapsto 1/z$, the logarithmic spiral is mapped to a curve congruent to the original spiral. To prove Theorem 4.2, we prepare the following lemma.

Lemma 4.3. *A Möbius transformation maps circles to circles. Moreover, the region to the left-hand side of a circle will be mapped to the left-hand side of the image circle. Here the term "circles" includes straight lines, which can be regarded as circles with infinite radius.*

We used the terms "left-hand side" and "right-hand side" of the circle, instead of the interior and exterior, since a Möbius transformation may map the interior of the circle to the exterior of the image circle.

Proof. The conclusion is evident for the first three transformations of (4.3). So we shall prove the lemma for the transformation $z \mapsto 1/z$. A circle on the xy-plane is expressed as

$$(4.4)\quad a(x^2 + y^2) + 2bx + 2cy + d = 0 \quad (a, b, c, d \in \boldsymbol{R},\ b^2 + c^2 - ad > 0)$$

in general. In particular, when $a = 0$, (4.4) represents a straight line. It is sufficient to prove that this circle is mapped to a circle by the transformation $z \mapsto 1/z$. Letting $z = x + iy$, (4.4) is rewritten as

$$a|z|^2 + (b - ic)z + (b + ic)\bar{z} + d = 0 \qquad (\bar{z} := x - iy).$$

Hence if one sets $w = 1/z$, it becomes

$$A|w|^2 + (B - iC)w + (B + iC)\bar{w} + D = 0,$$
$$(A := d, B := b, C := -c, D := a).$$

In particular, since $B^2 + C^2 - AD = b^2 + c^2 - ad > 0$, the image is also a circle. Note that the center of the original circle may not be mapped to the center of the image circle.

Now we prove the latter conclusion. Assume a point P is mapped to a point Q by a given Möbius transformation T. Then it is sufficient to show that the determinant of the Jacobi matrix (cf. page 18) is positive. In fact, if this is true, T is a positive coordinate change of a neighborhood of P to a neighborhood of Q, and then the left side of the circle is mapped to the left side of the image circle. Since translations, rotations, and similarity expansions and reductions have this property, one can accomplish the proof by checking this property for the transformation $z \mapsto 1/z$. In fact, the determinant of the Jacobi matrix of the coordinate change $z \mapsto 1/z$, i.e., $\xi(x, y) = x/(x^2 + y^2)$, $\eta(x, y) = -y/(x^2 + y^2)$, is positive. □

Proof of Theorem 4.2. Let $\gamma(s)$ be a positive spiral parametrized by the arc-length s. Then the curvature function κ satisfies $\kappa'(s) > 0$. By a Möbius transformation $z \mapsto T(z)$, the osculating circle C_s of γ at $\gamma(s)$ is mapped to a circle $T(C_s)$ by Lemma 4.3.

By Theorem 2.4, C_s and γ have second order contact at $\gamma(s)$. Here, T is a diffeomorphism in a neighborhood of the point. Then the order of contact is preserved by T, because of Lemma 2.6. Hence the circle $T(C_s)$ and the curve $T \circ \gamma(s)$ obtained by a composition of T and γ have second order contact. In particular, $T(C_s)$ is the osculating circle of $T \circ \gamma(s)$.

A point where $\kappa'(s) = 0$ is where C_s and γ have third order contact (cf. Problem **9**-(3) in Section 2). The property of third order contact is preserved by Möbius transformations, as well as the first and second order cases (cf. Problem **9**-(2) in Section 2).

Hence if there exists a point on the curve $T \circ \gamma$ where the derivative of the curvature function vanishes, the circle $T(C_s)$ and $T \circ \gamma$ have third order contact at s. Then, by composing with the inverse transformation of T, one can see that C_s and γ have third order contact at s, which implies $\kappa'(s) = 0$, a contradiction to the assumption that $\kappa'(s) \neq 0$. Therefore $T \circ \gamma$ is a spiral.

Moreover, as we assumed that γ is a positive spiral, γ crosses tangentially to C_s from the right-hand side to the left-hand side at s, because of

Theorem 4.1. This implies that $T \circ \gamma$ is also a positive spiral. □

Using these properties of spirals, we can prove the four-vertex theorem for simple closed curves (not necessarily convex).

Theorem 4.4. *There exist at least four vertices on any simple closed regular curve that is not a circle.*

Proof. Let $\gamma(s)$ be a simple closed regular curve parametrized by arc-length. Similar to the proof of Theorem 2.11 in Section 2, we show a contradiction assuming the number of vertices to be 2.

Let P and Q be the vertices. Without loss of generality, we may assume that the curvature function of γ attains its minimum at P. We denote by γ_1 (resp. γ_2) the subarc from P to Q (resp. Q to P). Then the curvature function is non-decreasing (resp. non-increasing) on γ_1 (resp. γ_2).

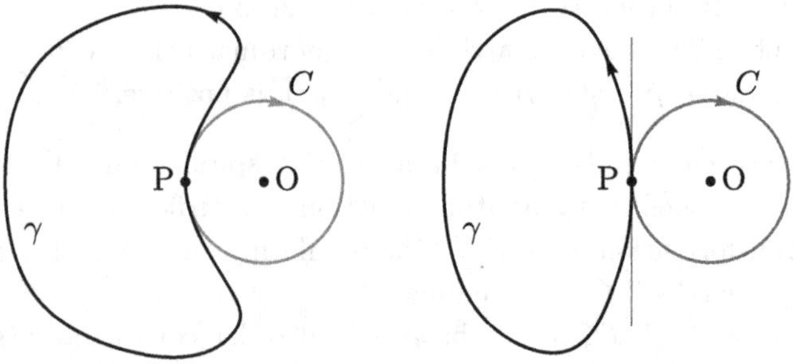

Fig. 4.4

If the curvature function κ of γ is positive at P, then the curve is strictly convex (cf. Corollary B.2.2 in Appendix B.2), and the conclusion follows by Theorem 2.11. So we consider the case that the curvature is non-positive at P. If $\kappa < 0$ at P, we let C be the osculating circle of γ at P (cf. Fig. 4.4, left). On the other hand, if κ vanishes at P, we choose C to be a circle tangent to the osculating line lying in the exterior domain of the curve γ. In this situation, if we give C the orientation that is compatible with the orientation of the curve, then C is clockwise oriented (cf. Fig. 4.4, right). Let O be the center of the circle C. By a homothetic transformation with respect to O, we may assume that C is a unit circle.

Applying the Möbius transformation $T(z) = 1/z$, the image $T(C)(= C)$ is the unit circle with anti-clockwise orientation. If $\kappa < 0$ at P, $T(C)$ itself is the osculating circle of $T \circ \gamma$. On the other hand, if $\kappa = 0$ at P, then the osculating circle lies in the interior of C. Thus, the curvature of $T \circ \gamma$ at $T(\mathrm{P})$ is positive.

Even when the curve has a part of a straight line or a circular arc, where the curvature function is constant (Corollary 2.10 in Section 2), this remains so under Möbius transformations. By Theorem 4.2 and this fact, the property of having a non-decreasing and non-increasing curvature function is preserved by Möbius transformations. In particular, $T(\mathrm{P})$ is the minimum of the curvature function of the curve $T \circ \gamma$. Therefore the curvature of $T \circ \gamma$ is positive. Then $T \circ \gamma$ is a strictly convex curve (cf. Corollary B.2.2 in Appendix B.2), contradicting the four-vertex theorem for strictly convex curves (Theorem 2.11 in Section 2). □

A global property of spirals. Consider a car driving along a positive spiral. If the driver would stop turning the wheel, the orbit of the car would become a circle, which is the osculating circle at that point on the spiral. Then by turning the wheel to the left, the car travels to the inside of the osculating circle. So, one can see that the future osculating circles lie inside the present osculating circle. The following theorem describes this fact rigorously:

Theorem 4.5. *Let $\gamma(s)$ $(a \leq s \leq b)$ be a positive (resp. negative) spiral. For each value of s, let the osculating circle $C(s)$ of γ at s have the same orientation as $\gamma(s)$. Denote by D_s the open domain which lies to the left (resp. right) of $C(s)$. Then, when $t < s$ (resp. $t > s$), it holds that*

$$\overline{D}_s \subset D_t,$$

where $\overline{D}_s$ is the closure of D_s, that is, the union of D_s and its boundary ∂D_s.

Remark 4.6. When the osculating circle is counterclockwise (resp. clockwise) oriented, the interior (resp. exterior) of the circle lies on the left. In particular, if the curvature of γ is positive, D_s is the interior of the osculating circle of γ at $\gamma(s)$ (see Fig. 4.5).

Proof of Theorem 4.5. We shall prove the case that $\gamma(s)$ $(a \leq s \leq b)$ is a positive spiral, and $t = a$, $s = b$. (By reflecting the curve across a line, the conclusion for negative spirals is obtained.) Without loss of generality, the parameter is arc-length and $\kappa(a) > 0$. In fact, when $\kappa(a) \leq 0$, the osculating circle C_a of $\gamma(s)$ at $s = a$ is a clockwise-turning circle. In this case, choose the point O in the right-hand domain of C_a and apply the

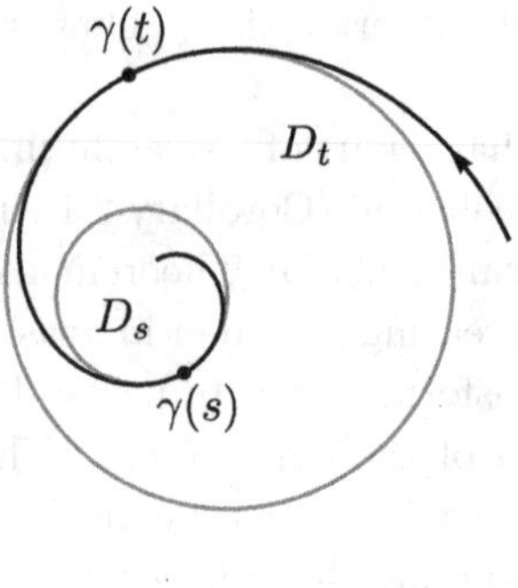

Fig. 4.5 Theorem 4.5.

Möbius transformation $T(z) = 1/z$ with respect to the origin O. Then $T(C_a)$ is a counterclockwise-turning circle and the curvature of $T \circ \gamma(s)$ at $s = a$ is positive.

Thus, we may assume $\kappa(a) > 0$ and prove the case of a positive spiral. Denote the evolute of γ (i.e., the locus of the centers of the osculating circles of $\gamma(s)$, see Appendix B.1) by

$$\sigma(s) := \gamma(s) + \frac{1}{\kappa(s)}\boldsymbol{n}(s)$$

(cf. Theorem B.1.1 of Appendix B.1). Differentiating this, we have $\sigma'(s) = (1/\kappa(s))'\boldsymbol{n}(s)$ by the Frenet formula (2.15).

On the other hand, let r_a and r_b be the radii of the osculating circles C_a and C_b at $s = a$ and $s = b$, respectively. Then, by the triangle inequality for integrals (Theorem A.1.4 in Appendix A.1), we have

$$\begin{aligned}|\sigma(b) - \sigma(a)| &= \left|\int_a^b \sigma'(s)\,ds\right| = \left|\int_a^b \left(\frac{1}{\kappa(s)}\right)' \boldsymbol{n}(s)\,ds\right| \\ &\leq \int_a^b \left|\left(\frac{1}{\kappa(s)}\right)' \boldsymbol{n}(s)\right| ds = -\int_a^b \left(\frac{1}{\kappa(s)}\right)' ds \\ &= \frac{1}{\kappa(a)} - \frac{1}{\kappa(b)} = r_a - r_b.\end{aligned}$$

The condition for equality in the above inequality is that $\boldsymbol{n}(s)$ must be a constant vector, which contradicts that the curvature of $\gamma(s)$ is increasing. Hence we have $|\sigma(b) - \sigma(a)| < r_a - r_b$, which implies that the distance $|\sigma(b) - \sigma(a)|$ between the center of the osculating circles is less than the difference of the radii $r_a - r_b$. Thus C_b lies to the inside of C_a, proving $\overline{D}_b \subset D_a$. □

Corollary 4.7. *A spiral does not have self-intersections.*

Proof. Without loss of generality, we may assume $\gamma(s)$ $(a \leq s \leq b)$ is a positive spiral. If there would be a self-intersection, there exist t, s with $a \leq t < s \leq b$ such that $\gamma(s) = \gamma(t)$. Then by Theorem 4.5, it holds that $\gamma(s) \in \overline{D}_s \subset D_t$, a contradiction to $\gamma(t) \notin D_t$. □

An application of Theorem 4.5 to a generalization of the four-vertex theorem is given by Pinkall [30]. One of the deepest results is about the arrangement of vertices, stated as follows: For each simple closed curve $\gamma : [a, b] \to \boldsymbol{R}^2$, there exist four points $(a \leq)t_1 < t_2 < t_3 < t_4(< b)$ such that the osculating circles at t_1 and t_3 (resp. t_2 and t_4) are inscribed (resp. circumscribed), see [38], and also [20, 39].

Exercises 4

1 Prove that the area bounded by the Archimedean spiral $r = a\theta$ and the half-lines $\theta = \alpha$, $\theta = \beta$ $(0 < \beta - \alpha < 2\pi)$ is $a^2(\beta^3 - \alpha^3)/6$. This fact was discovered by Archimedes himself.

2 Show that a similarity expansion of the logarithmic spiral with respect to the origin coincides with the original spiral up to the rotation about the origin. Moreover, any half-line starting at the origin will make a single constant angle with the logarithmic spiral, independent of choice of direction of the half-line and also of choice of intersection point (see Fig. 4.1).

3 Explain, using the analogy of driving a car, that the positivity and negativity of spirals do not depend on the orientation of curves.

4 Find the curvature functions of the Archimedean spirals and the logarithmic spirals.

5 (1) Show that the number of vertices of the hyperbola $x^2 - y^2 = 1/a^2$ is 2. Here a is a positive constant.

(2) Applying the Möbius transformation $z \mapsto 1/z$ $(z = x + iy)$, show that the lemniscate in Examples 1.1 and 1.3 is mapped to the hyperbola $x^2 - y^2 = 1/a^2$. (In this case, the origin of the lemniscate is mapped to infinity.)

(3) Show that the lemniscate has exactly two vertices.

6 Show that any Möbius transformation can be expressed as a composition of the four transformations in (4.3).

7* Show that the locus of the centers of the osculating circles of the logarithmic spiral (i.e., the evolute, cf. Appendix B.1) is again the logarithmic spiral.

5. Space curves

In this section, we consider curves in 3-dimensional Euclidean space, or more briefly "space curves". To do this, we will need the vector product (or cross product) of two vectors in space (cf. Appendix A.3).

Space curves. Using a parameter t, we can describe a curve in space as

$$\gamma(t) = (x(t), y(t), z(t)).$$

Just like in the case of planar curves, a curve satisfying $\dot{\gamma}(t) \neq \mathbf{0}$ for all t is called a *regular curve.* In this section, we consider only regular space curves.

Similar to the case of planar curves, the length of the part of the curve corresponding to an interval $[a, t]$ is given by

$$s(t) = \int_a^t |\dot{\gamma}(u)|\, du = \int_a^t \sqrt{\dot{x}(u)^2 + \dot{y}(u)^2 + \dot{z}(u)^2}\, du.$$

Like the case of planar curves, we can use this to get an arc-length parameter s.

Now let us consider the space curve $\gamma(s)$ with arc-length parameter s. We will denote the derivative with respect to s with a " $'$ ". Then the same equation as (2.2) in the case of planar curves holds, and

$$\boldsymbol{e}(s) := \gamma'(s) = \frac{\dot{\gamma}}{|\dot{\gamma}|}$$

is a unit vector. We call this the *unit tangent vector* of the curve. Since $\boldsymbol{e}(s)$ is a unit vector, the inner product $\boldsymbol{e}(s) \cdot \boldsymbol{e}(s)$ is identically equal to 1. Differentiating both sides of the equation $\boldsymbol{e}(s) \cdot \boldsymbol{e}(s) = 1$ with respect to s, we have (cf. Corollary A.3.10 in Appendix A.3)

$$\boldsymbol{e}'(s) \cdot \boldsymbol{e}(s) = 0,$$

that is, $\boldsymbol{e}'(s)$ is perpendicular to the unit tangent vector $\boldsymbol{e}(s)$. In the case of planar curves, the direction perpendicular to the curve (more precisely, to the tangent direction of the curve) is uniquely determined. However, this is not the case for space curves. Now, when $\boldsymbol{e}'(s) \neq \mathbf{0}$, we can define

$$\boldsymbol{n}(s) := \frac{\boldsymbol{e}'(s)}{|\boldsymbol{e}'(s)|} \left(= \frac{\gamma''(s)}{|\gamma''(s)|} \right),$$

called the *principal normal vector* to the curve γ. The vector $\boldsymbol{n}(s)$ is a unit vector perpendicular to $\boldsymbol{e}(s)$.

We say that

$$\kappa(s) := |\boldsymbol{e}'(s)| \quad (\text{when } \boldsymbol{e}'(s) = \mathbf{0}, \text{ we set } \kappa(s) = 0),$$

is the *curvature* of the space curve $\gamma(s)$. When $\gamma''(s) = \boldsymbol{e}'(s) = \boldsymbol{0}$, we cannot define the principal normal vector, so we now consider the case that $\gamma''(s)$ never vanishes. Then we have

$$(\gamma''(s) =)\boldsymbol{e}'(s) = \kappa(s)\boldsymbol{n}(s) \qquad (\kappa(s) > 0).$$

Since $\boldsymbol{e}$ and $\boldsymbol{n}$ are a pair of perpendicular unit vectors, using the properties of vector products (see Appendix A.3, page 205), the vector product

$$\boldsymbol{b}(s) := \boldsymbol{e}(s) \times \boldsymbol{n}(s)$$

is a unit vector perpendicular to both $\boldsymbol{e}(s)$ and $\boldsymbol{n}(s)$, and is called the *binormal vector*. For each value of s,

$$\{\boldsymbol{e}(s), \boldsymbol{n}(s), \boldsymbol{b}(s)\}$$

is a positively oriented orthonormal basis for $\boldsymbol{R}^3$, that is, the determinant of the 3×3-matrix $(\boldsymbol{e}(s), \boldsymbol{n}(s), \boldsymbol{b}(s))$ is equal to 1 (cf. Appendix A.3).

For each s, the plane containing the point $\gamma(s)$ and spanned by $\boldsymbol{e}(s)$ and $\boldsymbol{n}(s)$ is called the *osculating plane* of the regular curve γ at s (Fig. 5.1). In addition, the plane containing the point $\gamma(s)$ and spanned by $\boldsymbol{n}(s)$ and $\boldsymbol{b}(s)$ (resp. $\boldsymbol{e}(s)$ and $\boldsymbol{b}(s)$) is called the *normal plane* (resp. the *rectifying plane*) of the curve γ at s (Fig. 5.2).

Here, we define the *torsion* of the curve as

$$\tau(s) := -\boldsymbol{b}'(s) \cdot \boldsymbol{n}(s), \tag{5.1}$$

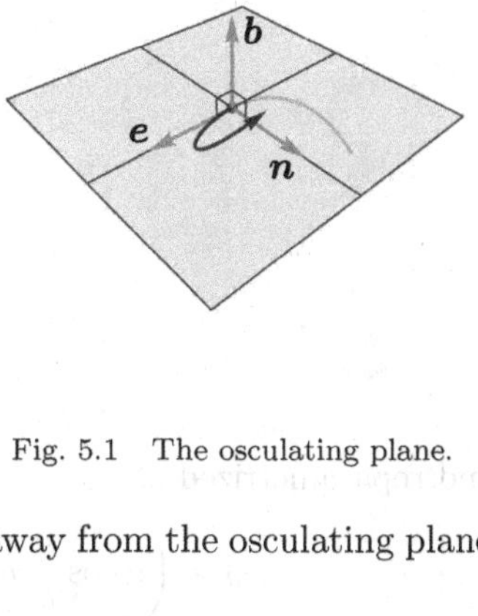

Fig. 5.1 The osculating plane.

which measures how quickly the curve pulls away from the osculating plane (cf. Problem **6** at the end of this section).

If two regular curves differ only by orientation preserving congruences (cf. page 203 in Appendix A.3), then their curvatures and torsions coincide (cf. Problem **1**).

Example 5.1. The curve parametrized by

$$\gamma(t) = (a\cos t, a\sin t, bt)$$

is called a *helix* (Fig. 5.3). Here a and b are both non-zero constants. Because $|\dot\gamma(t)| = \sqrt{a^2+b^2}$, this curve has arc-length parameter

$$s = t\sqrt{a^2+b^2},$$

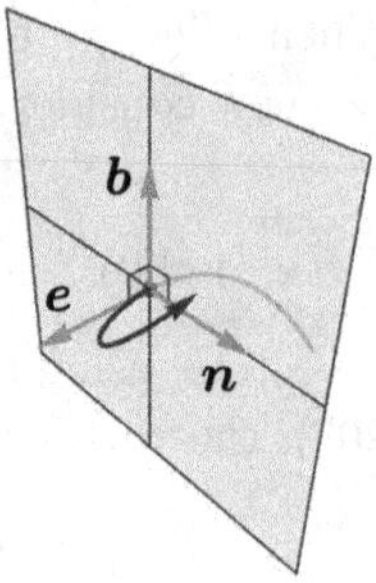

The normal plane.

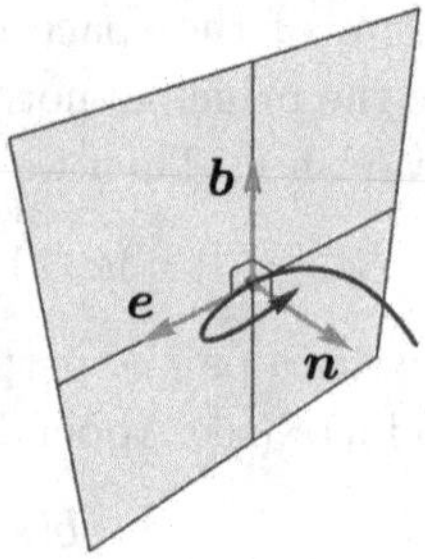

The rectifying plane.

Fig. 5.2 The normal and rectifying planes.

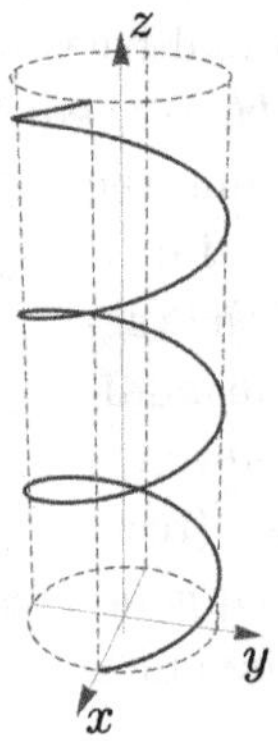

Fig. 5.3 The helix.

and reparametrized as

$$\gamma(s) = \left(a\cos\frac{s}{c}, a\sin\frac{s}{c}, \frac{bs}{c}\right) \qquad (c := \sqrt{a^2+b^2})$$

by the arc-length parameter s. Then we have

$$\boldsymbol{e}(s) = \gamma'(s) = \frac{1}{c}\left(-a\sin\frac{s}{c}, a\cos\frac{s}{c}, b\right),$$
$$\boldsymbol{e}'(s) = \gamma''(s) = \frac{1}{c^2}\left(-a\cos\frac{s}{c}, -a\sin\frac{s}{c}, 0\right)$$

which imply that the curvature κ and normal vector $\boldsymbol{n}$ are

$$\kappa = |\gamma''(s)| = \frac{|a|}{c^2} = \frac{|a|}{a^2+b^2}, \qquad \boldsymbol{n} = \frac{\boldsymbol{e}'(s)}{|\boldsymbol{e}'(s)|} = \frac{a}{|a|}\left(-\cos\frac{s}{c}, -\sin\frac{s}{c}, 0\right),$$

respectively. We then have that the binormal vector $\boldsymbol{b}$ and torsion τ are

$$\boldsymbol{b} = \boldsymbol{e} \times \boldsymbol{n} = \frac{a}{|a|}\frac{1}{c}\left(b\sin\frac{s}{c}, -b\cos\frac{s}{c}, a\right),$$

$$\tau = -\boldsymbol{b}' \cdot \boldsymbol{n} = \frac{b}{c^2} = \frac{b}{a^2+b^2}.$$

In particular, both the curvature and torsion of a helix are constant.

The Frenet-Serret formula. Let $\gamma(s)$ be a regular curve in $\boldsymbol{R}^3$ parametrized by arc-length such that $\kappa(s) > 0$ for each s. Since the unit tangent vector $\boldsymbol{e}(s)$, the principal normal vector $\boldsymbol{n}(s)$, and the binormal vector $\boldsymbol{b}(s)$ form an orthonormal basis for $\boldsymbol{R}^3$, any vector in $\boldsymbol{R}^3$ can be written as a linear combination of them. Therefore $\boldsymbol{n}'$ and $\boldsymbol{b}'$ can be written as

$$\boldsymbol{n}' = \frac{d\boldsymbol{n}}{ds} = \alpha_1\boldsymbol{e} + \alpha_2\boldsymbol{n} + \alpha_3\boldsymbol{b}, \tag{5.2}$$

$$\boldsymbol{b}' = \frac{d\boldsymbol{b}}{ds} = \beta_1\boldsymbol{e} + \beta_2\boldsymbol{n} + \beta_3\boldsymbol{b}, \tag{5.3}$$

where α_j, β_j $(j = 1, 2, 3)$ are real-valued functions depending on s.

Taking inner products of both sides of equation (5.2) with $\boldsymbol{e}$, $\boldsymbol{n}$ and $\boldsymbol{b}$, we find that

$$\begin{aligned}
\alpha_1 &= \boldsymbol{n}' \cdot \boldsymbol{e} = (\boldsymbol{n} \cdot \boldsymbol{e})' - \boldsymbol{n} \cdot \boldsymbol{e}' = -\kappa\boldsymbol{n} \cdot \boldsymbol{n} = -\kappa,\\
\alpha_2 &= \boldsymbol{n}' \cdot \boldsymbol{n} = \frac{1}{2}(\boldsymbol{n} \cdot \boldsymbol{n})' = 0,\\
\alpha_3 &= \boldsymbol{n}' \cdot \boldsymbol{b} = (\boldsymbol{n} \cdot \boldsymbol{b})' - \boldsymbol{n} \cdot \boldsymbol{b}' = \tau.
\end{aligned}$$

Again, taking inner products of both sides of equation (5.3) with $\boldsymbol{e}$, $\boldsymbol{n}$ and $\boldsymbol{b}$, we have

$$\begin{aligned}
\beta_1 &= \boldsymbol{b}' \cdot \boldsymbol{e} = (\boldsymbol{b} \cdot \boldsymbol{e})' - \boldsymbol{b} \cdot \boldsymbol{e}' = -\kappa\boldsymbol{b} \cdot \boldsymbol{n} = 0,\\
\beta_2 &= \boldsymbol{b}' \cdot \boldsymbol{n} = -\tau,\\
\beta_3 &= \boldsymbol{b}' \cdot \boldsymbol{b} = \frac{1}{2}(\boldsymbol{b} \cdot \boldsymbol{b})' = 0.
\end{aligned}$$

Substituting these into (5.2), (5.3) (together with $\boldsymbol{e}' = \kappa\boldsymbol{n}$), we have

$$\begin{cases}
\boldsymbol{e}'(s) = & \kappa(s)\,\boldsymbol{n}(s),\\
\boldsymbol{n}'(s) = -\kappa(s)\,\boldsymbol{e}(s) & \qquad +\tau(s)\,\boldsymbol{b}(s),\\
\boldsymbol{b}'(s) = & -\tau(s)\,\boldsymbol{n}(s).
\end{cases} \tag{5.4}$$

We call these the *Frenet-Serret*[1] *formula.*

[1]Serret, Joseph Alfred (1819–1885). See page 21 for Frenet.

Take the 3×3-matrix $\mathcal{F}(s) := (\boldsymbol{e}(s), \boldsymbol{n}(s), \boldsymbol{b}(s))$, where $\boldsymbol{e}(s)$, $\boldsymbol{n}(s)$ and $\boldsymbol{b}(s)$ are considered as column vectors. Then $\mathcal{F}(s)$ is an orthogonal matrix with determinant 1, since $\{\boldsymbol{e}(s), \boldsymbol{n}(s), \boldsymbol{b}(s)\}$ forms a positive orthonormal basis (a right-handed system, see Appendix A.3) for each s. Using this, the Frenet-Serret formula (5.4) is reformulated as

$$\frac{d\mathcal{F}}{ds} = \mathcal{F}\Omega, \qquad \Omega := \begin{pmatrix} 0 & -\kappa & 0 \\ \kappa & 0 & -\tau \\ 0 & \tau & 0 \end{pmatrix}. \tag{5.5}$$

In Example 5.1, we found the curvature and the torsion by reparametrizing the curve by an arc-length parameter. However, to compute curvatures and torsions for various examples, it is convenient to prepare formulas to compute these quantities for general parameters. Let $\gamma(t)$ be a regular curve parametrized by (not necessarily arc-length) parameter t. Then its unit tangent vector $\boldsymbol{e}(t)$, the principal normal vector $\boldsymbol{n}(t)$ and the binormal vector $\boldsymbol{b}(t)$ are given as

$$\boldsymbol{e} = \frac{\dot\gamma}{|\dot\gamma|}, \qquad \boldsymbol{n} = \frac{(\dot\gamma \times \ddot\gamma) \times \dot\gamma}{|(\dot\gamma \times \ddot\gamma) \times \dot\gamma|}, \qquad \boldsymbol{b} = \frac{\dot\gamma \times \ddot\gamma}{|\dot\gamma \times \ddot\gamma|}. \tag{5.6}$$

Moreover, the curvature and the torsion are expressed as

$$\kappa(t) = \frac{|\dot\gamma(t) \times \ddot\gamma(t)|}{|\dot\gamma(t)|^3}, \qquad \tau(t) = \frac{\det(\dot\gamma(t), \ddot\gamma(t), \dddot\gamma(t))}{|\dot\gamma(t) \times \ddot\gamma(t)|^2} \tag{5.7}$$

(cf. Problem **2**). In particular, when γ is parametrized by arc-length s, (5.7) is rewritten as

$$\kappa(s) = |\gamma''(s)|, \qquad \tau(s) = \frac{\det(\gamma'(s), \gamma''(s), \gamma'''(s))}{\kappa(s)^2}.$$

The fundamental theorem for space curves. We have already seen in Theorem 2.8 in Section 2 that a plane curve is determined uniquely by its curvature function. A similar result is true for space curves.

Theorem 5.2 (The fundamental theorem for space curves). *Let $\kappa(s)$ and $\tau(s)$ be two smooth functions defined on an interval $[a, b]$ such that $\kappa(s) > 0$ holds on $[a, b]$. Then there exists a regular curve $\gamma(s)$ $(a \leq s \leq b)$ parametrized by arc-length s whose curvature and torsion are $\kappa(s)$ and $\tau(s)$, respectively. Moreover, such a curve is unique up to orientation preserving congruences.*

Proof. The theorem can be proven in a similar way as we proved Theorem 2.8 for plane curves. Consider the initial value problem of the ordinary differential equation

$$(5.8) \qquad \mathcal{F}' = \mathcal{F}\Omega, \qquad \mathcal{F}(a) = I, \qquad \Omega = \begin{pmatrix} 0 & -\kappa & 0 \\ \kappa & 0 & -\tau \\ 0 & \tau & 0 \end{pmatrix}$$

whose unknown is the function $\mathcal{F}(s)$ taking values in the set of 3×3-matrices, where I is the 3×3-identity matrix. Since this equation is linear with respect to the nine components of $\mathcal{F}$, by the fundamental theorem for linear ordinary differential equations (Theorem A.2.2 in Appendix A.1), there exists a unique solution $\mathcal{F}(s)$ of (5.8) defined on the interval $[a, b]$. Here, since $\Omega(s)$ is a skew-symmetric matrix, in a similar way to the case for plane curves (cf. (2.17)), we know that $\mathcal{F}(s)$ is an orthogonal matrix with determinant 1. Thus, we write $\mathcal{F}(s) = (\boldsymbol{e}(s), \boldsymbol{n}(s), \boldsymbol{b}(s))$, and let

$$\gamma(s) = \int_{s_0}^{s} \boldsymbol{e}(u)\, du.$$

Then $\gamma(s)$ is the desired curve. The uniqueness part and final part of the proof are left as the exercises for readers. □

A simple closed space curve is called a *knot*. In contrast to the planar curve case that any simple closed planar curve can be deformed continuously to a circle, the knot as in Fig. 5.4 cannot be deformed to a circle. One of the important problems for space curves is to investigate properties of knots. Relationships between knots and their total curvatures (defined in a similar way as that for planar curves, see page 28) are discussed in [26].

Fig. 5.4 The trefoil knot.

Exercises 5

1 Show that the curvature and torsion of a space curve are invariant under orientation preserving congruences. Namely, let Φ be an orientation preserving congruence (cf. page 203 in Appendix A.3). Then the curvature and torsion for a curve $\gamma(s)$ and those for $\Phi \circ \gamma(s)$ coincide. Note that whereas the curvature is invariant under general congruence, the torsion may change sign.

2 Verify (5.6) and (5.7).

3 Let $\gamma(s) = (x(s), y(s))$ be a planar curve parametrized by arc-length s, and assume that the curvature of γ does not vanish. Prove that the curvature of the space curve $\tilde{\gamma}(s) := (x(s), y(s), 0)$ coincides with the absolute value of the curvature of $\gamma(s)$, and the torsion of $\tilde{\gamma}$ is identically zero.

4 Let $\gamma(s)$ be a space curve parametrized by arc-length s whose curvature and torsion are $\kappa(s)$ and $\tau(s)$, respectively. Show that, near $s = 0$, it holds that

$$\gamma(s) = \gamma(0) + s\boldsymbol{e}(0) + \frac{s^2}{2}\kappa(0)\boldsymbol{n}(0) + \frac{s^3}{6}(-\kappa(0)^2\boldsymbol{e}(0) + \kappa'(0)\boldsymbol{n}(0) + \kappa(0)\tau(0)\boldsymbol{b}(0)) + o(s^3),$$

where $o(s^3)$ represents a higher order term (cf. Appendix A.1). This formula is known as *Bouquet's formula*[2]. Compare with Problem **8** in Section 2.

5 Show that a space regular curve whose curvature is a positive constant, and whose torsion is a non-zero constant, coincides with a helix by an orientation preserving congruence.

6* Show that a space regular curve whose torsion vanishes identically (and having positive curvature) coincides with a curve obtained as in Problem **3** from a planar curve, up to an orientation preserving congruence.

7* Show that a regular curve that lies on a sphere and has constant curvature is a part of a circle.

8* Let $\gamma_1(t)$ be a planar regular curve with curvature $\kappa_1(t)$, and $\gamma_2(t)$ a curve obtained by reflecting $\gamma_1(t)$ across a line. Show that the curvature of $\gamma_2(t)$ is $-\kappa_1(t)$ (cf. Problem **4** in Section 2). Let $\tilde{\gamma}_1(t)$ and $\tilde{\gamma}_2(t)$ be space curves obtained from γ_1 and γ_2, respectively, as in Problem **3**. Then show that the curvature functions of these curves are $|\kappa_1|$, and the torsion functions vanish identically.

Here, since γ_1 and γ_2 have different curvature functions as planar curves, they cannot coincide by rotations and translations. In contrast, the space curves $\tilde{\gamma}_1$ and $\tilde{\gamma}_2$ have the same curvature and torsion, and hence by the fundamental theorem for space curves (Theorem 5.2), they coincide by a rotation and translation (an orientation preserving congruence). *Explain the reason for this difference between planar curves and space curves.*

[2]Bouquet, Jean Claude (1819–1885).

Chapter II

Surfaces

We introduce parametrizations of surfaces like we did for curves, and define the first and second fundamental forms. Using these, the Gaussian curvature and mean curvature, which do not depend on choice of parametrization, are introduced, and geometric meanings of these curvatures are explained. Moreover, the principal and asymptotic directions of surfaces, which play important roles in analyzing the shapes of surfaces, are introduced. As an application, we introduce the Gauss-Bonnet theorem, which measures the difference between π and the sum of interior angles of a triangle on a surface.

6. What exactly is a "surface"?

Things like the surface of an uncracked egg and the body of a car are what we ordinarily imagine to be surfaces. Here we introduce a way to represent such surfaces mathematically.

Graphs of functions of two variables. A function $f(x, y)$ is called *smooth* if it is C^∞-differentiable with respect to the variables x, y. For a function $f(x, y)$ defined on a region D lying in the plane $\boldsymbol{R}^2$ (we call f a *function of two variables*), the set of points $(x, y, f(x, y))$ for (x, y) in D forms a surface. This is the *graph* of $z = f(x, y)$.

Example 6.1. (1) The graph of the function $f(x, y) = ax + by + c$ is a plane, where a, b, c are real constants.

(2) The graph of the function

$$z = \frac{x^2}{a^2} + \frac{y^2}{b^2} \qquad (a,\ b \text{ are positive constants}) \tag{6.1}$$

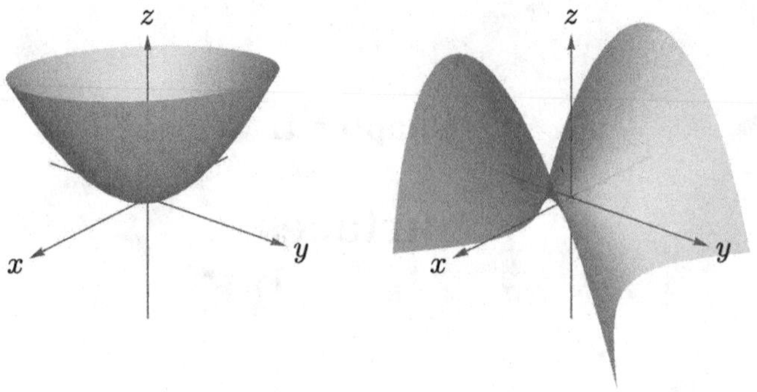

An elliptic paraboloid. A hyperbolic paraboloid.

Fig. 6.1

is called an *elliptic paraboloid* (see the left side of Fig. 6.1). Cutting this elliptic paraboloid with a plane parallel (resp. perpendicular) to the z-axis, one gets a parabola (resp. ellipse). In the case that $a = b$, we have a *paraboloid of revolution*, that is, the surface of revolution created by rotating the parabola $z = x^2/a^2$ in the xz-plane about the z-axis. Using a rotationally symmetric paraboloid as a mirror, all light coming in parallel to the axis of rotation is reflected so that it passes through, and can all be collected at, the point $(0, 0, a^2/4)$. A parabolic reflector makes use of this shape.

(3) The graph of the function

$$z = \frac{x^2}{a^2} - \frac{y^2}{b^2} \qquad (a,\ b \text{ are positive constants}) \tag{6.2}$$

is called a *hyperbolic paraboloid* (see the right side of Fig. 6.1). Intersecting this surface with a plane parallel to the z-axis, the intersection is a parabola. Intersecting it with a plane perpendicular to the z-axis, the intersection is a hyperbola, in the particular case that this plane is the xy-plane, the intersection becomes two intersecting lines.

Surfaces expressed by the implicit function $F(x, y, z) = 0$. For a function $F(x, y, z)$ of three variables, the set of points (x, y, z) in space that satisfy $F(x, y, z) = 0$ can form a surface. For example, the set of points satisfying $ax + by + cz + d = 0$ form a plane, where $(a, b, c) \neq (0, 0, 0)$.

Example 6.2. (1) The surface determined by the equation

$$\frac{x^2}{a^2} + \frac{y^2}{b^2} + \frac{z^2}{c^2} = 1 \qquad (a,\ b \text{ and } c \text{ are positive reals}) \tag{6.3}$$

is called an *ellipsoid* (see the upper-left of Fig. 6.2). In the case that $a = b$, this surface is symmetric with respect to the z-axis, called a *rotationally symmetric ellipsoid.* In the case that $a = b = c$, the surface becomes a *sphere* of radius a.

(2) The surface determined by

$$(6.4)\qquad \frac{x^2}{a^2} + \frac{y^2}{b^2} - \frac{z^2}{c^2} = 1 \qquad (a,\ b \text{ and } c \text{ are positive reals})$$

is called a *hyperboloid of one sheet* (see the upper-right of Fig. 6.2). Intersecting this surface with a plane parallel to (resp. perpendicular to) the z-axis, the intersection set is a hyperbola (resp. ellipse).
In the case of

$$(6.5)\qquad \frac{x^2}{a^2} + \frac{y^2}{b^2} - \frac{z^2}{c^2} = -1 \qquad (a,\ b \text{ and } c \text{ are positive reals})\ ,$$

one has a *hyperboloid of two sheets* (see the lower-left of Fig. 6.2). Just like for the case of a hyperboloid of one sheet, intersecting the hyperboloid of two sheets with a plane parallel to (resp. perpendicular to) the z-axis, the intersection set is a hyperbola (resp. ellipse).

(3) The surface determined by

$$(6.6)\quad (x^2 + y^2 + z^2)^2 - a^2(z^2 - x^2 - y^2) = 0 \qquad (a \text{ is a positive real})\ ,$$

as shown in the lower-right of Fig. 6.2, is a surface that is the image of the lemniscate rotated about its axis (cf. Section 1). The figure looks like an hourglass.

Like in the previous case of a function of two variables, a function $F(x, y, z)$ of three variables is called *smooth* if it is C^∞-differentiable with respect to the variables x, y, z. When a surface is given as the solution set of the equation $F(x, y, z) = 0$, if

$$F_x(x_0, y_0, z_0) = F_y(x_0, y_0, z_0) = F_z(x_0, y_0, z_0) = 0$$
$$\left(F_x = \frac{\partial F}{\partial x},\quad F_y = \frac{\partial F}{\partial y},\quad F_z = \frac{\partial F}{\partial z}\right),$$

then the point (x_0, y_0, z_0) is called a *singular point* of the surface. It is possible that the surface will not look smooth at a singular point, in spite of the fact that it is represented by a function of class C^∞. In fact, the rotated lemniscate given by (6.6) has a singular point at the origin. In general, there are numerous possibilities for the shape of a surface near a singular point.

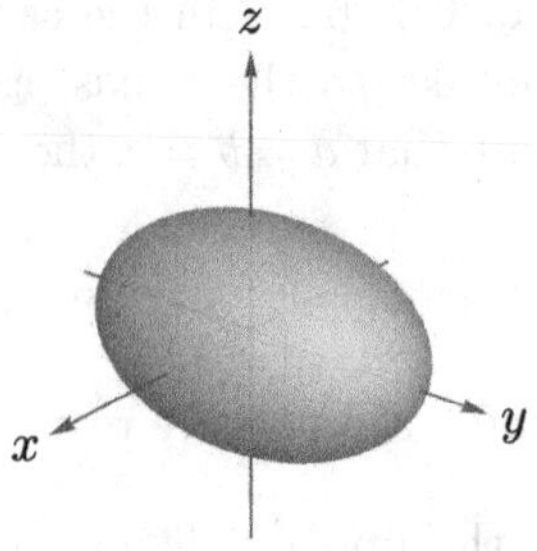

An ellipsoid.

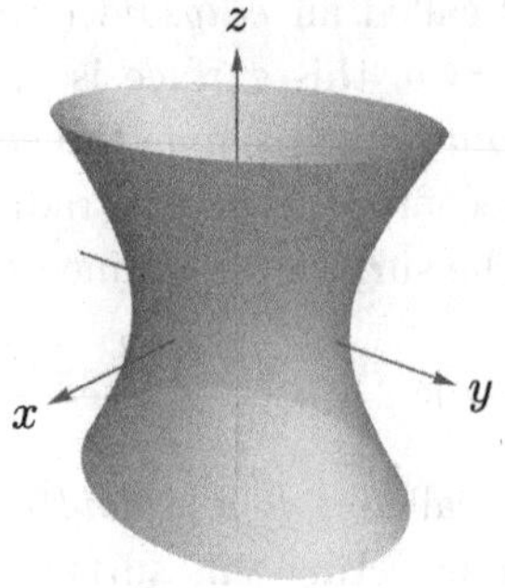

A hyperboloid of one sheet.

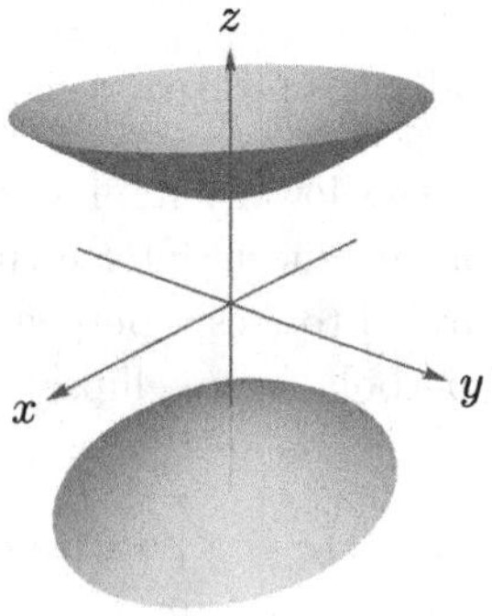

A hyperboloid of two sheets.

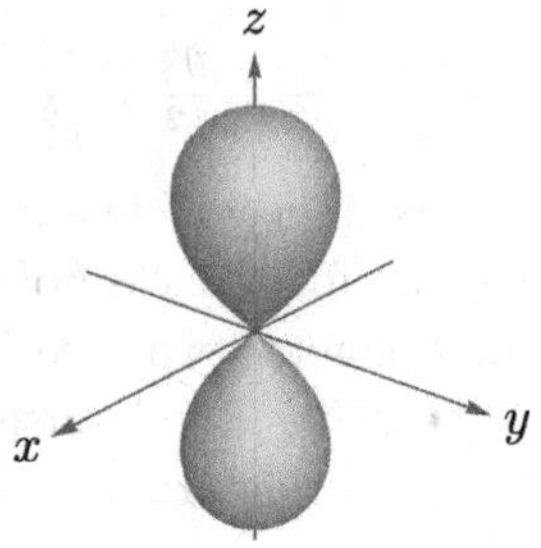

The rotated lemniscate.

Fig. 6.2

On the other hand, away from singular points, the surface has a smooth shape without self-intersections. For example, if $F_z(x_0, y_0, z_0) \neq 0$, then the surface can be written as a smooth graph $z = f(x, y)$ near the point (x_0, y_0, z_0) by the implicit function theorem (cf. Theorem A.1.6 of Appendix A.1).

Surfaces via parametrizations. Though a curve is expressible with one variable, a parametrization of a surface generally needs two variables. Collecting three functions $(x(u, v), y(u, v), z(u, v))$, each of two variables, each pair (u, v) gives a point in the space $\boldsymbol{R}^3$. Moving (u, v) in a region D in the uv-plane, the points $(x(u, v), y(u, v), z(u, v))$ might likely carve out a surface in $\boldsymbol{R}^3$. Such a representation is called the *parametrization* of a surface, and the pair (u, v) of variables is called *parameters.* The examples of surfaces given in Example 6.2 can be represented in this way as follows:

Ellipsoid:

$$x := a\cos u\cos v,\ y := b\cos u\sin v,\ z := c\sin u;$$
$$D := \left\{(u,v)\ \middle|\ |u| < \frac{\pi}{2},\ 0 \le v < 2\pi\right\}.$$

Hyperboloid of one sheet:

$$x := a\cosh u\cos v,\quad y := b\cosh u\sin v,\quad z := c\sinh u;$$
$$D := \{(u,v)\,|\,u \in \boldsymbol{R},\ 0 \le v < 2\pi\}.$$

Upper half of the hyperboloid of two sheets:

$$x := a\sinh u\cos v,\quad y := b\sinh u\sin v,\quad z := c\cosh u;$$
$$D := \{(u,v)\,|\,u \in \boldsymbol{R},\ 0 \le v < 2\pi\}.$$

The rotated lemniscate:

$$x := \frac{a\sin u\cos u\cos v}{1+\sin^2 u},\quad y := \frac{a\sin u\cos u\sin v}{1+\sin^2 u},\quad z := \frac{a\cos u}{1+\sin^2 u};$$
$$D := \{(u,v)\,|\,0 \le u \le \pi,\ 0 \le v < 2\pi\}.$$

Here we can describe the surfaces by collecting the three functions of two variables into a position vector

$$\boldsymbol{p}(u,v) = (x(u,v), y(u,v), z(u,v)),$$

where (u,v) is a point in a region D in the uv-plane $\boldsymbol{R}^2$. We now assume that $\boldsymbol{p}(u,v)$ is smooth, that is, the three functions $x(u,v), y(u,v), z(u,v)$ are smooth functions on D. A point where the derivative vectors

$$\boldsymbol{p}_u = \frac{\partial \boldsymbol{p}}{\partial u} = (x_u, y_u, z_u), \qquad \boldsymbol{p}_v = \frac{\partial \boldsymbol{p}}{\partial v} = (x_v, y_v, z_v)$$

are linearly independent (resp. linearly dependent) is called a *regular point* (resp. *singular point*). Similar to a surface determined by an implicit function $F(x,y,z) = 0$, a parametrization of a surface might not form an apparently smooth surface near a singular point. For example, the points corresponding to $u = \pi/2$ satisfy $\boldsymbol{p}_v = \boldsymbol{0}$. These points correspond to the hourglass point in Fig. 6.2. There are numerous possibilities of shapes near singular points. See the paragraph "Singularities on surfaces" at the end of this section.

A parametrization $\boldsymbol{p}(u,v)$ defined on a region D of the uv-plane is said to be a *regular surface* if

(6.7) $\boldsymbol{p}_u(u,v)$ and $\boldsymbol{p}_v(u,v)$ are linearly independent for each $(u,v) \in D$.

The region D is called the *region determined by the parametrization.* In the surfaces in Fig. 6.2, the ellipsoid, the hyperboloid of one sheet, and the hyperboloid of two sheets are regular, but the rotated lemniscate admits a cone-like singular point. In this book, a *surface* means a regular surface unless otherwise noted.

For a smooth function $f(x, y)$ of two variables,

$$\boldsymbol{p}(u, v) := (u, v, f(u, v)) \tag{6.8}$$

represents the graph of a function $z = f(x, y)$. Since $\boldsymbol{p}_u = (1, 0, f_u)$ and $\boldsymbol{p}_v = (0, 1, f_v)$ are linearly independent, this is a special case of a regular surface.

Given a regular parametrization $\boldsymbol{p}(u, v)$, and fixing the value for v, we have a map $u \mapsto \boldsymbol{p}(u, v)$ that gives a curve lying on the surface. We call this a *u-curve*. Likewise, if we instead fix u, the curve $v \mapsto \boldsymbol{p}(u, v)$ is a *v-curve* of the surface. These two families of curves give what can be thought of as a net over the surface (see Fig. 6.3). At each point in each u-curve, the vector $\boldsymbol{p}_u(u, v)$ is the velocity vector of the u-curve through that point, and likewise $\boldsymbol{p}_v(u, v)$ is the velocity vector of each v-curve at any given point.

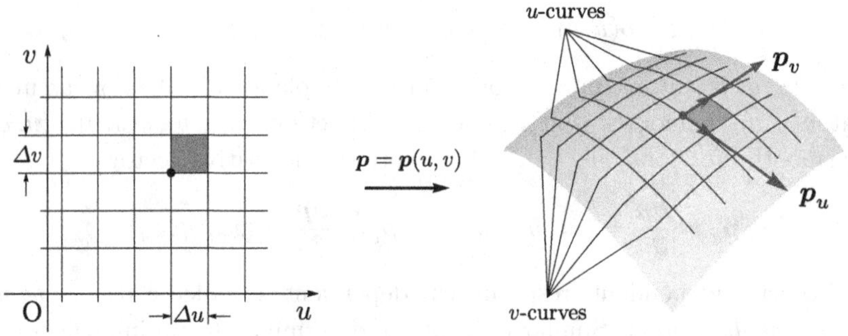

Fig. 6.3 The u-curves and the v-curves; a small quadrilateral on a surface.

At each point $\boldsymbol{p} = \boldsymbol{p}(u, v)$ on the surface, the tangent vectors can be written as linear combinations of $\boldsymbol{p}_u(u, v)$ and $\boldsymbol{p}_v(u, v)$. Hence the tangent plane of the surface at $\boldsymbol{p}(u, v)$ is expressed as

$$\{\boldsymbol{p}(u, v) + s\boldsymbol{p}_u(u, v) + t\boldsymbol{p}_v(u, v) \mid s, t \in \boldsymbol{R}\},$$

that is, the plane passing through the point $\boldsymbol{p}(u, v)$ and parallel to both $\boldsymbol{p}_u(u, v)$ and $\boldsymbol{p}_v(u, v)$. There is a $\pm$-ambiguity for the choice of the unit vector perpendicular to both $\boldsymbol{p}_u(u,v)$ and $\boldsymbol{p}_v(u,v)$, called the *unit normal vector*. One choice of this vector is given by

$$\nu := \frac{\boldsymbol{p}_u \times \boldsymbol{p}_v}{|\boldsymbol{p}_u \times \boldsymbol{p}_v|}. \tag{6.9}$$

Here the symbol "×" denotes the vector product, or outer product (see Appendix A.3, page 205). For example, the graph of the function $f(x,y)$ given by (6.8) has the unit normal vector

$$\nu = \frac{1}{\sqrt{1+f_x^2+f_y^2}}(-f_x, -f_y, 1). \tag{6.10}$$

A given surface might not be represented by a single parametrization in general. For example, in the parametrization of an ellipsoid in page 61, $\boldsymbol{p}_v = \boldsymbol{0}$ at the points where $u = \pm\pi/2$, that is, the condition (6.7) does not hold. In other words, the points $(0,0,\pm c)$ cannot be treated with this parametrization. However, writing this ellipsoid as two graphs

$$z = \pm c\sqrt{1 - \frac{x^2}{a^2} - \frac{y^2}{b^2}}, \tag{6.11}$$

we obtain parametrizations of the ellipsoid on a neighborhood of $(0,0,\pm c)$, which are excluded in the parametrization on page 61.

In this way, more than one parametrization may be needed to express the entirety of the surface. In this case, a pair (u,v) of each parametrization $\boldsymbol{p}(u,v)$ of the surface is called a *local coordinate system*, and a *coordinate change* is a change of parametrization to another one (cf. Problems **4** and **7**). The unit normal vector given in (6.9) is invariant up to $\pm$-ambiguity under coordinate changes, because it gives the direction perpendicular to the tangent space of the surface (cf. Problem **4**).

The area of a surface. We now show how to determine the area of a surface $\boldsymbol{p}(u,v)$ given as a map from a bounded closed region[1] $\overline{D}$, that is, the map $\boldsymbol{p}$ can be extended to an open domain containing $\overline{D}$. In the uv-plane, where $\overline{D}$ lies, let Δu and Δv denote small distances in the u and v directions, respectively. Then $[u, u+\Delta u] \times [v, v+\Delta v]$ is a small rectangle in the uv-plane and we now approximate the area of its image under $\boldsymbol{p}$ (see Fig. 6.3). By Taylor's formula (see Theorem A.1.2 in Appendix A.1), we have

$$\begin{aligned}
\boldsymbol{p}(u+\Delta u, v) - \boldsymbol{p}(u,v) &\approx \boldsymbol{p}_u(u,v)\Delta u, \\
\boldsymbol{p}(u, v+\Delta v) - \boldsymbol{p}(u,v) &\approx \boldsymbol{p}_v(u,v)\Delta v, \\
\boldsymbol{p}(u+\Delta u, v+\Delta v) - \boldsymbol{p}(u,v) &\approx \boldsymbol{p}_u(u,v)\Delta u + \boldsymbol{p}_v(u,v)\Delta v\,.
\end{aligned}$$

[1]For a region (a connected open set, cf. page 194 in Appendix A.1) D, the union $\overline{D}$ of D and its boundary ∂D is called a *closed region* or a *closed domain*.

Here, the symbol $\approx$ means that the left-hand side and the right-hand side are approximately equal.

Hence the region of the surface under consideration is approximately a quadrilateral with vertices

$$\begin{aligned} &\boldsymbol{p}(u,v), && \boldsymbol{p}(u,v)+\boldsymbol{p}_u(u,v)\Delta u,\\ &\boldsymbol{p}(u,v)+\boldsymbol{p}_v(u,v)\Delta v, && \boldsymbol{p}(u,v)+\boldsymbol{p}_u(u,v)\Delta u+\boldsymbol{p}_v(u,v)\Delta v. \end{aligned}$$

The area of this quadrilateral is

$$|(\boldsymbol{p}_u(u,v)\Delta u)\times(\boldsymbol{p}_v(u,v)\Delta v)| = |\boldsymbol{p}_u(u,v)\times\boldsymbol{p}_v(u,v)|\,\Delta u\,\Delta v,$$

so, by adding up the areas of such quadrilaterals and taking the limit as Δu and Δv approach zero, we find that the area of the image of $\overline{D}$ under $\boldsymbol{p}$ is[2]

$$\iint_{\overline{D}} |\boldsymbol{p}_u(u,v)\times\boldsymbol{p}_v(u,v)|\,du\,dv. \tag{6.12}$$

We set the inside of the integral (6.12) as

$$dA := |\boldsymbol{p}_u(u,v)\times\boldsymbol{p}_v(u,v)|\,du\,dv \tag{6.13}$$

and call it the *area element* of the surface. It can be easily proved that this area does not depend on parameters, using the formula for changing variables of double integrals (Theorem A.1.7 in Appendix A.1). In particular, the area of the surface given by a graph $z = f(x,y)$ $\big((x,y)\in\overline{D}\big)$ is given by

$$\iint_{\overline{D}} \sqrt{1+f_x^2+f_y^2}\,dx\,dy\,, \tag{6.14}$$

see Problem **3**.

For a given closed regular surface, we can divide it into a finite union of closed parametrized regions so that any two adjacent regions have no common interior points. Summing up the areas of each of these closed regions, we can define the (total) area of the surface.

Remark. In this text, we mainly consider regular surfaces without self-intersections. However, like with curves having self-intersections, there are important examples of regular surfaces with self-intersections as follows:

The *Klein*[3] *bottle* as in the top figure of Fig. 6.4 having self-intersections is an example of a closed surface for which one cannot distinguish between the front and back. When a surface has self-intersections, for a person

[2]We do not give a rigorous proof of this formula. For a proof, see [24].

[3]Klein, Felix C. (1849–1925).

traveling on the surface who encounters a self-intersection, we consider the other local piece of the surface as a "wall" which he can pass through but cannot transfer onto.

Surprisingly, there is a closed regular surface with constant mean curvature (the mean curvature will be defined in Section 8) that is different from the sphere. In fact, the bottom figure of Fig. 6.4, called the *Wente torus*, which was given by Wente[4] in 1986, is a closed surface with self-intersections, and topologically a torus (cf. Problem **1**).

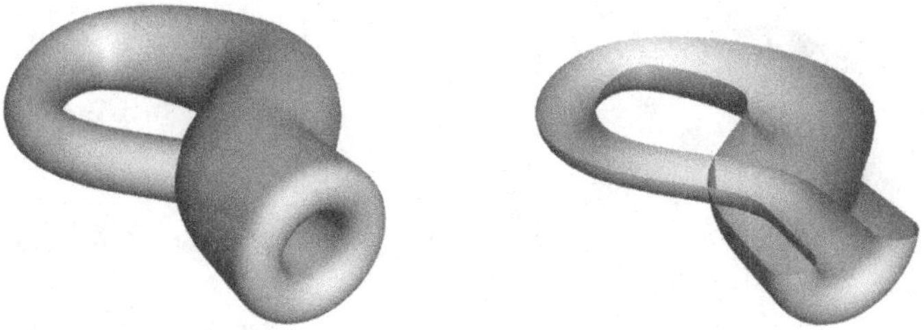

A Klein bottle.

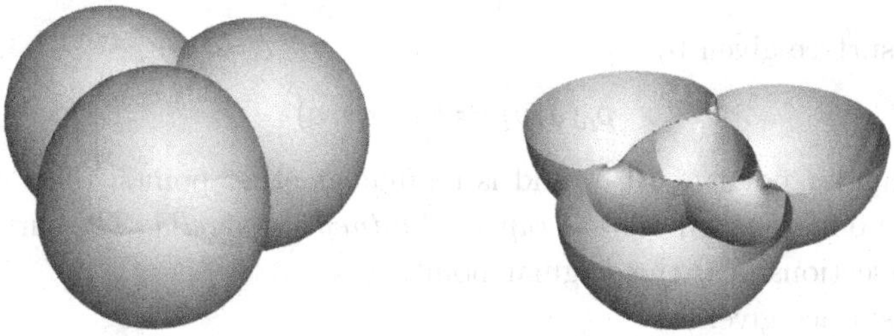

The Wente torus.

Fig. 6.4 Surfaces having self-intersections.

Singularities on surfaces. Various kinds of singularities can appear on surfaces. As explained below, self-intersections frequently appear in surfaces with singularities. We introduce some typical singularities here (Fig. 6.5).

---

[4]Wente, Henry C.

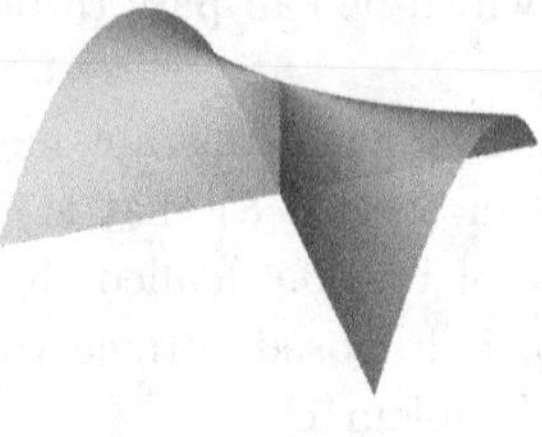

The cross cap.

The cuspidal edges.

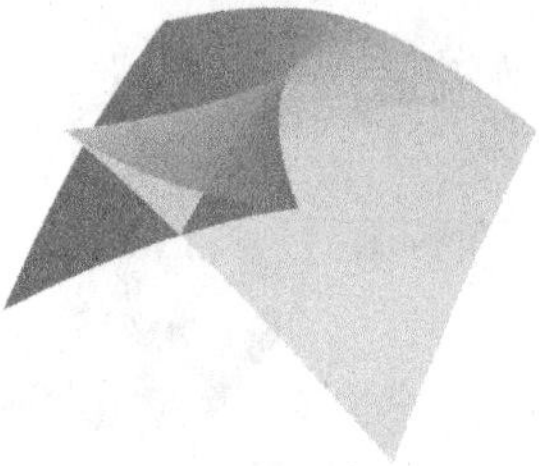

The swallowtail.

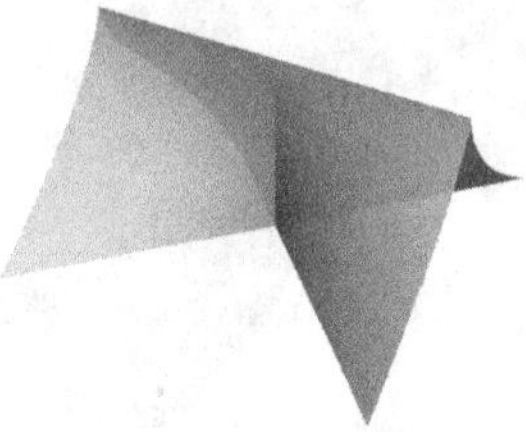

The cuspidal cross cap.

Fig. 6.5 Singularities on surfaces.

The surface given by

$$\boldsymbol{p}_0(u, v) := (u^2, uv, v) \tag{6.15}$$

has a singular point at $(0, 0)$ and is regular at other points. This isolated singular point is called a *cross cap* or *Whitney's umbrella*. The surface has self-intersections near the singular point.

The surface given by

$$\boldsymbol{p}_1(u, v) := (u, v^2, v^3) \tag{6.16}$$

has singular points on the u-axis in the uv-plane. These singular points of p_1 are called *cuspidal edges*.

The surface

$$\boldsymbol{p}_2(u, v) := (3u^4 + u^2 v, 4u^3 + 2uv, v) \tag{6.17}$$

has singular points along the parabola $6u^2 + v = 0$ in the uv-plane. In particular, the singular point corresponding the origin is called a *swallowtail*. Singularities other than at the origin are all cuspidal edges. The swallowtail looks like a cusp of cuspidal edges, and the surface has self-intersections.

The set of singular points of the surface

$$\boldsymbol{p}_3(u,v) := (u, v^2, uv^3)$$

is the u-axis in the uv-plane. The singular point corresponding to the origin is called a *cuspidal cross cap*, which looks like a hybrid of a cross cap and a cuspidal edge. The singular points other than at the origin are all cuspidal edges.

Among the four types of singularities here, criteria for cross caps, cuspidal edges and swallowtails are introduced in Appendix B.9.

Exercises 6

1 Rotating a circle in the xz-plane centered at $(a,0,0)$ with radius b about the z-axis, a torus as in the following figure is obtained. Here the constants a, b satisfy $0 < b < a$. Find a parametrization of such a surface.

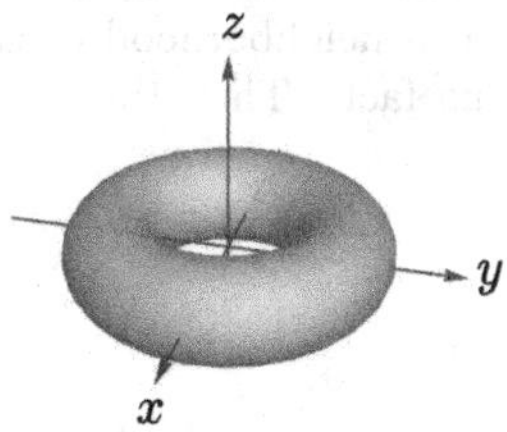

Torus.

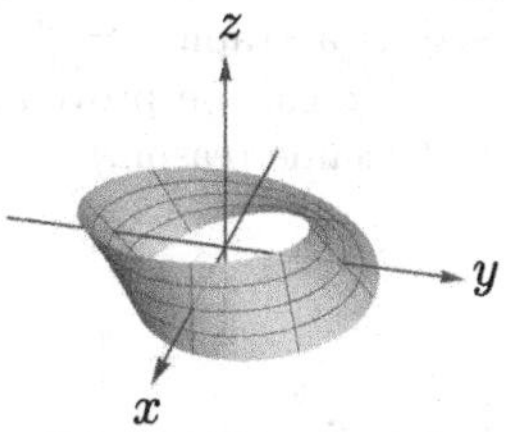

Möbius strip.

2 Find a parametrization of a Möbius strip as in the figure above, whose centerline C is a circle in the xy-plane centered at the origin with radius 2, and is foliated by line segments perpendicular to C and turning at constant speed along C.

3 Derive (6.14) from (6.12). Using this, prove that the area of the sphere with radius R is $4\pi R^2$.

4 Consider a diffeomorphism (cf. page 18 in Section 2)

$$(\xi, \eta) \longmapsto (u(\xi,\eta), v(\xi,\eta))$$

from a region $\widetilde{D}$ in the $\xi\eta$-plane to a region D in the uv-plane. For a parametrization $\boldsymbol{p}(u,v)$ of a surface defined on D, $\tilde{\boldsymbol{p}}(\xi,\eta) = \boldsymbol{p}(u(\xi,\eta), v(\xi,\eta))$ gives another parametrization of the same surface as $\boldsymbol{p}$. When $\tilde{\boldsymbol{p}}(\xi,\eta)$ is obtained by a coordinate change from $\boldsymbol{p}(u,v)$, we often denote it by $\boldsymbol{p}(\xi,\eta)$ instead of $\boldsymbol{p}(u,v)$, as long as there is no risk of confusion. In the situation above, prove that

$$\frac{\boldsymbol{p}_u \times \boldsymbol{p}_v}{|\boldsymbol{p}_u \times \boldsymbol{p}_v|} = \pm \frac{\boldsymbol{p}_\xi \times \boldsymbol{p}_\eta}{|\boldsymbol{p}_\xi \times \boldsymbol{p}_\eta|},$$

where the sign coincides with that of the determinant of the Jacobi matrix of the coordinate change.

5 Show that the area of a surface does not depend on choice of parametrization, that is, prove that, under a parameter change as in Problem **4**, the areas computed by (6.12) in each parametrization coincide, using the substitution rule for double integrals.

6 Show that the area of a surface is invariant under congruence. That is, verify that a parametrized surface $\boldsymbol{p}(u,v)$ and the surface given by $\boldsymbol{p}(u,v) + \boldsymbol{q}$ have the same area, where T is an orthogonal matrix and $\boldsymbol{q}$ is a constant vector.

7* Assume a parametrization $\boldsymbol{p}(u,v)$ of a regular surface defined on a region D of the uv-plane and a parametrization $\tilde{\boldsymbol{p}}(\xi,\eta)$ on a region $\widetilde{D}$ of the $\xi\eta$-plane give the same surface (image surface). Then for each point (ξ,η) in $\widetilde{D}$, there exists the unique point (u,v) in D satisfying $\tilde{\boldsymbol{p}}(\xi,\eta) = \boldsymbol{p}(u,v)$. Prove that the correspondence $\varphi\colon (\xi,\eta) \mapsto (u(\xi,\eta), v(\xi,\eta))$ is a diffeomorphism from $\widetilde{D}$ to D. (Hint: As shown in the proof in Theorem 8.7 on page 82, a surface can be expressed as a graph $z = f(x,y)$ in a neighborhood of each point. The smoothness of φ can be proved by this fact. Then the inverse φ^{-1} is also smooth by the same reason.)

7. The first fundamental form

In this section, we introduce the first fundamental form, which encodes the information needed to determine quantities such as infinitesimal length and angles on the surface. In preparation for this, we first describe the exterior derivative (or the total differential) of functions.

The exterior derivative of functions. Consider a function $f(u,v)$ depending on two variables. Moving a point (u,v) to another nearby point $(u+\Delta u, v+\Delta v)$, the change in the value of f is

$$\Delta f = f(u+\Delta u, v+\Delta v) - f(u,v),$$

and by Taylor's formula (Theorem A.1.2 in Appendix A.1), when Δu, Δv are sufficiently small, we have the approximation

$$\Delta f \approx f_u(u,v)\,\Delta u + f_v(u,v)\,\Delta v.$$

We write this approximation symbolically as

$$df(u,v) = f_u(u,v)\,du + f_v(u,v)\,dv \tag{7.1}$$

and we call df the *exterior derivative* or the *total differential* of f. Here we are thinking of du, dv simply as symbols.

Having defined this exterior derivative df, we will see below why it is independent of the choice of coordinates. Considering the change of coordinates (cf. page 18 in Section 2)

$$u = u(\xi, \eta), \qquad v = v(\xi, \eta),$$

since the u, v are functions of ξ, η, the definition (7.1) of the exterior derivative gives

$$du(\xi,\eta) = u_\xi(\xi,\eta)\, d\xi + u_\eta(\xi,\eta)\, d\eta,$$
$$dv(\xi,\eta) = v_\xi(\xi,\eta)\, d\xi + v_\eta(\xi,\eta)\, d\eta.$$

Substituting these into (7.1) and applying the chain rule for partial derivatives of functions of several variables, we have

$$\begin{aligned} df &= f_u\, du + f_v\, dv \\ &= f_u(u_\xi\, d\xi + u_\eta\, d\eta) + f_v(v_\xi\, d\xi + v_\eta\, d\eta) \\ &= (f_u u_\xi + f_v v_\xi)\, d\xi + (f_u u_\eta + f_v v_\eta)\, d\eta \\ &= f_\xi\, d\xi + f_\eta\, d\eta. \end{aligned}$$

Thus we have now checked that the exterior derivative $f_\xi\, d\xi + f_\eta\, d\eta$ of f with respect to the $\xi\eta$-plane is the same as df in (7.1).

Like the properties of derivatives of functions of a single variable, the exterior derivative satisfies the following (see Problem **3** at the end of this section):

(7.2) $$d(f+g) = df + dg,$$

(7.3) $$d(fg) = g\, df + f\, dg,$$

(7.4) $$d(\varphi(f)) = \dot{\varphi}(f)\, df.$$

Here $\varphi = \varphi(t)$ is a smooth function of one variable, and $\dot{\varphi}(t)$ is its derivative. For specific calculations, one may use these formulas.

Now we will check, for an explicit example, that the exterior derivative is independent of change of coordinates.

Example 7.1. The exterior derivative of the function $f(u,v) := u^2 - v^2$ is given by $df(u,v) = 2(u\, du - v\, dv)$. Setting

(7.5) $$\xi := u + v, \qquad \eta := u - v,$$

f is represented as $f(\xi,\eta) = \xi\eta$ with respect to (ξ,η), and so we have

(7.6) $$df = \eta\, d\xi + \xi\, d\eta.$$

By (7.5),

$$u = \frac{\xi + \eta}{2}, \qquad v = \frac{\xi - \eta}{2}. \tag{7.7}$$

The exterior derivatives of these are

$$du = \frac{d\xi + d\eta}{2}, \qquad dv = \frac{d\xi - d\eta}{2}.$$

Inserting these into $df = 2(u\,du - v\,dv)$, we arrive at (7.6), so we have checked by direct computation that df is invariant under this particular change of coordinates.

The first fundamental form. Let $\boldsymbol{p}(u,v)$ be a regular surface defined on a domain, and Δs the distance between the points $\boldsymbol{p}(u,v)$ and $\boldsymbol{p}(u+\Delta u, v+\Delta v)$. Then for sufficiently small Δu and Δv, the square $(\Delta s)^2$ is approximated as

$$\begin{aligned}(\Delta s)^2 &= |\boldsymbol{p}(u+\Delta u, v+\Delta v) - \boldsymbol{p}(u,v)|^2 \\ &\approx |\boldsymbol{p}_u(u,v)\Delta u + \boldsymbol{p}_v(u,v)\Delta v|^2 \\ &= (\boldsymbol{p}_u(u,v)\cdot\boldsymbol{p}_u(u,v))(\Delta u)^2 + 2(\boldsymbol{p}_u(u,v)\cdot\boldsymbol{p}_v(u,v))\Delta u\Delta v \\ &\quad + (\boldsymbol{p}_v(u,v)\cdot\boldsymbol{p}_v(u,v))(\Delta v)^2,\end{aligned}$$

by Taylor's formula (Theorem A.1.2 in Appendix A.1), where " $\cdot$ " is the inner product of $\boldsymbol{R}^3$ (see Appendix A.3).

Similar to the case of exterior derivatives, we define the first fundamental form as a formal expression of the approximation above, as follows: For three functions defined by the inner products of tangent vectors $\boldsymbol{p}_u(u,v)$ and $\boldsymbol{p}_v(u,v)$ as

$$E := \boldsymbol{p}_u \cdot \boldsymbol{p}_u, \quad F := \boldsymbol{p}_u \cdot \boldsymbol{p}_v, \quad G := \boldsymbol{p}_v \cdot \boldsymbol{p}_v, \tag{7.8}$$

the formal sum

$$ds^2 = E\,du^2 + 2F\,du\,dv + G\,dv^2 \tag{7.9}$$

is called the *first fundamental form*, and the functions E, F and G are called the *coefficients of the first fundamental form*. Replacing du and dv by Δu and Δv, respectively, in (7.9), we obtain the original approximation. The first fundamental form can be written formally using matrix multiplication as

$$ds^2 = (du\ dv)\begin{pmatrix} E & F \\ F & G \end{pmatrix}\begin{pmatrix} du \\ dv \end{pmatrix}. \tag{7.10}$$

Here, the matrix

$$\widehat{I} := \begin{pmatrix} E & F \\ F & G \end{pmatrix}$$

is a symmetric matrix whose eigenvalues are positive (cf. Problem **2**). In particular, $\widehat{I}$ is a regular matrix. We call this matrix the *first fundamental matrix*.

Invariance of the first fundamental form. Regarding a parametrization $\boldsymbol{p}(u,v)$ of a surface as a vector-valued function

$$\boldsymbol{p}(u,v) = (x(u,v), y(u,v), z(u,v)),$$

we define the exterior derivative as

$$d\boldsymbol{p} := \boldsymbol{p}_u\, du + \boldsymbol{p}_v\, dv = (dx, dy, dz).$$

So we have another expression of the first fundamental form:

$$ds^2 = d\boldsymbol{p} \cdot d\boldsymbol{p}. \tag{7.11}$$

The exterior derivative $d\boldsymbol{p} = (dx, dy, dz)$ is invariant under coordinate changes as seen in page 69, so *the first fundamental form is also invariant under coordinate changes.* Let us find the transformation formula of the coefficients of the first fundamental form under coordinate changes, using this fact. Let $\boldsymbol{p}(\xi,\eta)$ be a parametrization obtained by changing coordinates $u = u(\xi,\eta)$, $v = v(\xi,\eta)$ of the parametrization $\boldsymbol{p}(u,v)$, and let us find the coefficients of the first fundamental form of $\boldsymbol{p}(\xi,\eta)$.

Substituting the exterior derivatives

$$du = u_\xi\, d\xi + u_\eta\, d\eta, \qquad dv = v_\xi\, d\xi + v_\eta\, d\eta$$

into the definition (7.9) of ds^2, we have

$$\begin{aligned} ds^2 &= E(u_\xi\, d\xi + u_\eta\, d\eta)^2 + 2F(u_\xi\, d\xi + u_\eta\, d\eta)(v_\xi\, d\xi + v_\eta\, d\eta) \\ &\quad + G(v_\xi\, d\xi + v_\eta\, d\eta)^2 \\ &= (Eu_\xi^2 + 2Fu_\xi v_\xi + Gv_\xi^2)\, d\xi^2 \\ &\quad + 2(Eu_\xi u_\eta + F(u_\xi v_\eta + u_\eta v_\xi) + Gv_\xi v_\eta)\, d\xi\, d\eta \\ &\quad + (Eu_\eta^2 + 2Fu_\eta v_\eta + Gv_\eta^2)\, d\eta^2. \end{aligned}$$

Then the coefficients $\widetilde{E}$, $\widetilde{F}$, $\widetilde{G}$ of the first fundamental form of $\boldsymbol{p}(\xi,\eta)$ are

$$\begin{cases} \widetilde{E} = \boldsymbol{p}_\xi \cdot \boldsymbol{p}_\xi = Eu_\xi^2 + 2Fu_\xi v_\xi + Gv_\xi^2, \\ \widetilde{F} = \boldsymbol{p}_\xi \cdot \boldsymbol{p}_\eta = Eu_\xi u_\eta + F(u_\xi v_\eta + u_\eta v_\xi) + Gv_\xi v_\eta, \\ \widetilde{G} = \boldsymbol{p}_\eta \cdot \boldsymbol{p}_\eta = Eu_\eta^2 + 2Fu_\eta v_\eta + Gv_\eta^2. \end{cases} \tag{7.12}$$

One can rewrite these relations as

$$\begin{pmatrix} \widetilde{E} & \widetilde{F} \\ \widetilde{F} & \widetilde{G} \end{pmatrix} = \begin{pmatrix} u_\xi & u_\eta \\ v_\xi & v_\eta \end{pmatrix}^T \begin{pmatrix} E & F \\ F & G \end{pmatrix} \begin{pmatrix} u_\xi & u_\eta \\ v_\xi & v_\eta \end{pmatrix}. \tag{7.13}$$

Here, $\left(\begin{smallmatrix} u_\xi & u_\eta \\ v_\xi & v_\eta \end{smallmatrix}\right)$ is the Jacobi matrix of the coordinate change (cf. page 18), which is a regular matrix. The relation is used in the next section to show the invariance of the Gaussian curvature and mean curvature under coordinate changes.

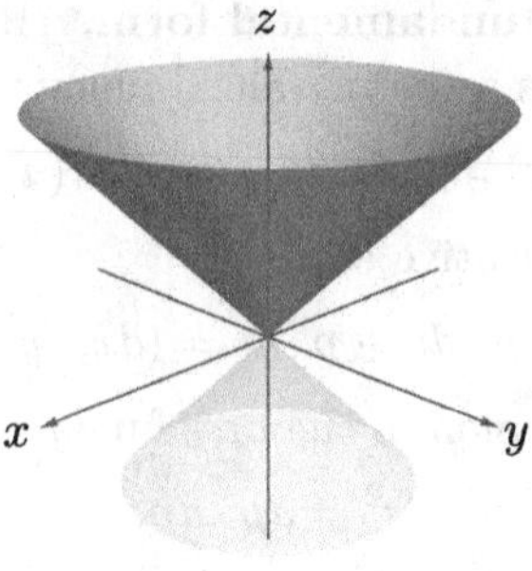

Fig. 7.1 The cone and its extension.

Example 7.2. Consider a surface which is the graph of the function

$$f(x,y) := \sqrt{x^2+y^2} \qquad ((x,y) \neq (0,0)),$$

and which is obtained by rotating a half-line $z = x$ $(x > 0)$ in the xz-plane about the z-axis, that is, the cone having vertex at the origin (Fig. 7.1). Using the polar coordinates (r, θ) given by

$$x = r\cos\theta, \quad y = r\sin\theta,$$

the surface is reparametrized as

$$\boldsymbol{p}(r,\theta) = (r\cos\theta, r\sin\theta, r) \qquad (r > 0)$$

in which no square root appears, and the map $\boldsymbol{p}$ can be extended smoothly to $r \leq 0$ and the image of the singular set $\{r = 0\}$ consists of a point, which is a typical example of a *cone-like singularity* (see Fig. 7.1).

Differentiating this, it holds that

$$d\boldsymbol{p} = (\cos\theta, \sin\theta, 1)\,dr + r(-\sin\theta, \cos\theta, 0)\,d\theta,$$

and then the first fundamental form is written as

$$ds^2 = d\boldsymbol{p} \cdot d\boldsymbol{p} = 2dr^2 + r^2 d\theta^2.$$

In particular, the coefficients of the first fundamental form with respect to the polar coordinates (r, θ) are $E = 2$, $F = 0$ and $G = r^2$. One can check that the first fundamental form is also expressed using the (x, y) coordinate system as

$$ds^2 = \left(1 + \frac{x^2}{x^2+y^2}\right) dx^2 + \frac{2xy}{x^2+y^2}\, dx\, dy + \left(1 + \frac{y^2}{x^2+y^2}\right) dy^2.$$

Area and the first fundamental form. Given a parametrization $\boldsymbol{p}(u,v)$ of a surface, Lagrange's identity ((A.3.15) on page 206 of Appendix A.3) tells us that

$$|\boldsymbol{p}_u \times \boldsymbol{p}_v|^2 = (\boldsymbol{p}_u \cdot \boldsymbol{p}_u)(\boldsymbol{p}_v \cdot \boldsymbol{p}_v) - (\boldsymbol{p}_u \cdot \boldsymbol{p}_v)^2 = EG - F^2$$

is satisfied, so the area element defined in the previous section ((6.13) on page 64) can be written as

$$dA = \sqrt{EG - F^2}\, du\, dv. \tag{7.14}$$

In particular, the area of the surface can be computed from the first fundamental form.

Example 7.3. Let us compute the area of the torus

$$\boldsymbol{p}(u,v) := ((a + b\cos u)\cos v, (a + b\cos u)\sin v, b\sin u) \quad (0 \le u, v \le 2\pi),$$

where we assume $0 < b < a$ (cf. Problem **1** in Section 6). The exterior derivative is

$$\begin{aligned} d\boldsymbol{p} = b\,&(-\sin u\cos v, -\sin u\sin v, \cos u)\, du \\ &+ (a + b\cos u)(-\sin v, \cos v, 0)\, dv, \end{aligned}$$

so it follows that

$$ds^2 = b^2\, du^2 + (a + b\cos u)^2\, dv^2$$

and the components of the first fundamental form are

$$E = b^2, \qquad F = 0, \qquad G = (a + b\cos u)^2.$$

In particular, the area element becomes

$$dA = \sqrt{EG - F^2}\, du\, dv = b(a + b\cos u)\, du\, dv$$

and the area is

$$\iint_{\{0 \le u,v \le 2\pi\}} b(a + b\cos u)\, du\, dv = (2\pi)^2 ab.$$

This is the same as the area of the surface of a cylinder of radius b and height $2\pi a$.

Lengths and angles. For a parametrization $\boldsymbol{p}(u,v)$ of a surface, points on the surface correspond to points in the uv-plane, and in this way we can consider the uv-plane to be giving us a *map* of the surface. Here we introduce a method for finding lengths and angles on that map.

For a surface $\boldsymbol{p}(u,v)$ parametrized by the variables (u,v), let us fix a point $\mathrm{P} = \boldsymbol{p}(u_0,v_0)$. As we saw in Section 6, the vectors tangent to the surface at P can all be written as linear combinations of $\boldsymbol{p}_u(u_0,v_0)$ and $\boldsymbol{p}_v(u_0,v_0)$. Thus, for a tangent vector $\boldsymbol{x}$, at the point $\mathrm{P} = \boldsymbol{p}(u_0,v_0)$, we can give the following correspondence

$$\boldsymbol{x} = x_1\,\boldsymbol{p}_u(u_0,v_0) + x_2\,\boldsymbol{p}_v(u_0,v_0) \longleftrightarrow \tilde{\boldsymbol{x}} = (x_1,x_2) \tag{7.15}$$

to a vector $\tilde{\boldsymbol{x}}$ in the *plane*.

Similarly, with the tangent vector $\boldsymbol{y} = y_1\boldsymbol{p}_u(u_0,v_0)+y_2\boldsymbol{p}_v(u_0,v_0)$ corresponding to the planar vector $\tilde{\boldsymbol{y}} = (y_1,y_2)$, the inner product of the spacial vectors $\boldsymbol{x}$ and $\boldsymbol{y}$ can be written as

$$\boldsymbol{x}\cdot\boldsymbol{y} = x_1y_1E + (x_1y_2 + x_2y_1)F + x_2y_2G = (x_1,x_2)\,\widehat{I}\begin{pmatrix} y_1 \\ y_2 \end{pmatrix} \tag{7.16}$$

in terms of the first fundamental form and the corresponding planar vectors, where $E = E(u_0,v_0)$, $F = F(u_0,v_0)$, $G = G(u_0,v_0)$ are evaluated at $(u,v) = (u_0,v_0)$. In particular, the norm of the vector is

$$|\boldsymbol{x}| = \sqrt{E\,{x_1}^2 + 2F\,x_1\,x_2 + G\,{x_2}^2}. \tag{7.17}$$

In general, for a curve $(u(t),v(t))$ $(a \le t \le b)$ in the uv-plane, $\gamma(t) = \boldsymbol{p}(u(t),v(t))$ is a space curve which lies on the surface, whose velocity vector is

$$\dot\gamma(t) = \frac{d}{dt}\boldsymbol{p}(u(t),v(t)) = \dot u(t)\boldsymbol{p}_u(u(t),v(t)) + \dot v(t)\boldsymbol{p}_v(u(t),v(t)).$$

Hence, by (7.17), the length of the space curve γ is

$$\mathcal{L}(\gamma) := \int_a^b |\dot\gamma(t)|\,dt = \int_a^b \sqrt{E\dot u^2 + 2F\dot u\dot v + G\dot v^2}\,dt, \tag{7.18}$$

and so it turns out that we can write this length in terms of the first fundamental form. Also, for two curves $(u_1(t),v_1(t))$ and $(u_2(t),v_2(t))$ in the uv-plane that intersect at $t = t_0$, the velocity vectors of the curves $\boldsymbol{p}(u_1(t),v_1(t))$ and $\boldsymbol{p}(u_2(t),v_2(t))$ on the surface, at $t = t_0$, are

$$\dot u_1\boldsymbol{p}_u + \dot v_1\boldsymbol{p}_v \qquad \dot u_2\boldsymbol{p}_u + \dot v_2\boldsymbol{p}_v$$

respectively. Denoting the angle between these two vectors as θ, (7.16) and (7.17) give

$$\cos\theta = \frac{E\dot u_1\dot u_2 + F(\dot u_1\dot v_2 + \dot v_1\dot u_2) + G\dot v_1\dot v_2}{\sqrt{E\,{\dot u_1}^2 + 2F\,\dot u_1\,\dot v_1 + G\,{\dot v_1}^2}\sqrt{E\,{\dot u_2}^2 + 2F\,\dot u_2\,\dot v_2 + G\,{\dot v_2}^2}}, \tag{7.19}$$

and so this angle can also be described in terms of the first fundamental form. In particular, when $E = G$ and $F = 0$, we have

$$\cos\theta = \frac{\dot{u}_1\dot{u}_2 + \dot{v}_1\dot{v}_2}{\sqrt{\dot{u}_1{}^2 + \dot{v}_1{}^2}\sqrt{\dot{u}_2{}^2 + \dot{v}_2{}^2}},$$

and then the angle θ between these two vectors in space is the same as the angle between the two velocity vectors of the two curves in the uv-plane. Coordinate systems that have the property $E = G$ and $F = 0$ are called the *isothermal coordinate systems.* It is known that all surfaces have such coordinates on local neighborhoods (see Section 16).

When considering the uv-plane as a map of the parametrized surface $\mathbf{p}(u, v)$, by (7.18), (7.19), (7.14), the first fundamental form gives information about lengths, angles and areas on the map. Here, the length of the space curve computed by (7.18) in the uv-plane may not coincide with the length of the planar curve in the uv-plane. The following proposition holds.

Proposition 7.4. *Let $\mathbf{p}(u, v)$ be a regular surface, defined on a region D in the uv-plane. A necessary and sufficient condition for the length of an arbitrarily chosen curve $(u(t), v(t))$ $(a \le t \le b)$ in the region D to be the same as the length of the corresponding curve $\gamma(t) = \mathbf{p}(u(t), v(t))$ in the surface is that $\mathbf{p}(u, v)$ has first fundamental form with coefficients*

$$E = G = 1, \qquad F = 0$$

on D.

Proof. Let $E = G = 1$ and $F = 0$. Then by (7.18), the length of the curve γ is $\mathcal{L}(\gamma) = \int_a^b \sqrt{\dot{u}(t)^2 + \dot{v}(t)^2}\, dt$, which is the same as the length (cf. (1.5) in Section 1) of the planar curve $(u(t), v(t))$ $(a \le t \le b)$, which proves sufficiency.

Let us now prove necessity. Fix an angle α, and consider a curve $(u_0 + t\cos\alpha, v_0 + t\sin\alpha)$ $(0 \le t \le c)$ in D passing through an arbitrarily chosen point (u_0, v_0) in D and forming the angle α with the u-direction. This is a line segment of length c as a curve in the uv-plane, and if the corresponding spacial curve

$$\mathbf{p}(u_0 + t\cos\alpha, v_0 + t\sin\alpha) \qquad (0 \le t \le c)$$

has the same length, (7.18) yields

$$c = \int_0^c \sqrt{E\cos^2\alpha + 2F\cos\alpha\sin\alpha + G\sin^2\alpha}\, dt.$$

Differentiating this with respect to c and letting $c \to 0$, it holds that

$$E\cos^2\alpha + 2F\cos\alpha\sin\alpha + G\sin^2\alpha = 1$$

at (u_0, v_0). Since this equality holds for any α, we have $E = 1$ by letting $\alpha = 0$, and $G = 1$ by letting $\alpha = \pi/2$. Moreover, since α is arbitrary, we have $F = 0$. □

A map that satisfies the condition in Proposition 7.4 is referred to as a *distance-angle-preserving map*. Regarding maps of the earth that we live on, we refer the reader to Appendix B.4. Regarding the types of surfaces for which one can make a distance-angle-preserving map, see Section B.6, and Lemma 16.2 of Section 16.

Exercises 7

1 Prove that for any 2×2 symmetric matrix $A = \left(\begin{smallmatrix} a & b \\ b & c \end{smallmatrix}\right)$ with real components, the eigenvalues are reals, and it can be diagonalized by an orthogonal matrix of determinant one (Problem **1** in Section 2) as follows:

(1) Show that the eigenvalues of A are reals, using the discriminant of quadratic equations.

(2) Let λ be an eigenvalue of A, and $\boldsymbol{e} := \left(\begin{smallmatrix} x \\ y \end{smallmatrix}\right)$ the corresponding eigenvector. Then show that the vector $\boldsymbol{n} := \left(\begin{smallmatrix} -y \\ x \end{smallmatrix}\right)$ obtained by 90°-counterclockwise rotation of $\boldsymbol{e}$ is the eigenvector corresponding to the eigenvalue $\mu := a + c - \lambda$.

(3) Choose $\boldsymbol{e}$ so that $|\boldsymbol{e}| = 1$ by applying a scalar multiplication, if necessary. Then show that the 2×2-matrix $P = (\boldsymbol{e}, \boldsymbol{n})$ is an orthogonal matrix of determinant 1. In addition, show that

$$AP = P\begin{pmatrix} \lambda & 0 \\ 0 & \mu \end{pmatrix}$$

holds. Hence $P^{-1}AP$ is a diagonal matrix.

2 Show that all eigenvalues of a 2×2-symmetric matrix $A = \left(\begin{smallmatrix} a & b \\ b & c \end{smallmatrix}\right)$ are positive if and only if $a > 0$, $ac - b^2 > 0$ hold. Moreover, show that the eigenvalues of the first fundamental matrix $\widehat{I}$ are positive.

3 Verify formulas (7.2), (7.3), (7.4).

4 Derive directly the transformation formula (7.13) without using the invariance under coordinate changes of the first fundamental form.

5 Find the coefficients of the first fundamental form of the graph $z = f(x, y)$ of the function f.

6 Compute the coefficients of the first fundamental forms of the elliptic paraboloids, hyperbolic paraboloids, ellipsoids, hyperboloids of one sheet, and hyperboloids of two sheets, as given in Section 6.

7 A surface obtained by rotating a curve $(x(t), z(t))$ in the xz-plane about the z-axis (a surface of revolution) can be parametrized as

$$\boldsymbol{p}(u,v) = (x(u)\cos v, x(u)\sin v, z(u)) \qquad (-\pi < v \le \pi).$$

Find the coefficients of the first fundamental form of this surface. In particular, simplify them when the curve is parametrized by arc-length, or given as a graph $x = f(z)$ of the function f.

8. The second fundamental form

In this section we define the second fundamental form, and use it to define the Gauss and mean curvatures of a surface, which are measures of how the surface is bending.

The second fundamental form. Given a regular surface $\boldsymbol{p}(u,v)$, the *second fundamental form* is defined as

$$I\!I = -d\boldsymbol{p}\cdot d\nu = -(\boldsymbol{p}_u\,du + \boldsymbol{p}_v\,dv)\cdot(\nu_u\,du + \nu_v\,dv),$$

where $\nu = \nu(u,v)$ is a unit normal vector. Expanding out the above equation, we have

$$I\!I = (-\boldsymbol{p}_u\cdot\nu_u)\,du^2 + (-\boldsymbol{p}_u\cdot\nu_v - \boldsymbol{p}_v\cdot\nu_u)\,du\,dv + (-\boldsymbol{p}_v\cdot\nu_v)\,dv^2.$$

Since the tangent vectors $\boldsymbol{p}_u$ and $\boldsymbol{p}_v$ are perpendicular to the unit normal vector ν, we have $\boldsymbol{p}_u\cdot\nu = 0$, $\boldsymbol{p}_v\cdot\nu = 0$, and we can differentiate these two equations to get

$$\begin{cases} \boldsymbol{p}_{uu}\cdot\nu = -\boldsymbol{p}_u\cdot\nu_u, & \boldsymbol{p}_{vv}\cdot\nu = -\boldsymbol{p}_v\cdot\nu_v, \\ \boldsymbol{p}_{uv}\cdot\nu = -\boldsymbol{p}_u\cdot\nu_v = -\boldsymbol{p}_v\cdot\nu_u. \end{cases}$$

So we can write the second fundamental form as

$$I\!I = L\,du^2 + 2M\,du\,dv + N\,dv^2, \tag{8.1}$$

where L, M, N are

$$\begin{cases} L := -\boldsymbol{p}_u\cdot\nu_u = \boldsymbol{p}_{uu}\cdot\nu, \\ M := -\boldsymbol{p}_u\cdot\nu_v = -\boldsymbol{p}_v\cdot\nu_u = \boldsymbol{p}_{uv}\cdot\nu, \\ N := -\boldsymbol{p}_v\cdot\nu_v = \boldsymbol{p}_{vv}\cdot\nu. \end{cases} \tag{8.2}$$

These three functions L, M, N of two variables are called the *coefficients of the second fundamental form*.

Just like for the first fundamental form, the second fundamental form can also be written in matrix form

$$II = (du\ dv)\ \widehat{II}\begin{pmatrix} du \\ dv \end{pmatrix}, \qquad \widehat{II} := \begin{pmatrix} L & M \\ M & N \end{pmatrix}.$$

We call the symmetric matrix $\widehat{II}$ the *second fundamental matrix.*

We now investigate how the second fundamental form changes under a change of coordinates. In this text, we shall fix a choice of unit normal vector independent of coordinates, and then define the second fundamental form using this chosen normal vector. Then, the second fundamental form is independent of coordinate changes, as is the first fundamental form.

Using the invariance of the second fundamental form, we can easily see how the coefficients of the second fundamental form transform under a change of coordinates. Applying a coordinate change $u = u(\xi, \eta)$, $v = v(\xi, \eta)$ to $\boldsymbol{p} = \boldsymbol{p}(u, v)$, and thinking of $\boldsymbol{p} = \boldsymbol{p}(\xi, \eta)$ as now depending on ξ and η, we can write the components of the second fundamental form of $\boldsymbol{p}(\xi, \eta)$ as $\widetilde{L}$, $\widetilde{M}$, $\widetilde{N}$. Like we had for the first fundamental form, we have

$$\begin{aligned} \widetilde{L} &= -\boldsymbol{p}_\xi \cdot \nu_\xi = Lu_\xi^2 + 2Mu_\xi v_\xi + Nv_\xi^2, \\ \widetilde{M} &= -\boldsymbol{p}_\xi \cdot \nu_\eta = Lu_\xi u_\eta + M(u_\xi v_\eta + u_\eta v_\xi) + Nv_\xi v_\eta, \\ \widetilde{N} &= -\boldsymbol{p}_\eta \cdot \nu_\eta = Lu_\eta^2 + 2Mu_\eta v_\eta + Nv_\eta^2. \end{aligned}$$

In terms of matrices, these relations become

$$\begin{pmatrix} \widetilde{L} & \widetilde{M} \\ \widetilde{M} & \widetilde{N} \end{pmatrix} = \begin{pmatrix} u_\xi & u_\eta \\ v_\xi & v_\eta \end{pmatrix}^T \begin{pmatrix} L & M \\ M & N \end{pmatrix} \begin{pmatrix} u_\xi & u_\eta \\ v_\xi & v_\eta \end{pmatrix}. \tag{8.3}$$

Since the first fundamental matrix is invertible for regular surfaces (cf. page 70), here we can define the new matrix

$$A := \begin{pmatrix} E & F \\ F & G \end{pmatrix}^{-1} \begin{pmatrix} L & M \\ M & N \end{pmatrix}, \tag{8.4}$$

that is,

$$A = \frac{1}{EG - F^2} \begin{pmatrix} GL - FM & GM - FN \\ -FL + EM & -FM + EN \end{pmatrix}. \tag{8.5}$$

By the transformation formulas (7.13), (8.3) and (8.4), we have that the matrix A transforms into

$$\widetilde{A} = \begin{pmatrix} u_\xi & u_\eta \\ v_\xi & v_\eta \end{pmatrix}^{-1} A \begin{pmatrix} u_\xi & u_\eta \\ v_\xi & v_\eta \end{pmatrix}. \tag{8.6}$$

Because of this, the eigenvalues λ_1, λ_2 of the matrix A are unchanged by any coordinate change. In this textbook, we call the matrix A the *Weingarten matrix*. Furthermore, these eigenvalues of A are real numbers (cf. Problem **1** or Proposition 9.3 of Section 9). We call these two reals the *principal curvatures* of the surface. The determinant and half the trace of the matrix A are, respectively,

$$
\begin{aligned}
K &:= \det A = \lambda_1\lambda_2 = \frac{LN - M^2}{EG - F^2}, \\
H &:= \frac{1}{2}\operatorname{tr} A = \frac{1}{2}(\lambda_1 + \lambda_2) = \frac{EN - 2FM + GL}{2(EG - F^2)},
\end{aligned} \tag{8.7}
$$

and K is called the *Gaussian curvature* and H the *mean curvature*, of the surface. Here, $\operatorname{tr} A$ denotes the sum of the diagonal components of the matrix A, that is, the *trace* of A.

The following facts provide us with geometric meanings of the Gaussian and mean curvatures:

- The sign of the Gaussian curvature is related closely to the local shape of the surface, as shown in Theorem 8.7 in this section.
- If both K and H vanish identically on the surface, $L = M = N = 0$ hold, and the surface is a part of a plane (Problem **6**).
- For a surface such that K is identically zero, one can produce a local distance-angle-preserving map (cf. Appendix B.6 and Lemma 16.2 in Section 16).
- A surface whose mean curvature H vanishes everywhere is called a *minimal surface* (cf. well-known examples of minimal surfaces are found in Problems **7**–**11**). Dipping a wire frame into a soap solution, and then pulling it out, a soap film spans the frame. It is known that the film takes the shape of a minimal surface. For example, see [23, 29]. The existence of minimal surfaces spanning arbitrary piecewise smooth simple closed curve in $\boldsymbol{R}^3$ is known. However, uniqueness of the surfaces does not necessarily hold.

Proposition 8.1. *We fix a unit normal vector $\nu(u,v)$ of a regular surface $\mathbf{p}(u,v)$. Then the Gaussian curvature K and mean curvature H are invariant under changes of coordinates.*

Proof. Since the principal curvatures are invariant under coordinate changes, the proposition follows. □

By Proposition 8.1, the values of the Gaussian and mean curvatures at each point of a surface are independent of coordinate changes, so, for the purpose of computation, it is allowed and is often advisable to find the most convenient choice of parametrization.

We introduce examples of surfaces whose Gaussian and mean curvatures are both constant:

Example 8.2 (The plane). A surface $\boldsymbol{p} = (u, v, 0)$ gives a parametrization of the horizontal plane in $\boldsymbol{R}^3$. One can take the unit normal vector to be $\nu := (0, 0, 1)$, which is constant. Then the first and second fundamental forms are

$$ds^2 = d\boldsymbol{p} \cdot d\boldsymbol{p} = du^2 + dv^2, \qquad I\!I = -d\boldsymbol{p} \cdot d\nu = 0.$$

Since the second fundamental form vanishes, the Weingarten matrix is identically zero, and thus the Gaussian and mean curvatures are identically zero. Conversely, a surface whose Gaussian and mean curvatures vanish simultaneously is a part of the plane (Problem **6**).

Example 8.3 (The circular cylinder of radius 1). A surface

$$\boldsymbol{p} = (\cos u, \sin u, v)$$

gives a parametrization of the cylinder of radius 1 in $\boldsymbol{R}^3$. The unit normal vector can be chosen as $\nu = (\cos u, \sin u, 0)$. Then the first and the second fundamental forms are

$$ds^2 = d\boldsymbol{p} \cdot d\boldsymbol{p} = du^2 + dv^2, \qquad I\!I = -d\boldsymbol{p} \cdot d\nu = -du^2.$$

In particular, $E = G = 1$, $F = 0$, $L = -1$ and $M = N = 0$, and then the Weingarten matrix A is $\left(\begin{smallmatrix} -1 & 0 \\ 0 & 0 \end{smallmatrix}\right)$. Hence the Gaussian curvature is 0, and the mean curvature is $-1/2$. The first fundamental form of this surface coincides with that of the plane in Example 8.2. This example and the previous example (Example 8.2) show that the first fundamental form alone does not determine the shape of the surface.

Example 8.4 (The sphere). For a positive number a,

$$\boldsymbol{p} = a(\cos u \cos v, \cos u \sin v, \sin u) \qquad \left(|u| < \frac{\pi}{2},\ 0 \leq v < 2\pi\right)$$

gives a parametrization of the sphere centered at the origin with radius a. The unit normal vector can be chosen as $\nu = -(1/a)\boldsymbol{p}$, and then

$$I\!I = -d\boldsymbol{p} \cdot d\nu = \frac{1}{a} d\boldsymbol{p} \cdot d\boldsymbol{p} = \frac{1}{a} ds^2,$$

that is, the second fundamental form is proportional to the first fundamental form. Thus, the Weingarten matrix A is $1/a$ times the identity matrix, and hence the Gaussian curvature is $1/a^2$, and the mean curvature is $1/a$.

It is known that a surface whose Gaussian and mean curvatures are both constant is (essentially) one of these examples; that is, a part of a plane, a circular cylinder or a sphere (see Proposition B.7.4 in Appendix B.7).

We return to the general theory. The Weingarten matrix A defined in (8.4) is closely related to the behavior of the unit normal vector ν:

Proposition 8.5 (The Weingarten formula). *For a regular surface $\boldsymbol{p}(u,v)$, it holds that*

$$(\nu_u, \nu_v) = -(\boldsymbol{p}_u, \boldsymbol{p}_v)A. \tag{8.8}$$

In particular, at each point on the surface, ν_u and ν_v are linearly dependent if and only if the Gaussian curvature vanishes.

Proof. Consider vectors $\boldsymbol{a}$, $\boldsymbol{b} \in \boldsymbol{R}^3$ as column vectors. Then the inner product $\boldsymbol{a} \cdot \boldsymbol{b}$ is written by a matrix product as $\boldsymbol{a} \cdot \boldsymbol{b} = \boldsymbol{a}^T \boldsymbol{b}$ (cf. (A.3.1) in Appendix A.3). Let $P_1 := (\boldsymbol{p}_u, \boldsymbol{p}_v, \nu)$ and $P_2 := (\nu_u, \nu_v, \nu)$ be the 3×3-matrices formed by column vectors. Then by the definition of the first and the second fundamental matrices, we have

$$P_1{}^T P_1 = \begin{pmatrix} \widehat{I} & \boldsymbol{0} \\ \boldsymbol{0}^T & 1 \end{pmatrix}, \qquad -P_1{}^T P_2 = \begin{pmatrix} \widehat{II} & \boldsymbol{0} \\ \boldsymbol{0}^T & -1 \end{pmatrix},$$

where $\boldsymbol{0} = \left(\begin{smallmatrix} 0 \\ 0 \end{smallmatrix}\right)$. Hence it holds that

$$-P_1^{-1} P_2 = -(P_1{}^T P_1)^{-1} P_1{}^T P_2 = \begin{pmatrix} \widehat{I}^{-1} \widehat{II} & \boldsymbol{0} \\ \boldsymbol{0}^T & -1 \end{pmatrix},$$

that is,

$$(\nu_u, \nu_v, \nu) = P_2 = -P_1 \begin{pmatrix} \widehat{I}^{-1} \widehat{II} & \boldsymbol{0} \\ \boldsymbol{0}^T & -1 \end{pmatrix} = -(\boldsymbol{p}_u, \boldsymbol{p}_v, \nu) \begin{pmatrix} A & \boldsymbol{0} \\ \boldsymbol{0}^T & -1 \end{pmatrix},$$

proving (8.8). Since $\boldsymbol{p}_u$ and $\boldsymbol{p}_v$ are linearly independent, ν_u and ν_v are linearly dependent if and only if the determinant of A is 0. □

Proposition 8.6. *Under a congruent transformation (cf. Appendix A.3), the Gaussian curvature K and the absolute value of H are invariant.*

Proof. For a parametrization $\boldsymbol{p}$ of a surface, we set $\tilde{\boldsymbol{p}} = T\boldsymbol{p} + \boldsymbol{c}$, where T is an orthogonal matrix and $\boldsymbol{c}$ is a constant vector. Then

$$\tilde{\boldsymbol{p}}_u = T\boldsymbol{p}_u, \quad \tilde{\boldsymbol{p}}_v = T\boldsymbol{p}_v, \quad \tilde{\boldsymbol{p}}_{uu} = T\boldsymbol{p}_{uu}, \quad \tilde{\boldsymbol{p}}_{uv} = T\boldsymbol{p}_{uv}, \quad \tilde{\boldsymbol{p}}_{vv} = T\boldsymbol{p}_{vv}.$$

Then, by (A.3.12) in Appendix A.3, $\tilde{\nu} := T\nu$ is a unit normal vector of $\tilde{\boldsymbol{p}}$. Then the coefficients of the first and second fundamental forms of $\tilde{\boldsymbol{p}}$ coincide with those of $\boldsymbol{p}$. Substituting these into (8.7), we have the conclusion. □

On the surface, points where the Gaussian curvature K is positive are called *elliptic points*. Similarly, points where $K = 0$, resp. $K < 0$, are called *parabolic points*, resp. *hyperbolic points*. The following theorem gives geometric properties of Gaussian curvatures.

Theorem 8.7. *The principal curvatures at each point of a regular surface are real numbers. In addition, near an elliptic point (a point where $K > 0$), the surface is convex. On the other hand, near a hyperbolic point (a point where $K < 0$), the surface takes the shape of a saddle, see Fig. 8.1.*

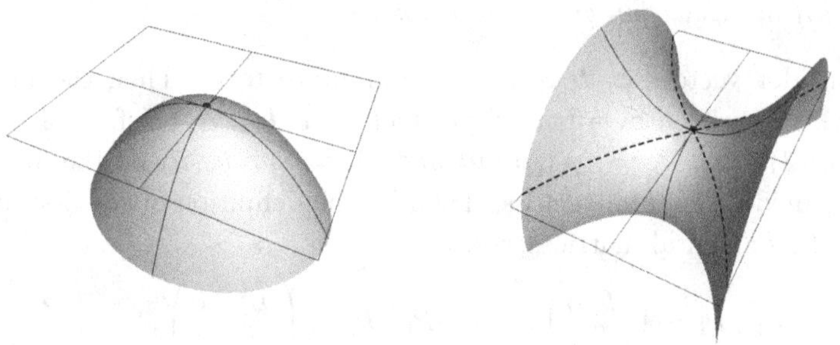

An elliptic point.

A hyperbolic point.
The dotted curves show the intersection with the tangent plane.

Fig. 8.1

Proof. Fix a point P on the surface. By a rotation and a translation, we may assume that P is the origin, and the tangent plane of the surface at P is the xy-plane, without loss of generality.

Let $\boldsymbol{p}(u,v) = (x(u,v), y(u,v), z(u,v))$ be the parametrization of the surface, and take a map $\varphi\colon (u,v) \mapsto (x(u,v), y(u,v))$ from the uv-plane to the xy-plane. Here, we set $\mathrm{P} = (0,0,0) = \boldsymbol{p}(u_0, v_0)$. Since

$$\boldsymbol{p}_u(u_0,v_0) = (x_u(u_0,v_0), y_u(u_0,v_0), 0) = (\varphi_u, 0),$$
$$\boldsymbol{p}_v(u_0,v_0) = (x_v(u_0,v_0), y_v(u_0,v_0), 0) = (\varphi_v, 0)$$

are linearly independent, the determinant of the Jacobi matrix of φ does not vanish at (u_0, v_0). Hence, by the inverse function theorem (Theorem A.1.5 in Appendix A.1), there exists an inverse ψ of φ defined on a neighborhood of the origin of the xy-plane. Thus, we can choose (x, y) as a coordinate system of the surface, and it can be expressed as a graph of a function

$z = f(x, y)$ near the origin P. By assumption, we have

$$f(0,0) = f_x(0,0) = f_y(0,0) = 0,$$

and the coefficients of the first fundamental form at origin are $E = G = 1$, $F = 0$ (Problem **5** in Section 7), and the Weingarten matrix A coincides with the second fundamental matrix at the origin by (8.4), that is,

$$A = \begin{pmatrix} L & M \\ M & N \end{pmatrix} \tag{8.9}$$

holds at the origin, where L, M, N are the coefficients of the second fundamental form with respect to the parameters (x, y). In particular, since A is a symmetric matrix in this case, the eigenvalues (i.e., the principal curvatures) are reals (Problem **1** in Section 7). Hence the first part of the theorem is proven.

Since $f(0,0) = f_x(0,0) = f_y(0,0) = 0$ holds, Taylor's formula (Theorem A.1.2 in Appendix A.1) implies that

$$f(x, y) = \frac{1}{2}\left(f_{xx}(0,0)x^2 + 2f_{xy}(0,0)xy + f_{yy}(0,0)y^2\right) + o(x^2 + y^2).$$

This shows that the surface is approximated by a quadric

$$z = \frac{1}{2}(ax^2 + 2bxy + cy^2) \tag{8.10}$$

$$(a := f_{xx}(0,0),\ b := f_{xy}(0,0),\ c := f_{yy}(0,0))$$

in a neighborhood of the origin. In the case that $a = c = 0$ and $b \neq 0$, (8.10) is the graph of $z = bxy$. This is a surface obtained by 45° rotation of the graph of $z = b(x^2 - y^2)/2$ (a special case of Example 6.1) about the z-axis, that is, a hyperbolic paraboloid. On the other hand, in the case that $a \neq 0$, (8.10) can be written as

$$z = \frac{1}{2}\left[a\left(x + \frac{b}{a}y\right)^2 + \left(\frac{ac - b^2}{a}\right)y^2\right].$$

In particular, when $ac - b^2 = f_{xx}f_{yy} - f_{xy}^2 > 0$, if $a > 0$ (resp. $a < 0$), the graph of $f(x, y)$ is contained in the upper half-space $\{z \geq 0\}$ (resp. the lower half-space $\{z \leq 0\}$), and the shape of the surface looks like a bowl, as seen on the left side of Fig. 8.1. When $ac - b^2 < 0$, the graph is saddle-shaped, as shown on the right side of Fig. 8.1. Since the Gaussian curvature K is given by

$$K = \frac{f_{xx}f_{yy} - f_{xy}^2}{(1 + f_x^2 + f_y^2)^2}$$

(cf. Problem **3** in this section), the conclusion follows. □

Surfaces of revolution with constant Gaussian curvature. Now we examine the structure of surfaces of revolution with constant Gaussian curvature K. We have already seen that the plane, circular cylinder and sphere all have constant Gaussian curvature, but we can make many other examples as well.

Taking a regular surface $\boldsymbol{p}(u, v)$ and changing its size by a factor of c, by replacing it with $c\boldsymbol{p}(u, v)$, the Gaussian curvature and mean curvature change to be $1/c^2$ times and $1/c$ times the Gaussian and mean curvatures of the original surface, respectively (see Problem **5** at the end of this section). So without loss of generality we may consider only the cases that the Gaussian curvature is 0, 1 or -1. Here, we classify the cases that $K = 0$ and 1. The case that $K = -1$ is considered in Appendix B.8.

We start with a curve in the xz-plane, with arc-length parameter s:

$$\gamma(s) = (x(s), z(s)) \qquad (x(s) > 0). \tag{8.11}$$

Rotating this about the z-axis, we have the surface of revolution

$$\boldsymbol{p}(u, v) := (x(u) \cos v, x(u) \sin v, z(u)), \tag{8.12}$$

whose Gaussian and mean curvatures we now find. Since $\gamma(s)$ is parametrized by arc-length,

$$(x')^2 + (z')^2 = 1 \tag{8.13}$$

holds, where " $'$ " denotes the derivative with respect to u. Differentiating both sides of this equation, we have

$$x'x'' + z'z'' = 0. \tag{8.14}$$

Using (7.11) and (8.13), the first fundamental form is

$$ds^2 = d\boldsymbol{p} \cdot d\boldsymbol{p} = \{(x')^2 + (z')^2\}\, du^2 + x^2\, dv^2 = du^2 + x^2\, dv^2,$$

and the components of the first fundamental form are $E = 1$, $F = 0$ and $G = x^2$.

Also, the unit normal vector of $\boldsymbol{p}(u, v)$ is

$$\nu(u, v) = (-z'(u) \cos v, -z'(u) \sin v, x'(u)),$$

and its exterior derivative is

$$d\nu = (-z'' \cos v, -z'' \sin v, x'')\, du + (z' \sin v, -z' \cos v, 0)\, dv,$$

so the second fundamental form becomes

$$I\!I = -d\boldsymbol{p} \cdot d\nu = (x'z'' - z'x'')\, du^2 + xz'\, dv^2,$$

and so we have $L = x'z'' - z'x''$, $M = 0$ and $N = xz'$. By (8.13) and (8.14),

$$x'z'' - z'x'' = -x''/z'$$

holds and so we have $L = -x''/z'$. (By (2.14) in Section 2, L is the curvature of the plane curve $(x(u), z(u))$.) Therefore, by (8.7), the Gaussian and mean curvatures of the surface are

$$K = \frac{LN}{EG} = -\frac{xx''}{x^2} = -\frac{x''}{x}, \tag{8.15}$$

$$H = \frac{EN + GL}{2EG} = \frac{xz' - (x^2x''/z')}{2x^2} = \frac{z'}{2x} - \frac{x''}{2z'}. \tag{8.16}$$

Using this, let us now find the constant Gaussian curvature surfaces of revolution.

(i) The case that $K = 0$: By (8.15), if $K = 0$, then $x'' = 0$ holds, so with (8.13) we have

$$x(u) = au+b, \quad z(u) = u\sqrt{1-a^2} \qquad (a \text{ and } b \text{ are constants with } |a| \leq 1).$$

In particular, if $a = 0$, then the surface is a *circular cylinder*, if $0 < |a| < 1$, it is a *cone*; and if $|a| = 1$, it is a *plane*.

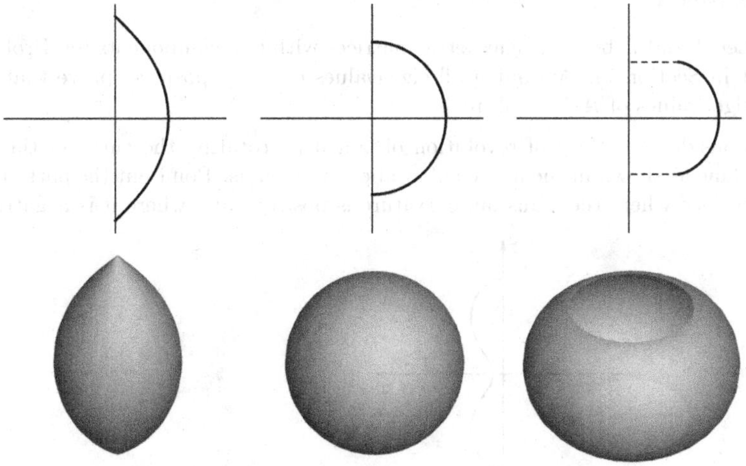

$a < 1$ $a = 1$ $a > 1$

Fig. 8.2 Surfaces of revolution with Gaussian curvature 1.

(ii) The case that $K = 1$: By (8.15), $x'' = -x$ holds. By (8.11), x can be written as $((\alpha, \beta) \neq (0,0))$

$$x(u) = \alpha \cos u + \beta \sin u = a \cos(u - \delta) \quad \left(a := \sqrt{\alpha^2 + \beta^2},\ \delta := \arctan \frac{\beta}{\alpha}\right).$$

This is a periodic function, and without loss of generality we can take $x(u) = a \cos u$ $(a > 0)$. Then by (8.13),

$$z(u) = \int_0^u \sqrt{1 - a^2 \sin^2 t}\, dt \tag{8.17}$$

holds. In particular, if $a = 1$, then the surface is a *sphere*; if $a < 1$, it is a surface of revolution shaped like rugby ball; and if $a > 1$, it is shaped like a barrel (see Fig. 8.2). Figure B.7.5 in Appendix B.7 shows the extensions of these surfaces as smooth maps. The relationship between these surfaces and surfaces of revolution with constant mean curvature is also discussed in Appendix B.7.

By the same method, one can determine the surfaces of revolution with constant mean curvature $H = 0$ (see Problem **7**). We will consider surfaces of revolution with H a non-zero constant in Section B.7. The surfaces of revolution with Gaussian curvature $K = -1$ are classified in Appendix B.8.

Exercises 8

1 Let A and B be 2×2 symmetric matrices with real components (cf. Problem **1** in Section 7). Assuming all eigenvalues of A are positive, prove that the eigenvalues of AB are all reals.

2 Consider a surface of revolution obtained by rotating the curve in the xz-plane as shown in the figure below about the z-axis. Point out the part of the surface where the Gaussian curvature is positive, and where it is negative.

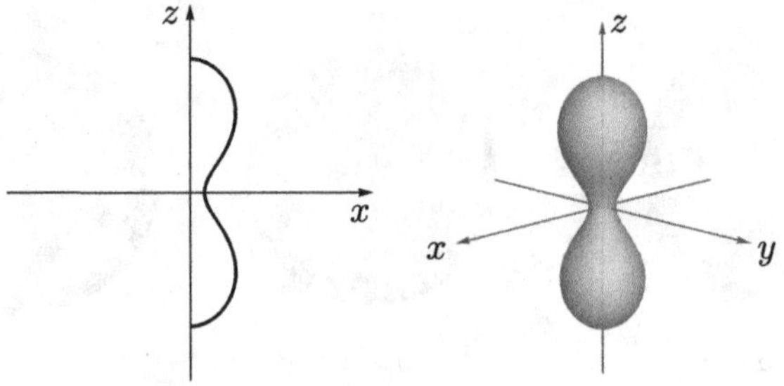

3 Compute the coefficients of the second fundamental form, the Gaussian curvature and the mean curvature of the graph of a function $z = f(x, y)$.

4 Compute the coefficients of the second fundamental forms, the Gaussian curvatures and the mean curvatures of the elliptic paraboloids, the hyperbolic paraboloids, the ellipsoids, the hyperboloids of one sheet, and the hyperboloids of two sheets, as given in Section 6.

5 Let K and H be the Gaussian curvature and the mean curvature of a surface $\boldsymbol{p}(u,v)$, respectively. Show that, for each positive constant c, the surface $c\boldsymbol{p}(u,v)$ obtained by c-factor scaling of the surface $\boldsymbol{p}(u,v)$ has the Gaussian curvature and mean curvature K/c^2, H/c, respectively.

6* Show that if the Gaussian curvature and the mean curvature of the surface vanish identically, the coefficients of the second fundamental form L, M, N also vanish identically. Moreover, prove that such a surface is a part of a plane.

7* Let us find minimal surfaces of revolution. Verify that if $H = 0$,

$$x(u)x''(u) = z'(u)^2 = 1 - x'(u)^2$$

holds, using (8.16). Assuming this equation, show that $(xx')' = 1$ and find $x(u)$ and $z(u)$. Moreover, if $z(u)$ is not constant, eliminating the parameter u, show that

$$x = a\cosh\left(\frac{z-b}{a}\right) \qquad (a \text{ and } b \text{ are constants and } a \neq 0).$$

This means that the surface is obtained by the catenary (Example 1.4 in Section 1). The surface is called a *catenoid*, one of the typical examples of a minimal surface (see the figure below).

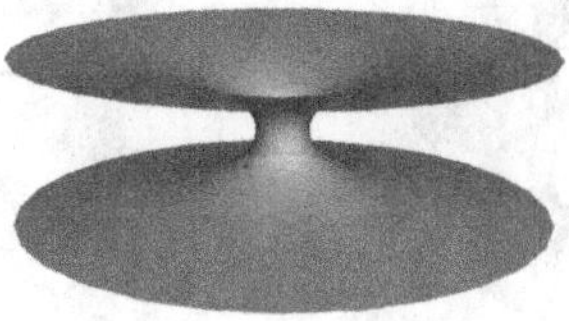

The catenoid.

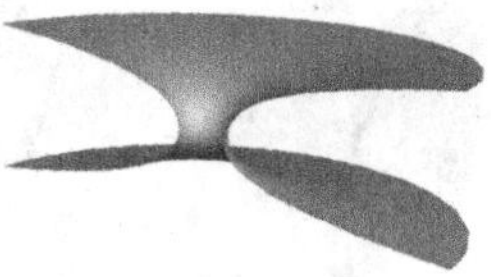

A half-cut of the catenoid.

8* A parametrized surface $\boldsymbol{p}(u,v) = (u\cos v, u\sin v, v)$ $(u, v \in \boldsymbol{R})$ satisfies that each u-curve is a helix as in Example 5.1 in Section 5, and each v-curve is a line parallel to the xy-plane. In particular, the surface is filled by a family of lines. Such a surface is said to be a *ruled surface* (see Appendix B.6). The surface $p(u,v)$ is called the *helicoid*, see the figure below. Show that the helicoid is a minimal surface.

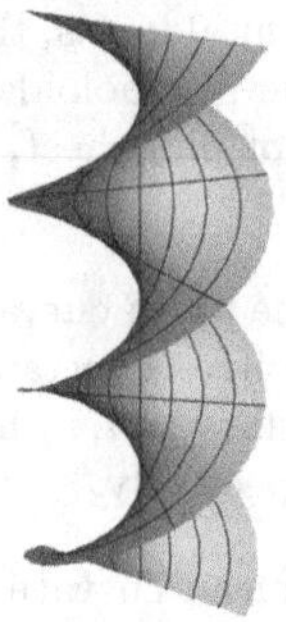

The helicoid.

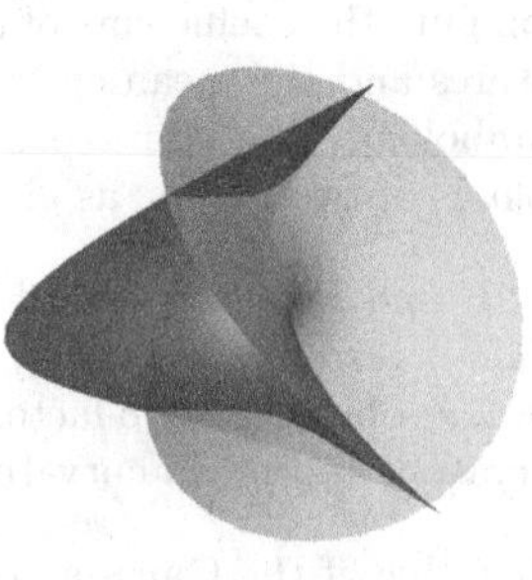

Enneper's surface.

9* Show that the parametrized surface

$$\boldsymbol{p}(u,v) = \left(u - \frac{u^3}{3} + uv^2, -v + \frac{v^3}{3} - u^2v, u^2 - v^2\right) \qquad (u, v \in \boldsymbol{R})$$

is minimal. This surface is called *Enneper's surface*[1] (see the figure above).

10* Let $f(x,y) := \log\left(\frac{\cos y}{\cos x}\right)$, which is a smooth function defined on the subset

$$\bigcup_{\substack{m,\, n\text{: integers},\\ m+n\text{: even}}} \left\{(x,y) \in \boldsymbol{R}^2 \;\middle|\; |x - m\pi| < \frac{\pi}{2},\ |y - n\pi| < \frac{\pi}{2}\right\}$$

of $\boldsymbol{R}^2$. Show that the graph of f is minimal surface, using Problem **3**. This surface is called the *Scherk*[2] *surface*, see the figure below.

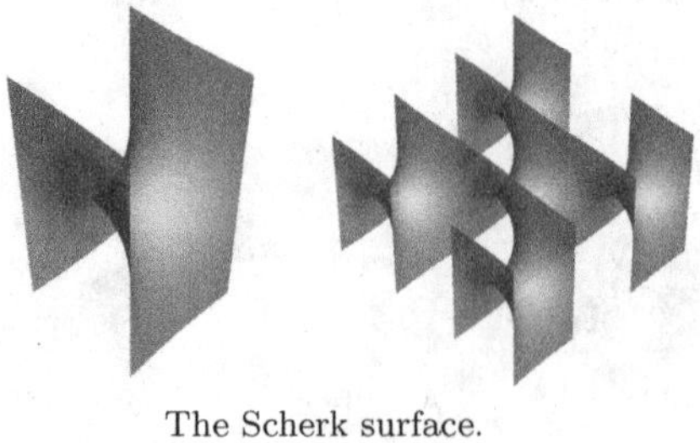

The Scherk surface.

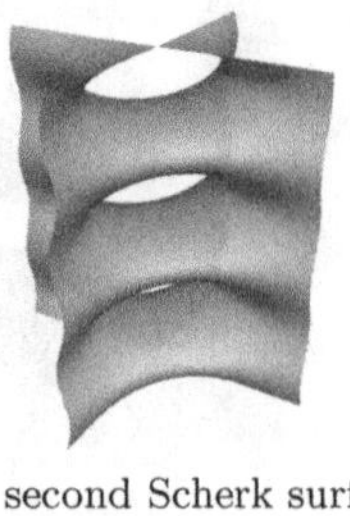

The second Scherk surface.

11* Show that a surface given by the implicit function

$$(\sinh x)(\sinh y) = \sin z$$

[1]Enneper, Alfred (1830–1885).
[2]Scherk, Heinrich Ferdinand (1798–1885).

is a minimal surface. (Hint: Rewrite the equation in the form $z = f(x, y)$ using the inverse sine function and calculate the mean curvature.) This surface is called the *second Scherk surface* or *Scherk's saddle tower*, see the figure above.

9. Principal and asymptotic directions

In this section, we introduce the geometric meaning of the principal curvatures defined in Section 8. Throughout this section, we consider a regular surface parametrized as $\boldsymbol{p}(u, v)$, and fix a point $\mathrm{P} = \boldsymbol{p}(u_0, v_0)$ on the surface.

Curves on a surface. Consider a spatial curve $\gamma(s)$ lying on a regular surface $\boldsymbol{p}(u, v)$, and suppose that s is an arc-length parameter for γ.

Because this curve $\gamma(s)$ lies on the surface $\boldsymbol{p}(u, v)$, we can write

$$\gamma(s) = \boldsymbol{p}(u(s), v(s)) \tag{9.1}$$

and there exists a corresponding curve $\bigl(u(s), v(s)\bigr)$ in the uv-plane. Using the chain rule for differentiation, the velocity vector for this curve becomes

$$\gamma'(s) = \boldsymbol{p}_u(u(s), v(s))\, u'(s) + \boldsymbol{p}_v(u(s), v(s))\, v'(s). \tag{9.2}$$

Because s is an arc-length parameter for $\gamma(s)$, we have

$$\gamma' \cdot \gamma' = E(u')^2 + 2Fu'v' + G(v')^2 = 1. \tag{9.3}$$

Any vector in $\boldsymbol{R}^3$ can be written uniquely as the sum of a vector tangent to the surface and a vector normal to the surface, so the acceleration vector of γ can be written as

$$\gamma''(s) = \boldsymbol{\kappa}_g(s) + \boldsymbol{\kappa}_n(s) \quad \begin{pmatrix} \boldsymbol{\kappa}_g(s) \text{ is tangent to the surface,} \\ \boldsymbol{\kappa}_n(s) \text{ is perpendicular to the surface} \end{pmatrix}. \tag{9.4}$$

In this decomposition, we call $\boldsymbol{\kappa}_n(s)$ the *normal curvature vector* of $\gamma(s)$, and $\boldsymbol{\kappa}_g(s)$ the *geodesic curvature vector* of $\gamma(s)$.

Normal curvature. Taking ν to be a unit normal vector to the surface, we denote by $\nu(s)$ the restriction of ν to the curve $\gamma(s)$. Then $\gamma(s)$ over the surface has normal curvature vector

$$\boldsymbol{\kappa}_n(s) = (\gamma''(s) \cdot \nu(s))\, \nu(s).$$

Defining the coefficient here as

$$\kappa_n(s) := \gamma''(s) \cdot \nu(s),$$

we call this the *normal curvature* of the curve.

Writing γ as in (9.1) in terms of the corresponding curve $(u(s), v(s))$ in the uv-plane, $\gamma'(s) \cdot \nu(s) = 0$ gives that the normal curvature is

$$(9.5) \quad \begin{aligned} \kappa_n &= \gamma''(s) \cdot \nu(s) = -\gamma'(s) \cdot \nu'(s) \\ &= -(\boldsymbol{p}_u u' + \boldsymbol{p}_v v') \cdot (\nu_u u' + \nu_v v') = L(u')^2 + 2Mu'v' + N(v')^2. \end{aligned}$$

Here L, M, N are the components of the second fundamental form as in (8.2). Because the tangent vector to $\gamma(s)$ is determined by u' and v', as in (9.2), (9.5) shows that the value of κ_n is determined by the tangent vector. In particular, because s is an arc-length parameter, the *normal curvature κ_n is determined by just the direction of the tangent vector.*

Now we consider a geometric meaning of the normal curvatures. Cutting the surface at some point by a plane containing the normal vector at that point, the plane and the surface intersect along a planar curve. That planar curve is called a *normal section* of the surface at that point (see Fig. 9.1).

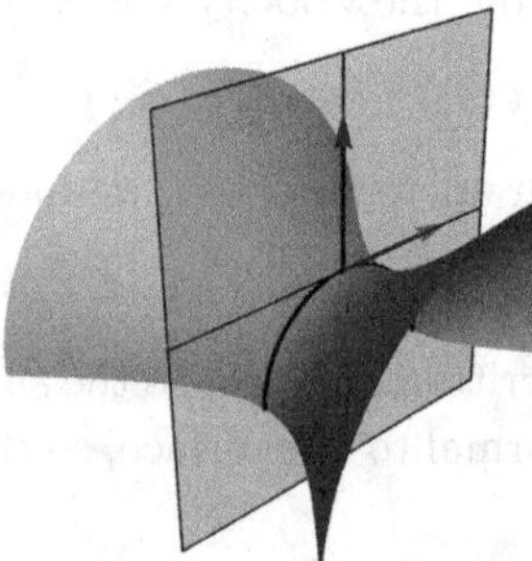

Fig. 9.1 The normal section.

Theorem 9.1. *Take a space curve $\gamma(s)$ on the regular surface going through the point $\mathrm{P} = \gamma(s_0)$ when $s = s_0$. Consider the normal section through P with the same tangent vector as $\gamma'(s_0)$ of γ at P. Regarding the normal section as lying in the xy-plane oriented so that the x-axis points in the direction of $\gamma'(s_0)$ and the y-axis points in the normal direction $\nu(s_0)$ (see Fig. 9.1), the normal curvature of γ at P is the same as the planar curvature of the normal section at P.*

Proof. Assume s is the arc-length parameter of γ, and take the normal section $\sigma(s)$ with an arc-length parametrization such that $\sigma(s_0) = \mathrm{P}$. The normal curvature κ_n of γ at P is determined by only the tangent direction, and $\gamma''(s_0) \cdot \nu(s_0) = \sigma''(s_0) \cdot \nu(s_0)$ holds, and in particular $\sigma''(s_0) \cdot \nu(s_0) = \kappa_n$. However, $\sigma''(s_0)$ is perpendicular to $\sigma'(s_0) = \gamma'(s_0)$, and $\sigma(s)$ always stays within the plane determined by the normal vector at s_0 and the vector $\sigma'(s_0) = \gamma'(s_0)$, so $\sigma''(s_0)$ is parallel to $\nu(s_0)$, and then $\sigma''(s_0) = \kappa_n(s_0)\nu(s_0)$. Thus we have shown that the planar curvature of the planar curve $\sigma(s)$ is κ_n at s_0. □

By this theorem, we see that the normal curvature is determined by merely the direction of the planar slice containing the normal vector. Hence, the normal curvature measures the bending of the surface in a given direction. The following formula is convenient for computing normal curvatures.

Proposition 9.2. *Let* P *be a point on the regular surface* $\boldsymbol{p}(u,v)$*. For the direction* $(\alpha,\beta) \in \boldsymbol{R}^2 \setminus \{(0,0)\}$ $(\beta = \alpha \tan\theta)$*, making angle* θ *with the* u*-axis in the* uv*-plane, the normal curvature* κ_n *at* P *is*

$$\kappa_n = \frac{L\alpha^2 + 2M\alpha\beta + N\beta^2}{E\alpha^2 + 2F\alpha\beta + G\beta^2} = \frac{L\cos^2\theta + 2M\cos\theta\sin\theta + N\sin^2\theta}{E\cos^2\theta + 2F\cos\theta\sin\theta + G\sin^2\theta}, \tag{9.6}$$

where E, F, G *are the coefficients of the first fundamental form of the surface at* P*, and* L, M, N *are those of the second fundamental form at* P.

We remark that the angle θ in this formula is the angle in the uv-plane, which might not correspond to the actual angle on the surface. Later we will give a formula (cf. Theorem 9.8) to find the normal curvature using the actual angle on the surface.

Proof. Let $\gamma(s) = \boldsymbol{p}(u(s), v(s))$ be a curve parametrized by arc-length s lying on the surface $\boldsymbol{p}(u,v)$, and assume $\gamma(0) = \mathrm{P}$. Let θ be the angle between the initial velocity vector $(u'(0), v'(0))$ in the uv-plane and the u-axis. Then there exists a positive number ρ satisfying

$$(u'(0), v'(0)) = \rho(\cos\theta, \sin\theta).$$

Since s is the arc-length, (9.3) yields

$$\begin{aligned} 1 &= Eu'(0)^2 + 2Fu'(0)v'(0) + Gv'(0)^2 \\ &= \rho^2(E\cos^2\theta + 2F\cos\theta\sin\theta + G\sin^2\theta). \end{aligned}$$

Using this and (9.5), we have

$$\begin{aligned}\kappa_n &= Lu'(0)^2 + 2Mu'(0)v'(0) + Nv'(0)^2 \\ &= \rho^2(L\cos^2\theta + 2M\cos\theta\sin\theta + N\sin^2\theta) \\ &= \frac{L\cos^2\theta + 2M\cos\theta\sin\theta + N\sin^2\theta}{E\cos^2\theta + 2F\cos\theta\sin\theta + G\sin^2\theta}. \qquad \square\end{aligned}$$

Fixing an arbitrary point $\mathrm{P} = \boldsymbol{p}(u_0, v_0)$ on the surface, and looking at arc-length parametrized curves $\gamma(s) = \boldsymbol{p}(u(s), v(s))$ on the surface going through P, we now find the maximum and minimum values of the normal curvatures at P. By Proposition 9.2, we see that κ_n is a continuous function of only one variable θ with period π and hence takes the maximum λ_1 and the minimum λ_2. By (9.6), these maximum and minimum values coincide with those of the function

$$\lambda(\alpha, \beta) = \frac{L\alpha^2 + 2M\alpha\beta + N\beta^2}{E\alpha^2 + 2F\alpha\beta + G\beta^2} \tag{9.7}$$

of variables (α, β) on $\boldsymbol{R}^2\backslash\{(0,0)\}$. At such an (α, β), it holds that $\partial\lambda/\partial\alpha = \partial\lambda/\partial\beta = 0$. Differentiating both sides of

$$L\alpha^2 + 2M\alpha\beta + N\beta^2 - (E\alpha^2 + 2F\alpha\beta + G\beta^2)\lambda = 0$$

with respect to α and β, we have

$$\begin{cases} (L - \lambda E)\alpha + (M - \lambda F)\beta = 0, \\ (M - \lambda F)\alpha + (N - \lambda G)\beta = 0 \end{cases} \tag{9.8}$$

holding for (α, β) which attain a maximum or minimum of the function λ. In order for this linear system to have a solution at some (α, β) not equal to $(0, 0)$, we need that

$$\det\begin{pmatrix} L - \lambda E & M - \lambda F \\ M - \lambda F & N - \lambda G \end{pmatrix} = 0$$

holds, that is, the maximum and the minimum of the normal curvature κ_n at P are the roots of the quadratic equation

$$(EG - F^2)\lambda^2 - (EN + GL - 2FM)\lambda + (LN - M^2) = 0 \tag{9.9}$$

with respect to λ. By the relationship between the roots and coefficients of quadratic equations,

$$\begin{cases} \lambda_1 + \lambda_2 = \dfrac{EN + GL - 2FM}{EG - F^2} = 2H, \\[2ex] \lambda_1\lambda_2 = \dfrac{LN - M^2}{EG - F^2} = K \end{cases} \tag{9.10}$$

hold. In particular, λ_1 and λ_2 are the eigenvalues of the Weingarten matrix

$$A = \begin{pmatrix} E & F \\ F & G \end{pmatrix}^{-1} \begin{pmatrix} L & M \\ M & N \end{pmatrix},$$

that is, the principal curvatures at P.

Summing up, we have the following proposition.

Proposition 9.3. *The maximum and minimum values of the normal curvature of a regular surface at a point* P *equal the principal curvatures at that point. Thus the product and average of these two values are the Gaussian and mean curvatures at that point.*

At a point P on the surface with principal curvatures λ_1, λ_2, we call the directions of the corresponding tangent vectors $\boldsymbol{v}_1$, $\boldsymbol{v}_2$ the *principal curvature directions.* Expressing tangent vectors of the surface at P in the form

$$\alpha \boldsymbol{p}_u + \beta \boldsymbol{p}_v,$$

a necessary and sufficient condition for a tangent vector $\alpha \boldsymbol{p}_u + \beta \boldsymbol{p}_v$ of the surface at P to point in a principal curvature direction is that, with λ being a principal curvature, α and β satisfy equation (9.8), that is, $(\alpha, \beta)^T$ is an eigenvector of A.

When the two principal curvatures λ_1 and λ_2 are equal, we say that this point on the surface is an *umbilic point.*The principal curvatures are the maximum and minimum of (9.6), so for an umbilic point, that function will be a constant independent of α and β, and every direction becomes a principal curvature direction. We can say that an umbilic point is a point where "the amount of bending is the same in every direction". By (9.10),

$$4(H^2 - K) = (\lambda_1 + \lambda_2)^2 - 4\lambda_1\lambda_2 = (\lambda_1 - \lambda_2)^2$$

holds, so we have the following proposition.

Proposition 9.4. *A point on a regular surface is an umbilic point if and only if* $H^2 - K = 0$ *at that point.*

Fix a point P on the surface. By a rotation and a translation of $\boldsymbol{R}^3$, we may assume that P is the origin, and the tangent plane of the surface at P is the xy-plane, without loss of generality. As seen in the proof of Theorem 8.7, the surface is parametrized as a graph of $z = f(x, y)$ in a neighborhood of P.

Lemma 9.5. *Under the setting above, by a suitable rotation within the xy-plane fixing the z-axis, the surface has a parametrization $\mathbf{p}(u,v) = (u,v,f(u,v))$ on a neighborhood of* P *such that the first and second fundamental matrices satisfy*

$$\widehat{I} = \begin{pmatrix} 1 & 0 \\ 0 & 1 \end{pmatrix} = I, \qquad \widehat{II} = \begin{pmatrix} \lambda_1 & 0 \\ 0 & \lambda_2 \end{pmatrix} \tag{9.11}$$

at P, *where λ_1 and λ_2 are the principal curvatures. In particular, when* P *is not an umbilic point, the u-direction and v-direction are orthogonal, and correspond to the principal directions.*

Proof. As seen in the proof of Theorem 8.7, the first fundamental matrix is the identity matrix, and the Weingarten matrix A is in the form as in (8.9). In particular, A is a symmetric matrix at P, and then it can be diagonalized by an orthogonal matrix of determinant 1 (cf. Problem **1** in Section 7). In other words, there exists an orthogonal matrix R with $\det R = 1$ such that

$$R^T A R = R^{-1} A R = \begin{pmatrix} \lambda_1 & 0 \\ 0 & \lambda_2 \end{pmatrix},$$

where λ_1 and λ_2 are the eigenvalues of A, i.e. the principal curvatures. Apply the isometric transformation

$$\begin{pmatrix} x \\ y \\ z \end{pmatrix} \longmapsto \begin{pmatrix} R^{-1} \begin{pmatrix} x \\ y \end{pmatrix} \\ z \end{pmatrix},$$

which gives a rotation of the xy-plane fixing the z-axis. Then the x-axis and the y-axis correspond to the eigendirections of A at P in the xy-plane. This implies that the Weingarten matrix with respect to the parametrization $(u,v) \mapsto (u,v,f(u,v))$ is diagonal at P, and $\widehat{II}$ satisfies (9.11). □

Using this lemma, we have the following proposition.

Proposition 9.6. *A point* P *on a regular surface is an umbilic point if and only if the Weingarten matrix A at the point is a scalar multiple of the identity matrix.*

Proof. Take the parametrization (u,v) as in Lemma 9.5. Then P is an umbilic point if and only if A is a scalar multiple of the identity matrix at P. This property is independent of choice of parameters, because of (8.6). Thus we have the conclusion. □

An elliptic paraboloid (6.1) for $a > b$ has two umbilic points at $(0, \pm bd, d^2)$, where $d := \sqrt{a^2 - b^2}/2$ (see Problem **2** in Section 15). On the other hand, an ellipsoid (6.3) satisfying $a > b > c$ has four umbilic points at $(\pm ap, 0, \pm cq)$, where $p := \sqrt{a^2 - b^2}/\sqrt{a^2 - c^2}$ and $q := \sqrt{1 - p^2}$ (see Example 15.8 in Section 15).

The Gaussian curvature and the mean curvature of a sphere of radius a are $1/a^2$ and $1/a$, respectively (cf. Example 8.4). So the sphere is a surface such that all points are umbilic. Conversely, the following holds. The proof is left for the exercises (cf. Problem **3**).

Proposition 9.7. *If every point of a regular surface is an umbilic point, then the surface is part of a plane or sphere.*

Here, we investigate the behavior of surfaces near non-umbilic points.

At any non-umbilic point P on a surface, we take the parametrization as in Lemma 9.5. Then, we obtain the following formula for the principal curvatures.

Theorem 9.8 (Euler's formula). *Let* P *be a non-umbilic point of a regular surface, and parametrize the surface as in Lemma 9.5 at* $\mathrm{P} = \boldsymbol{p}(0,0)$. *Let* λ_1 *and* λ_2 *be the principal curvatures of the surface at* P *with respect to the directions* $\boldsymbol{p}_u(0,0)$ *and* $\boldsymbol{p}_v(0,0)$, *respectively. Then the normal curvature* κ_n *of the direction in the tangent plane of the surface at* P *that makes angle* φ *with* $\boldsymbol{p}_u(0,0)$ *satisfies*

$$\kappa_n = \lambda_1 \cos^2 \varphi + \lambda_2 \sin^2 \varphi. \tag{9.12}$$

Proof. Since the first fundamental matrix with respect to (u, v) is the identity matrix at P, $\boldsymbol{p}_u(0,0)$ and $\boldsymbol{p}_v(0,0)$ are orthogonal unit vectors. Then the unit vector $\boldsymbol{v}$ on the tangent plane of the surface at P that makes angle φ with $\boldsymbol{p}_u(0,0)$ is written as

$$\boldsymbol{v} = (\cos\varphi)\boldsymbol{p}_u(0,0) + (\sin\varphi)\boldsymbol{p}_v(0,0),$$

which corresponds to the vector $(\cos\varphi, \sin\varphi)^T$ in the uv-plane. Hence we have the conclusion by (9.6) and (9.11). □

The angle θ in Proposition 9.2 is the angle in the uv-plane, which might not correspond to the actual angle on the surface. On the other hand, the angle φ in Euler's formula (9.12) is the actual angle measured on the surface from the principal direction. Thus we have now obtained a more geometrically natural formula than in Proposition 9.2.

A curve on the surface is called a *line of curvature* or a *curvature line* if the velocity vector of the curve is a principal curvature direction at each point of the curve. Near a non-umbilic point of a surface, one can choose a parameter (u, v) of the surface such that both the u-curves and v-curves are lines of curvature. Such a parameter is called a *curvature line coordinate system* or a *principal curvature coordinate system.* The existence of curvature line coordinates is shown in Appendix B.5. For example, the parametrization of a surface of revolution as (8.12) gives a curvature line coordinate system (cf. Problem **2** at the end of this section).

Asymptotic directions. Any direction for which the normal curvature is zero is called an *asymptotic direction.*

Proposition 9.9. *At a point on a regular surface where the Gaussian curvature K is positive, there do not exist asymptotic directions. At a non-umbilic point where $K = 0$, there is exactly one asymptotic direction. When $K < 0$, there exist two distinct asymptotic directions.*

Proof. Let λ_1 and λ_2 be the principal curvatures at a non-umbilic point. Then by Theorem 9.8, the normal curvature with respect to the direction which makes the angle φ with the λ_1-principal curvature direction is given by (9.12). If $K(= \lambda_1\lambda_2) > 0$, then λ_1 and λ_2 have the same sign, and then the normal curvature does not vanish. On the other hand, if $K < 0$, then λ_1 and λ_2 have opposite signs, and thus there is exactly one direction where the normal curvature vanishes, when φ moves from 0 to $\pi/2$ (resp. $\pi/2$ to π). Finally, if $K = 0$ at a non-umbilic point, exactly one of the principal curvatures λ_1, λ_2 is zero. Thus, the principal curvature direction for the principal curvature 0 is the unique asymptotic direction. □

The next theorem gives us a geometric meaning for asymptotic directions.

Theorem 9.10. *At a point* P *on a regular surface where $K < 0$, the intersection of the surface with the tangent plane at* P *is a pair of crossing curves near* P. *Furthermore, the tangent directions at* P *of those two curves are the same as the asymptotic directions at* P (*see the right-hand figure of Fig. 8.1*).

Proof. Take the parametrization as in Lemma 9.5 so that the surface is represented as a graph $z = f(x, y)$. Then the first fundamental matrix at

P is the identity matrix, and

$$L = f_{xx}(0,0) = \lambda_1, \quad M = f_{xy}(0,0) = 0, \quad N = f_{yy}(0,0) = \lambda_2$$

hold. By Taylor's formula (Theorem A.1.2 in Appendix A.1), we have

$$f(x,y) = \frac{1}{2}(Lx^2 + Ny^2) + o(x^2 + y^2),$$

where $o(\cdot)$ is Landau's symbol. By assumption, we have $LN = K < 0$, and interchanging the x and y-axes if necessary, we can write $L = a^2$ and $N = -b^2$, for $a, b > 0$. Then, near the origin, $f(x,y)$ becomes

$$f(x,y) = \frac{1}{2}(ax + by)(ax - by) + o\big(x^2 + y^2\big).$$

Because the tangent plane is the xy-plane, its intersection with the surface is the set of points (two curves) in the xy-plane where $f(x,y) = 0$ holds. The above formula for f tells us that these two curves have the tangent directions $ax \pm by = 0$.

On the other hand, Euler's formula (9.12) tells us that the normal curvature in the direction at angle φ from the x-axis satisfies

$$\kappa_n = \lambda_1 \cos^2\varphi + \lambda_2 \sin^2\varphi = L\cos^2\varphi + N\sin^2\varphi.$$

Choosing φ so that this normal curvature is 0, we have

$$0 = L\cos^2\varphi + N\sin^2\varphi = (a\cos\varphi + b\sin\varphi)(a\cos\varphi - b\sin\varphi).$$

This equation tells us that the directions given by the lines $ax \pm by = 0$ are the asymptotic directions, which are the same as the tangent directions created by cutting the surface with the tangent plane. □

At points where the Gaussian curvature is negative, the two angles for the two different asymptotic directions have the following property.

Proposition 9.11. *At a point of a regular surface where $K < 0$, the two asymptotic directions make the same angle with a principal curvature direction. Letting μ $(0 < \mu \le \pi/2)$ denote that angle to the principal curvature,*

$$\tan\frac{\mu}{2} = \sqrt{\frac{|\lambda_1|}{|\lambda_2|}}. \tag{9.13}$$

Here we assume that the principal curvatures λ_1 and λ_2 satisfy $|\lambda_1| \le |\lambda_2|$.

This can be easily proven using Euler's formula, so we leave it as an exercise for the reader (see Problem **4** at the end of this section).

A curve on the surface is called an *asymptotic curve* if the velocity vector of the curve give the asymptotic direction at each point of the curve. Near a point of a surface where the Gaussian curvature is negative, one can choose a parameter (u, v) of the surface such that both u-curves and v-curves are asymptotic curves. Such a parameter is called the *asymptotic line coordinate system.* The existence of such coordinates is shown in Appendix B.5.

Exercises 9

1 Let γ be a curve on a surface and P a point on it. Show that the normal curvature κ_n of γ at P and the curvature κ of the space curve γ at P satisfy $\kappa_n = \kappa \cos\theta$. Here, θ is the angle between the unit normal vector ν of the surface at P and the principal normal vector $\boldsymbol{n}$ of the space curve γ at P (Meusnier's[1] theorem).

2 Show that the principal curvature directions of a surface of revolution are the direction of the generating curve and the direction of the rotation.

3 Prove Proposition 9.7 as follows: Let $\boldsymbol{p}(u, v)$ be a surface defined on a region D of the uv-plane and assume that all points on the surface are umbilics.

(1) Prove that $\nu_u = -\lambda \boldsymbol{p}_u$, $\nu_v = -\lambda \boldsymbol{p}_v$, where ν is the unit normal vector of the surface.
(2) Using $\nu_{uv} = \nu_{vu}$, show that λ is a constant.
(3) When $\lambda = 0$, the surface is a part of a plane because of Problem **6** in Section 8. When $\lambda \neq 0$, show that $\boldsymbol{p} + (1/\lambda)\nu$ is constant, which implies that $\boldsymbol{p}(u, v)$ is a part of the sphere.

4 Prove Proposition 9.11.

5 Show that, at a point on a surface with $K < 0$ and $H = 0$, two asymptotic directions are mutually perpendicular. (In particular, a minimal surface is characterized a surfaces such that the intersection of the tangent plane and the surface consists of two orthogonal curves at each point.)

[1] Meusnier, Jean-Baptiste Marie (1754–1793).

10. Geodesics and the Gauss-Bonnet theorem

The sum of the interior angles of a triangle in the plane is π. In this section, we investigate the sum of interior angles of "triangles" on surfaces. A planar triangle is a figure bounded by line segments. So, to define triangles on the surface, we consider geodesics, which correspond to the lines on the plane.

Geodesics. A line segment on the plane is the shortest among the curves joining two points on the plane. So, what is the shortest curve on a surface joining two points on the surface? For example, Fig. 10.1 shows the shortest curve joining two points of an ellipsoid. To mathematically understand this concept, we first show that such a shortest curve is a "geodesic" defined below, if the curve is parametrized by the arc-length (Theorem 10.5).

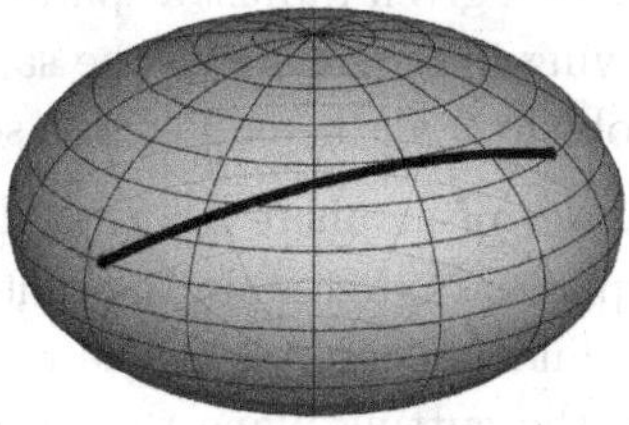

Fig. 10.1 The shortest curve joining two points of an ellipsoid.

Let $\gamma(t)$ be a space curve lying on a regular surface $\boldsymbol{p}(u,v)$; here, t is not necessarily the arc-length. Since $\gamma(t)$ lies on the surface $\boldsymbol{p}(u,v)$, there exists a curve $(u(t),v(t))$ on the uv-plane such that $\gamma(t) = \boldsymbol{p}(u(t),v(t))$. Any vector in $\boldsymbol{R}^3$ can be written uniquely as the sum of a vector tangent to the surface and a vector normal to the surface, so the acceleration vector of γ can be written as in[1] (9.4):

$$(10.1) \quad \ddot{\gamma}(t) = [\ddot{\gamma}(t)]^{\mathrm{H}} + [\ddot{\gamma}(t)]^{\mathrm{V}} \qquad \left(\begin{array}{l}[\ddot{\gamma}(t)]^{\mathrm{H}} \text{ is tangent to the surface,} \\ [\ddot{\gamma}(t)]^{\mathrm{V}} \text{ is perpendicular to the surface}\end{array}\right),$$

where the superscripts H and V stand for "horizontal" and "vertical", respectively.

The curve $\gamma(t)$ is said to be a *geodesic* if $[\ddot{\gamma}(t)]^{\mathrm{H}} = \mathbf{0}$ for all t. By definition, $\ddot{\gamma}(t)$ is orthogonal to $\dot{\gamma}(t)$ if $\gamma(t)$ is a geodesic. Then the derivative of $\dot{\gamma}\cdot\dot{\gamma}$ is identically zero, and hence we have the following proposition.

[1]The decomposition (9.4) in Section 9 corresponds to the case that t is the arc-length. Namely, if s is the arc-length parameter of γ, $\boldsymbol{\kappa}_g(s) = [\ddot{\gamma}(s)]^{\mathrm{H}}$, $\boldsymbol{\kappa}_n(s) = [\ddot{\gamma}(s)]^{\mathrm{V}}$.

Proposition 10.1. *If a curve $\gamma(t)$ on a regular surface is a geodesic, then $|\dot{\gamma}(t)|$ is constant. In particular, the parameter t is proportional to the arc-length.*

The definition tells us that, if one drives a car along a geodesic of the surface, the driver feels the acceleration in the normal direction. This means that a geodesic is a path along which one travels (on the surface) with the least amount of energy.

Example 10.2. Lines in the plane can be described as follows: taking a single point A with position vector $\boldsymbol{a}$, and taking a non-zero vector $\boldsymbol{v}$ parallel to the line, $\gamma(t) = \boldsymbol{a} + t\boldsymbol{v}$ is a parametrization of the line. Because $\ddot{\gamma}(t) = \mathbf{0}$, this is a geodesic in the plane.

Example 10.3. A helix with a given radius, as parametrized in Example 5.1 in Section 5 (page 51), lying in a cylinder of the same radius, is a geodesic in that cylinder (see Problem **2** at the end of this section).

Example 10.4. *Great circles* of a sphere are the curves obtained by intersecting the sphere with planes that contain the center of the sphere. When the sphere has radius 1 and is centered at the origin, choosing two unit vectors $\{\boldsymbol{e}_1, \boldsymbol{e}_2\}$ lying in the cutting plane that are perpendicular to each other, the resulting great circle has the parametrization

$$\gamma(s) = (\cos s)\boldsymbol{e}_1 + (\sin s)\boldsymbol{e}_2.$$

In particular, s is an arc-length parameter, and $\gamma''(s) = -\gamma(s)$ is perpendicular to the tangent plane of the surface at each point on the curve, so this is a geodesic in the sphere.

Geodesics and shortest-length paths. Regarding the shortest-length paths on surfaces, the following holds:

Theorem 10.5. *If, on a given regular surface, there exists a shortest-length regular curve between two given points* P *and* Q *in that surface, then that curve is a geodesic when it is parametrized by arc-length.*

Proof. Suppose that $\gamma(s)$ $(0 \leq s \leq l)$ is a parametrization of the curve on the surface with shortest length from P to Q. We prove the theorem using the fact that any deformation of this curve preserving the endpoints cannot shorten the length. First, we formulate the "deformation of a curve".

Consider a family of spacial curves $\{\gamma_w(s)\}_{|w|<\varepsilon}$ satisfying the following properties, where ε is a sufficiently small positive number, see Fig. 10.2.

- For each real number w, $\gamma_w(s)$ is a curve on the surface defined on $0 \leq s \leq l$ such that $\gamma_w(0) = \mathrm{P}$ and $\gamma_w(l) = \mathrm{Q}$.
- $\gamma_w(s)$ is a smooth function with respect to the two variables (w, s).
- $\gamma_0(s) = \gamma(s)$ holds for $0 \leq s \leq l$.

We call this type of family $\{\gamma_w\}_{|w|<\varepsilon}$ of curves an *endpoint-fixing variation* of the curve $\gamma(s)$. Since there are infinitely many ways to make such a family, there are infinitely many different endpoint-fixing variations of a single curve $\gamma(s)$.

For a given endpoint fixing variation $\{\gamma_w(s)\}_{|w|<\varepsilon}$ of a curve $\gamma(s)$, the vector $V(s)$ $(0 \leq s \leq l)$ defined as

$$V(s) := \left.\frac{\partial \gamma_w(s)}{\partial w}\right|_{w=0}$$

is a tangent vector of the surface at $\gamma(s)$. In general, if there are tangent vectors of the surface depending smoothly on s and placed at $\gamma(s)$, such a family of vectors is said to be a *vector field along the curve* $\gamma(s)$ on the surface. In particular, the vector field $V(s)$ along $\gamma(s)$ as above is called the *variational vector field* of the variation $\{\gamma_w(s)\}_{|w|<\varepsilon}$.

Assuming the shortest-length curve $\gamma(s)$ $(0 \leq s \leq l)$ joining P and Q is parametrized by the arc-length, we take an endpoint-fixing variation $\{\gamma_w(s)\}_{|w|<\varepsilon}$ of it. Though the parameter s is the arc-length of $\gamma = \gamma_0$, it may not be the arc-length for $\gamma_w(s)$ $(w \neq 0)$.

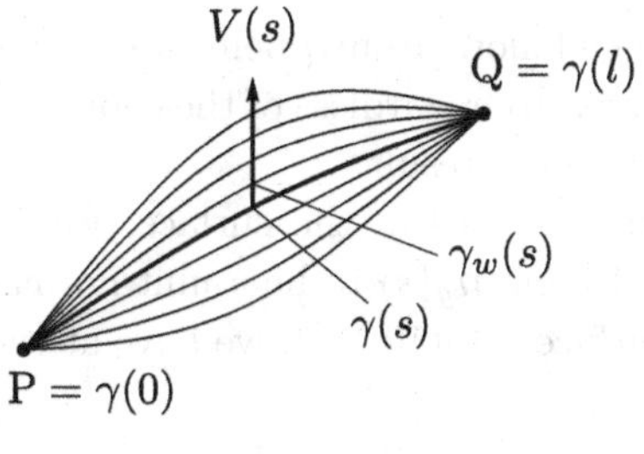

Fig. 10.2 An endpoint-fixing variation of a curve.

Because $\gamma(s)$ is the shortest path between its endpoints, the length

$$\mathcal{L}(\gamma_w) = \int_0^l |\gamma_w'(s)|\, ds \qquad \left({}' = \frac{d}{ds}\right)$$

of $\gamma_w(s)$, as a function of w, should have a minimum value at $w = 0$. In particular,

$$\left.\frac{d}{dw}\right|_{w=0} \mathcal{L}(\gamma_w) = 0 \tag{10.2}$$

holds. On the other hand, we have

$$\begin{aligned}\left.\frac{d}{dw}\right|_{w=0} \mathcal{L}(\gamma_w) &= \left.\frac{d}{dw}\right|_{w=0} \int_0^l |\gamma_w'(s)|\, ds = \int_0^l \left.\frac{\partial}{\partial w}\right|_{w=0} \sqrt{\gamma_w'(s)\cdot\gamma_w'(s)}\, ds \\ &= \int_0^l \gamma_0'(s)\cdot\left(\left.\frac{\partial}{\partial w}\right|_{w=0}\gamma_w'(s)\right)\frac{ds}{|\gamma_0'(s)|} \\ &= \int_0^l \gamma_0'(s)\cdot\frac{\partial}{\partial s}\left(\left.\frac{\partial}{\partial w}\right|_{w=0}\gamma_w(s)\right) ds = \int_0^l \gamma'(s)\cdot V'(s)\, ds \\ &= [\gamma'(s)\cdot V(s)]_{s=0}^{s=l} - \int_0^l \gamma''(s)\cdot V(s)\, ds.\end{aligned} \tag{10.3}$$

Here V is the variation vector field coming from the family $\{\gamma_w\}_{|w|<\varepsilon}$. Because the variation fixes the endpoints, we have $\gamma_w(0) = \gamma(0) = \mathrm{P}$ and $\gamma_w(l) = \gamma(l) = \mathrm{Q}$, and so

$$V(0) = \left.\frac{\partial \gamma_w(0)}{\partial w}\right|_{w=0} = \mathbf{0} \quad \text{and} \quad V(l) = \left.\frac{\partial \gamma_w(l)}{\partial w}\right|_{w=0} = \mathbf{0}$$

hold, giving

$$\left.\frac{d}{dw}\right|_{w=0} \mathcal{L}(\gamma_w) = -\int_0^l \gamma''(s)\cdot V(s)\, ds. \tag{10.4}$$

Because of (10.2), which follows from γ being a shortest-length path, the integral in (10.4) must be zero, regardless of the choice of variation $\{\gamma_w\}_{|w|<\varepsilon}$. Using this, we will prove the result.

Because the curve $\gamma(s)$ lies in the surface, we can parametrize it as $\gamma(s) = \boldsymbol{p}(u(s), v(s))$. Taking $\boldsymbol{n}_g(s)$ to be a unit normal vector to the curve and tangent to the surface (cf. (10.15)), we have at each point of the curve that

$$\boldsymbol{n}_g(s) = \alpha(s)\,\boldsymbol{p}_u(u(s), v(s)) + \beta(s)\,\boldsymbol{p}_v(u(s), v(s))$$

for some C^∞-functions $\alpha(s)$ and $\beta(s)$.

Then the curve γ is a geodesic if and only if

$$\kappa_g(s) := \gamma''(s)\cdot\boldsymbol{n}_g(s)$$

vanishes identically. Here, $\kappa_g(s)$ coincides with the geodesic curvature which will be defined in (10.13), up to the sign.

Assuming there exists c $(0 < c < l)$ satisfying $\kappa_g(c) \neq 0$, we will prove the existence of an endpoint-fixing variation such that (10.4) is not zero. Take a sufficiently small positive number δ such that $\kappa_g(s)$ does not change its sign on the interval $(c-\delta, c+\delta)$, and define a function ρ by

$$\rho(u) = \begin{cases} \exp\left(\dfrac{1}{(u-c)^2-\delta^2}\right) & (|u-c| < \delta), \\ 0 & (|u-c| \geq \delta), \end{cases} \tag{10.5}$$

see Fig. 10.3. This is a C^∞-function which is positive on $|u-c| < \delta$ and vanishes on $|u-c| \geq \delta$.

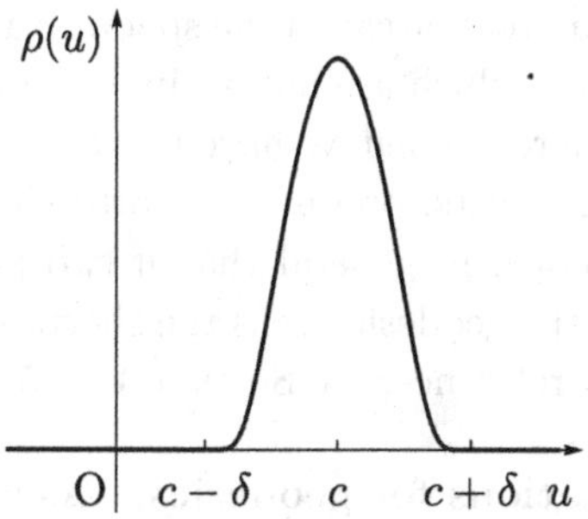

Fig. 10.3 The function ρ.

Consider an endpoint-fixing variation

$$\gamma_w(s) = \boldsymbol{p}(u(s) + w\rho(s)\alpha(s), v(s) + w\rho(s)\beta(s))$$

of $\gamma(s)$, whose variational vector field $V(s)$ is

$$\begin{aligned} V(s) &= \left.\frac{\partial \gamma_w(s)}{\partial w}\right|_{w=0} \\ &= \rho(s)(\alpha(s)\,\boldsymbol{p}_u(u(s), v(s)) + \beta(s)\,\boldsymbol{p}_v(u(s), v(s))) = \rho(s)\,\boldsymbol{n}_g(s). \end{aligned}$$

Substituting this into (10.4), we have

$$\begin{aligned} \left.\frac{d\mathcal{L}(\gamma_w)}{dw}\right|_{w=0} &= -\int_0^l \gamma''(s) \cdot V(s)\,ds \\ &= -\int_0^l \rho(s)\,(\gamma''(s) \cdot \boldsymbol{n}_g(s))\,ds = -\int_0^l \rho(s)\kappa_g(s)\,ds. \end{aligned}$$

Since $\rho(s)$ is positive on the interval $(c-\delta, c+\delta)$, $\rho(s)\kappa_g(s)$ does not change its sign on this interval. On the other hand, since $\rho(s) = 0$ outside this interval, we have

$$\left.\frac{d\mathcal{L}(\gamma_w)}{dw}\right|_{w=0} = -\int_{c-\delta}^{c+\delta} \rho(s)\kappa_g(s)\,ds \neq 0.$$

Hence if (10.2) holds for all endpoint-fixing variations, then $\kappa_g(c) = 0$, that is, the curve must be a geodesic. □

A shortest-length curve parametrized by a parameter proportional to the arc-length is a geodesic. However, the converse is not true in general. For example, consider a great circle on the sphere is a geodesic and take two points on the great circle which are not antipodal (symmetric with respect to the center of the sphere). Then we have two arcs of the great circle with given two points as endpoints, one is the shortest-length but the other is not. However, it is known in general that if two points on a geodesic is sufficiently close, then the geodesic arc is the shortest-length curves joining these two points (for a reference, see Section 3 of [5]).

The differential equations for geodesics. As in Section 7, the length of a curve in a surface is determined by the first fundamental form. So it is expected that geodesics are also determined from the first fundamental form alone. Letting E, F and G be the components of the first fundamental form of a surface $\boldsymbol{p}(u,v)$, we first formally define the following eight functions, and consider their meaning later,

$$(10.6)\qquad \begin{cases} \Gamma_{11}^1 := \dfrac{GE_u - 2FF_u + FE_v}{2(EG-F^2)}, \\ \Gamma_{11}^2 := \dfrac{2EF_u - EE_v - FE_u}{2(EG-F^2)}, \\ \Gamma_{12}^1 = \Gamma_{21}^1 := \dfrac{GE_v - FG_u}{2(EG-F^2)}, \\ \Gamma_{12}^2 = \Gamma_{21}^2 := \dfrac{EG_u - FE_v}{2(EG-F^2)}, \\ \Gamma_{22}^1 := \dfrac{2GF_v - GG_u - FG_v}{2(EG-F^2)}, \\ \Gamma_{22}^2 := \dfrac{EG_v - 2FF_v + FG_u}{2(EG-F^2)}. \end{cases}$$

These smooth functions are called the *Christoffel*[2] *symbols.* By definition, these functions are expressed in terms of the coefficients of the first fundamental form and their first derivatives. Using the Christoffel symbols, the second derivatives of $\boldsymbol{p}$ become

$$\begin{cases} \boldsymbol{p}_{uu} = \Gamma_{11}^1 \boldsymbol{p}_u + \Gamma_{11}^2 \boldsymbol{p}_v + L\nu, \\ \boldsymbol{p}_{uv} = \Gamma_{12}^1 \boldsymbol{p}_u + \Gamma_{12}^2 \boldsymbol{p}_v + M\nu, \\ \boldsymbol{p}_{vv} = \Gamma_{22}^1 \boldsymbol{p}_u + \Gamma_{22}^2 \boldsymbol{p}_v + N\nu \end{cases} \tag{10.7}$$

(see Proposition 11.1 in Section 11), where L, M, N are the coefficients of the second fundamental form and ν is the unit normal vector of the surface $\boldsymbol{p}$. Using this fact, we will soon show that geodesics can be characterized by a system of ordinary differential equations that depend only on the first fundamental form (cf. (10.8)).

For a curve $\gamma(t) = \boldsymbol{p}(u(t), v(t))$ lying on the surface p, the velocity and the acceleration vectors are computed as

$$\begin{aligned} \dot\gamma &= \dot u\, \boldsymbol{p}_u + \dot v\, \boldsymbol{p}_v\,, \\ \ddot\gamma &= \ddot u\, \boldsymbol{p}_u + \ddot v\, \boldsymbol{p}_v + \dot u^2 \boldsymbol{p}_{uu} + 2\dot u\, \dot v \boldsymbol{p}_{uv} + \dot v^2 \boldsymbol{p}_{vv} \\ &= \left(\ddot u + \dot u^2 \Gamma_{11}^1 + 2\dot u\, \dot v \Gamma_{12}^1 + \dot v^2 \Gamma_{22}^1\right) \boldsymbol{p}_u \\ &\quad + \left(\ddot v + \dot u^2 \Gamma_{11}^2 + 2\dot u\, \dot v \Gamma_{12}^2 + \dot v^2 \Gamma_{22}^2\right) \boldsymbol{p}_v + \left(\dot u^2 L + 2\dot u\, \dot v M + \dot v^2 N\right) \nu \end{aligned}$$

by the chain rule, then using the notation as in (10.1), we have

$$\begin{aligned} [\ddot\gamma]^{\mathrm{H}} &= \left(\ddot u + \dot u^2 \Gamma_{11}^1 + 2\dot u\, \dot v \Gamma_{12}^1 + \dot v^2 \Gamma_{22}^1\right) \boldsymbol{p}_u \\ &\quad + \left(\ddot v + \dot u^2 \Gamma_{11}^2 + 2\dot u\, \dot v \Gamma_{12}^2 + \dot v^2 \Gamma_{22}^2\right) \boldsymbol{p}_v. \end{aligned}$$

Here, since $\boldsymbol{p}_u$ and $\boldsymbol{p}_v$ are linearly independent, the necessary and sufficient condition for $\gamma(t) = \boldsymbol{p}(u(t), v(t))$ to be a geodesic is that $u(t)$, $v(t)$ satisfy the following differential equations:

$$\begin{cases} \ddot u + \dot u^2 \Gamma_{11}^1 + 2\dot u\, \dot v \Gamma_{12}^1 + \dot v^2 \Gamma_{22}^1 = 0, \\ \ddot v + \dot u^2 \Gamma_{11}^2 + 2\dot u\, \dot v \Gamma_{12}^2 + \dot v^2 \Gamma_{22}^2 = 0. \end{cases} \tag{10.8}$$

We call these the *equations for geodesics.*

Using new variables $\tilde u$, $\tilde v$, we can write the geodesic equations (10.8) as the following first-order differential equation system:

$$\begin{cases} \dfrac{du}{dt} = \tilde u, & \dfrac{d\tilde u}{dt} = -\Gamma_{11}^1 \tilde u^2 - 2\Gamma_{12}^1 \tilde u \tilde v - \Gamma_{22}^1 \tilde v^2, \\ \dfrac{dv}{dt} = \tilde v, & \dfrac{d\tilde v}{dt} = -\Gamma_{11}^2 \tilde u^2 - 2\Gamma_{12}^2 \tilde u \tilde v - \Gamma_{22}^2 \tilde v^2. \end{cases} \tag{10.9}$$

[2]Christoffel, Erwin Bruno (1829–1900).

Then we can now apply the fundamental theorem for ordinary differential equations (Theorem A.2.1 in Appendix A.2). Take a point $\mathrm{P} = \boldsymbol{p}(u_0, v_0)$ in the surface and a non-zero vector

$$\boldsymbol{w} = \xi\, \boldsymbol{p}_u(u_0, v_0) + \eta\, \boldsymbol{p}_v(u_0, v_0)$$

in the tangent plane at P, which is tangent to the surface at P. Then the initial conditions

$$u(0) = u_0, \quad v(0) = v_0, \quad \tilde{u}(0) = \frac{du}{dt}(0) = \xi, \quad \tilde{v}(0) = \frac{dv}{dt}(0) = \eta$$

for the system of differential equations (10.9) give existence of a unique solution for t sufficiently close to 0. For such a solution $(u(t), v(t))$, the curve $\gamma(t) = \boldsymbol{p}(u(t), v(t))$ is the unique geodesic on the surface with $\gamma(0) = \mathrm{P}$ and $\dot{\gamma}(0) = \boldsymbol{w}$.

This observation will be used in Section 11 for the proof of existence of geodesic polar coordinates.

The Gauss-Bonnet theorem. Geodesics are regarded as "straight" curves on the surface. A closed region bounded by three geodesic segments joining three distinct points on the surface is called a *geodesic triangle*, or a *triangle* for short, if the region is simply connected (see page 137), or equivalently, the region is homeomorphic to the triangular region

$$\{(x, y) \in \boldsymbol{R}^2 \mid x + y \leq 1, x \geq 0, y \geq 0\}$$

The three points are called *vertices*, and the geodesic segment joining them are called *edges*. The *interior angle* of a vertex of a geodesic triangle is the angle between two tangent vectors of the edges at the point, measured from the interior of the triangle. The following theorem is a generalization of the fact that the sum of interior angles of a planar triangle is π:

Theorem 10.6 (The Gauss-Bonnet[3] theorem). *Let $\angle\mathrm{A}$, $\angle\mathrm{B}$, $\angle\mathrm{C}$ be the interior angles of a geodesic triangle $\triangle\mathrm{ABC}$ on a regular surface.*

Then it holds that

$$\angle\mathrm{A} + \angle\mathrm{B} + \angle\mathrm{C} = \pi + \iint_{\triangle\mathrm{ABC}} K\, dA,$$

where dA is the area element (cf. (6.13) in Section 6) and K is the Gaussian curvature of the surface.

[3]Bonnet, Pierre Ossian (1819–1892). See page 30 for Gauss.

In particular, by the Gauss-Bonnet theorem, the sum of the interior angles of a triangle is greater than π (resp. less than π) if the Gaussian curvature is positive (resp. negative). See Problem **6** in this section.

The proof of Theorem 10.6 will be given in Section 11. The Gauss-Bonnet theorem can be applied not only to surfaces in the Euclidean 3-space, but also to general 2-dimensional Riemannian manifolds. Also, for triangles with non-geodesic edges, a formula of this type holds with a small modification. The statement and proof will be given in Section 14.

By the way, as pointed out in Section 7, the area element and the interior angles of triangles are determined by the coefficients of the first fundamental form E, F, G (see pages 73, 74). Also the geodesic equation (10.8) is written in terms of the coefficients of the first fundamental form. On the other hand, by the Gauss-Bonnet theorem, if one draws a geodesic triangle and measures its interior angles, the mean of the Gaussian curvature in the triangle can be determined. By these facts, it can be expected that the Gaussian curvature defined as (8.6) can be written in terms of the coefficients of the first fundamental form. In fact, the Gaussian curvature is written as

$$(10.10)\qquad \begin{aligned} K = {} & \frac{E(E_vG_v - 2F_uG_v + {G_u}^2)}{4(EG-F^2)^2} \\ & + \frac{F(E_uG_v - E_vG_u - 2E_vF_v - 2F_uG_u + 4F_uF_v)}{4(EG-F^2)^2} \\ & + \frac{G(E_uG_u - 2E_uF_v + {E_v}^2)}{4(EG-F^2)^2} - \frac{E_{vv} - 2F_{uv} + G_{uu}}{2(EG-F^2)} \end{aligned}$$

in terms of E, F, G (Theorem 11.2 in Section 11). This formula means that the Gaussian curvature, originally defined by the first and second fundamental forms, is expressed by the coefficients E, F, G of the first fundamental form and their first and second derivatives. Gauss named this fact as *Theorema Egregium.* In contrast to this, the mean curvature cannot be written in terms of the first fundamental form. In fact, the plane in Example 8.2 and the circular cylinder in Example 8.3 have the same first fundamental forms, but the mean curvatures are distinct.

Remark. As an application, we can prove the impossibility of creating a distance-angle-preserving map of the earth. Such a map would be a correspondence between a region of the earth (the sphere) and a region of the plane *preserving the length.*[4] If there were such a correspondence,

[4]For a precise definition, see page 76 in Section 7.

then there exists a coordinate system such that the coefficients of the first fundamental form satisfy $E = G = 1$, $F = 0$, because of Proposition 7.4. Then the Gaussian curvature must be identically zero, a contradiction.

After Gauss' discovery of these facts in the 19th century, the following idea was considered:

> *Can one consider surfaces independently of the ambient space? In this case, the first fundamental form can make sense. In this situation, not only lengths and angles, but also the notion of "measure of bending" is still available, and a meaningful theory of surfaces can be developed.*

This is one of the motivations that hatched the concept of manifolds [32].

An application to closed surfaces. A *closed surface* is a compact regular surface in $\boldsymbol{R}^3$ without boundary. The sphere and the torus (Problem **1** in Section 6) are examples of closed surfaces. In this section, we consider closed surfaces possibly with self-intersections, such as the Klein bottle and Wente tori (cf. page 65 in Section 6). By a *geodesic triangulation*, or a *triangulation* for short, we mean a decomposition of a closed surface into small geodesic triangles (Fig. 10.4, left). The following theorem is an important result that shows a relationship between the Gaussian curvature and topological properties of closed surfaces. It is also called the *Gauss-Bonnet theorem* because it is a direct consequence of Theorem 10.6.

Theorem 10.7 (The Gauss-Bonnet theorem for closed surfaces)**.** *Let S be a closed surface with a triangulation, and define the* Euler number $\chi(S)$ *of S as follows:*

$$\chi(S) := (\textit{number of vertices}) - (\textit{number of edges}) + (\textit{number of faces}). \tag{10.11}$$

Then

$$\iint_S K\, dA = 2\pi\chi(S)$$

holds. Here, K is the Gaussian curvature and dA is the area element.

Proof. Consider a triangulation of the surface S, and let $v_1, \ldots, v_n$ be the n vertices. We may assume that, each vertex v_j is shared by m_j-geodesic triangles $T_{j,1}, \ldots, T_{j,m_j}$ (Fig. 10.4, right). (The precise definition of triangulation and the existence of such a triangulation is given in Section 19. However, the reader can study the following proof without referring to that section.)

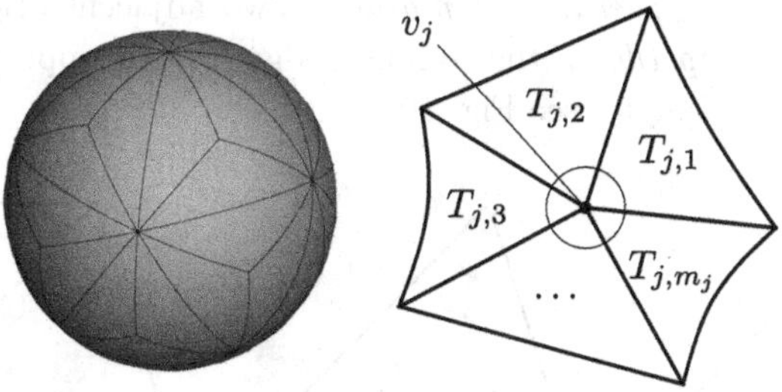

Fig. 10.4 A triangulation of the sphere and proof of Theorem 10.7.

Since the sum of the interior angles at the vertex v_j is 2π, by the Gauss-Bonnet theorem (Theorem 10.6) for geodesic triangles,

$$\begin{aligned} 2\pi n &= \sum_{i=1}^{n}\sum_{j=1}^{m_i}(\text{the interior angle of } T_{i,j} \text{ at } v_i) \\ &= (\text{the sum of the all interior angles of the triangles}) \\ &= \iint_S K\,dA + \pi \cdot (\text{the number of triangles}). \end{aligned}$$

Here, noticing that an edge is shared by two triangles and a triangle is bounded by three edges, we have

$$3 \cdot (\text{number of triangles}) = 2 \cdot (\text{number of edges}).$$

Then, recalling that the number of vertices is n, we have

$$\begin{aligned} \iint_S K\,dA &= 2\pi n - \pi \cdot (\text{number of triangles}) \\ &= 2\pi \cdot (\text{number of vertices}) \\ &\quad - 2\pi \cdot (\text{number of edges}) + 2\pi \cdot (\text{number of triangles}) \\ &= 2\pi\chi(S), \end{aligned}$$

which yields the conclusion. □

The left-hand side of the conclusion of Theorem 10.7 does not depend on geodesic triangulations of the surface. Moreover, it is known that this attains the same value for (not necessarily geodesic) triangulations. That is, the Euler number $\chi(S)$ does not depend on general triangulations of the surface.

For a geodesic triangulation, an *orientation* of a single triangle is a choice of the way rounding the triangle. If a triangle is oriented, such a choice is called the *positive orientation.* Two adjacent oriented triangles are said to be *compatible* if the common edge has the opposite direction as an edge of each triangle (see Fig. 10.5).

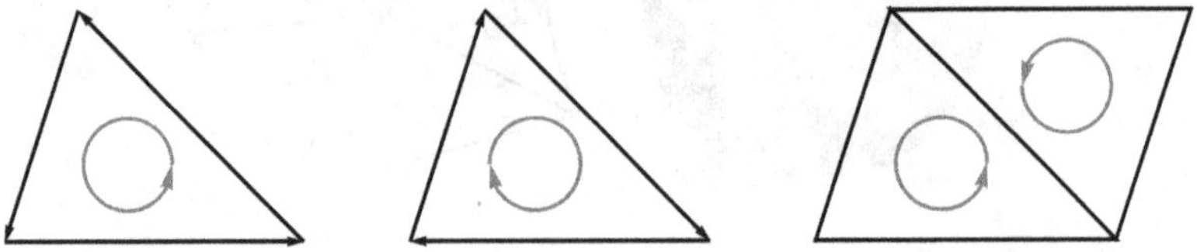

Orientation of a triangle. Two compatible triangles.

Fig. 10.5

For a triangulated surface, fix an orientation of one triangle, and choose compatible orientations of adjacent triangles. Continuing this procedure, if one can choose orientations of all triangles, the surface is said to be *orientable.* For example, the Möbius strip (Problem **2** in Section 6) is non-orientable, and the sphere is orientable.

We remark that *the proof of Theorem 10.7 works not only for orientable surfaces but also for non-orientable surfaces.*

Orientability of given regular surface is equivalent to existence of smooth unit normal vector ν defined on the whole surface. Refer to the definition of orientability of manifolds in Section 12, (13.18), and the first part of Section 19.

In this textbook, we treat mainly closed surfaces without self-intersections. It is known that such surfaces surround bounded domains (see [8, Section 18]), and it can be proven that they are orientable, since one can take inward unit normal vector fields defined on the entireties of the surfaces. On the other hand, Klein's bottle (cf. page 65 in Section 6) is an example of a non-orientable closed surface with self-intersections.

In general, a closed regular surface without self-intersection is orientable and diffeomorphic (namely, there is a smooth bijection whose inverse is also smooth) to one of the "doughnut with g holes" as in Fig. 10.6. The number g is called the *genus* of the surface.

The genus of a surface and the Euler number relate as

$$\chi(S) = 2 - 2g. \tag{10.12}$$

Then, by integrating the Gaussian curvature, one can compute the invariant describing the topology of the surface.

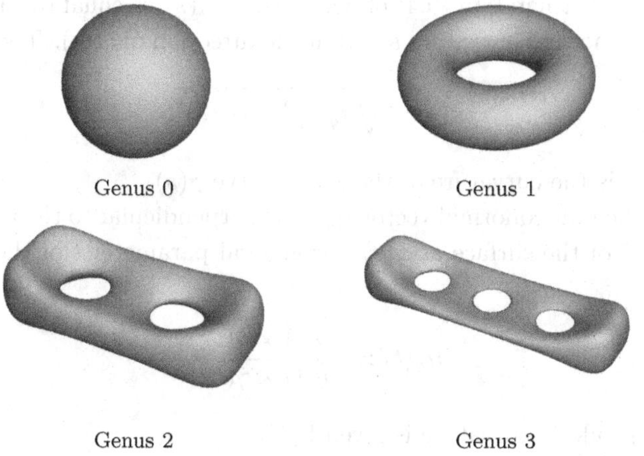

Fig. 10.6 The genus of surfaces.

Geodesic curvature. At the end of the section, we introduce the geodesic curvature which describes how a curve bends on a surface.

Let $\gamma(s) = \boldsymbol{p}(u(s), v(s))$ be a curve on the surface parametrized by the arc-length, and let $\nu(s) := \nu(u(s), v(s))$ be the unit normal vector of the surface at $\gamma(s)$. Then

$$\boldsymbol{n}_g(s) := \nu(s) \times \gamma'(s)$$

is the unit vector tangent to the surface at $\gamma(s)$ and perpendicular to $\gamma'(s)$. By definition, a triple $\{\gamma'(s), \boldsymbol{n}_g(s), \nu(s)\}$ gives a positive orthonormal basis of $\boldsymbol{R}^3$ for each fixed s (Fig. 10.7). We call this $\boldsymbol{n}_g(s)$ the *unit conormal vector* of the curve $\gamma(s)$ on the surface.

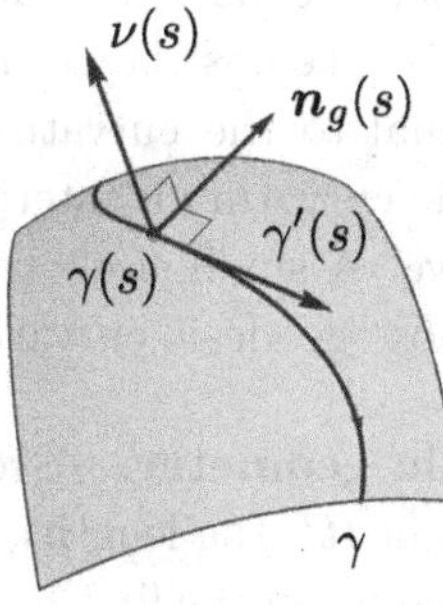

Fig. 10.7 The unit conormal vector.

Using the decomposition (9.4) in Section 9 and the formula of the scalar triple product (Proposition A.3.5), we define the value $\kappa_g(s)$ given by

$$\text{(10.13)} \qquad \kappa_g(s) = \boldsymbol{\kappa}_g(s) \cdot \boldsymbol{n}_g(s) = \gamma''(s) \cdot \boldsymbol{n}_g(s) = \det(\nu(s), \gamma'(s), \gamma''(s))$$

called the *geodesic curvature* of $\gamma(s)$. By definition, the absolute value of the geodesic curvature is equal to the norm $|\boldsymbol{\kappa}_g(s)|$ of the geodesic curvature vector given in (9.4).

On the other hand, in (9.4) of Section 9, $|\boldsymbol{\kappa}_n(s)|$ is equal to the absolute value of the normal curvature $\kappa_n(s)$ in the direction of $\gamma'(s)$. Then we have

$$\kappa(s) = \sqrt{\kappa_g(s)^2 + \kappa_n(s)^2}, \tag{10.14}$$

where $\kappa(s)$ is the curvature of the space curve $\gamma(s)$.

Since the unit conormal vector $\boldsymbol{n}_g(s)$ is perpendicular to the unit normal vector $\nu(s)$ of the surface at $\gamma(s)$, for general parameter t of the curve we have that

$$\boldsymbol{n}_g(t) := \frac{\nu(t) \times \dot{\gamma}(t)}{|\nu(t) \times \dot{\gamma}(t)|}. \tag{10.15}$$

Also, the geodesic curvature is given by

$$\kappa_g(t) = \frac{\det(\nu(t), \dot{\gamma}(t), \ddot{\gamma}(t))}{|\dot{\gamma}(t)|^3} = \frac{\ddot{\gamma}(t) \cdot \boldsymbol{n}_g(t)}{|\dot{\gamma}(t)|^2}, \tag{10.16}$$

see Problem **8** in this section.

The geodesic curvature of a geodesic is identically zero. Conversely, a curve whose geodesic curvature vanishes identically is a geodesic if the parameter is taken as the arc-length (Problem **9**). If the surface is a plane, $\kappa_g(t)$ is equal to the curvature as a planar curve (Problem **10**). Hence the geodesic curvature is interpreted as a generalization of the curvature of planar curve. A small circle (the intersection of the sphere and a plane) is a curve whose geodesic curvature is constant (Problem **11**).

Hyperbolic geometry. Even if one forgets that the surface is realized as a subset in $\boldsymbol{R}^3$, the lengths, the angles, and the Gaussian curvatures can be computed when the first fundamental form is given. In this way, one can investigate geometry on a region of the uv-plane if the first fundamental form is given on the region.

[1] Euclidean geometry. The uv-plane can be considered as a surface in $\boldsymbol{R}^3$, and the first fundamental form is given as $ds_{\mathrm{E}}^2 = du^2 + dv^2$ by Example 8.2 in Section 8.

Let $\mathrm{P} = (u_1, v_1)$ and $\mathrm{Q} = (u_2, v_2)$ be points of $\boldsymbol{R}^2$ and define the distance $d_{\mathrm{E}}(\mathrm{P}, \mathrm{Q})$ by the infimum of the lengths of piecewise smooth curves joining P and Q. Then it holds that

$$d_{\mathrm{E}}(\mathrm{P}, \mathrm{Q}) := \sqrt{(u_1 - u_2)^2 + (v_1 - v_2)^2}.$$

The geometry of planar figures with respect to such a distance is called *Euclidean geometry.* In Euclidean geometry, the shortest curve joining two points is the line segment. This distance is invariant under the transformation

$$\boldsymbol{R}^2 \ni \begin{pmatrix} u \\ v \end{pmatrix} \longmapsto A\begin{pmatrix} u \\ v \end{pmatrix} + \begin{pmatrix} b \\ c \end{pmatrix} \in \boldsymbol{R}^2,$$

where A is an orthogonal matrix and b, $c \in \boldsymbol{R}$, which plays the role of the congruent transformation, see (A.3.13) in Appendix A.3. Since the Gaussian curvature of the plane is 0, The Gauss-Bonnet theorem (Theorem 10.6) means that "the sum of interior angles of a triangle is π".

The term "Euclidean geometry" comes from Euclid, who developed the geometric theory of planar figures assuming the propositions (the five postulates and the five common notions) in his famous book *The Elements* [6]. Though Postulates and Common Notions were regarded as propositions for which the proofs are not needed, the fifth postulate

> *That, if a straight line falling on two straight lines make the interior angles on the same side less than two right angles, the two straight lines, if produced indefinitely, meet on that side on which are the angles less than the two right angles,*

was considered possibly to be able to be proven from other Postulates and Common Notions, until discovery of hyperbolic geometry.

[2] Hyperbolic geometry. The upper half-region

$$D := \{(u, v) \in \boldsymbol{R}^2 \mid v > 0\}$$

of the uv-plane is called the *upper half-plane.* Defining the first fundamental form

$$ds^2_{\mathrm{H}} = \frac{du^2 + dv^2}{v^2}$$

on the upper half-plane the length of the curves of the upper half-plane can be measured. The upper half-plane endowed with the first fundamental form ds^2_{H} is called the *hyperbolic plane.* For two given distinct points P, Q on D, there exists a unique Euclidean circle passing through P, Q and centered on the u-axis. (Here, a line parallel to the v-axis is also considered as a circle centered at the u-axis.) We call such a circle a *line*, and consider the circular arc $\overset{\frown}{\mathrm{PQ}}$ of such line as a *line segment* joining P and Q. Then it is the shortest curve joining P and Q (Fig. 10.8). We define the distance $d_{\mathrm{H}}(\mathrm{P}, \mathrm{Q})$ of two points $\mathrm{P} = (u_1, v_1)$, $\mathrm{Q} = (u_2, v_2)$ on D as the infimum of the length of piecewise smooth curves joining P and Q. By identifying $\boldsymbol{R}^2$ with $\boldsymbol{C}$ and by setting

$$z := u_1 + iv_1, \qquad w := u_2 + iv_2,$$

it holds that

$$d_{\mathrm{H}}(\mathrm{P},\mathrm{Q}) := \log \frac{|z-\bar{w}| + |z-w|}{|z-\bar{w}| - |z-w|},$$

which coincides with the length of the line segment $\overset{\frown}{\mathrm{PQ}}$. This distance is invariant under arbitrary finite compositions of the following transformations

$$D \ni z \longmapsto \frac{az+b}{cz+d} \in D, \qquad D \ni z \longmapsto -\bar{z} \in D,$$

where a, b, c, $d \in \boldsymbol{R}$ with $ad - bc = 1$, which give congruent transformations of the hyperbolic plane (see Problem **3** in Section 16).

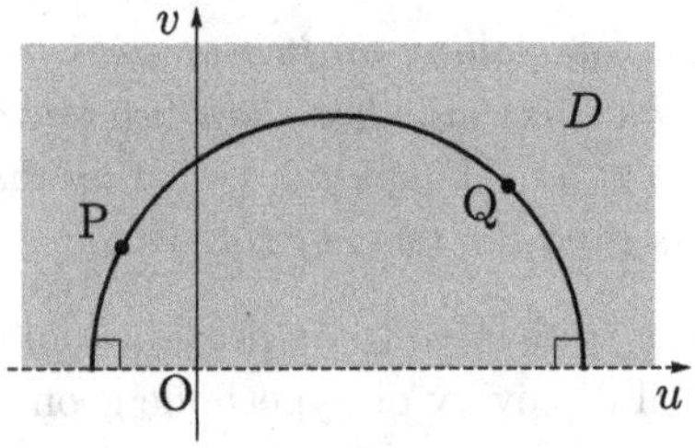

Fig. 10.8 The hyperbolic plane.

Moreover, the *angle* of an intersection of two lines is defined as the angle between two tangent vectors of the circles at the point. The postulates of Euclid are satisfied except the fifth postulate.

Such a geometry not satisfying the fifth postulate (*hyperbolic geometry*) was discovered by Lobachevski[5] and Bolyai[6] in the 19th century.

Since the Gaussian curvature of the first fundamental form ds^2_{H} is -1 (see Section 16), the sum of the interior angles of a geodesic triangle is less than π. For a right triangle $\triangle$ABC on the hyperbolic plane such that the angle $\angle$C is the right angle, it holds that

$$(*) \qquad \cosh \mathrm{AB} = \cosh \mathrm{BC} \cdot \cosh \mathrm{AC},$$

which is the hyperbolic correspondent of the Pythagorean theorem.

[5]Lobachevski, Nikolai Ivanovich (1793–1856).
[6]Bolyai, Janos (1802–1860).

The simple upper-half space model introduced here for hyperbolic geometry was discovered in the early 20th century. It is impossible for the entire upper half-plane D to be realized as a surface in $\boldsymbol{R}^3$ with first fundamental form ds_{H}^2 (Hilbert's theorem — see, for example, the final chapter of [13].) For more advanced reading on Euclidean and hyperbolic geometries, the book [10] is recommended.

Exercises 10

1 If a regular surface contains a line, show that the line is a geodesic if it is parametrized by the arc-length.

2 Show that a geodesic of the circular cylinder is one of the generating lines, the circles perpendicular to the generating lines, or the helices with the same radius of the cylinder, parametrized with the arc-length. Demonstrate this fact using the development of the cylinder.

3 Show that a geodesic of the sphere is a great circle parametrized by a parameter proportional to the arc-length.

4 Show that the generating curve of a surface of revolution is a geodesic if it is parametrized by the arc-length.

5 When a surface is symmetric with respect to reflection across a plane, show that the intersection of the surface and the plane is a geodesic.

6 Prove the special case of the Gauss-Bonnet theorem for geodesic triangles of the sphere of radius 1, as follows (see the figure below):

(1) Two great circles which meet at the A on the sphere with angle $\angle$A intersect again at the antipodal point (the point symmetric with respect to the center) of A. Then verify that the crescent shaped region bounded by two half circles has the area 2$\angle$A.

(2) Let $\angle$A, $\angle$B, $\angle$C be the interior angles of the geodesic triangle $\triangle$ABC on the sphere, and consider two crescent shaped regions bounded by two geodesics obtained by extending two edges at A, two regions bounded by geodesics extended two edges at B, and two regions bounded by geodesics extended two edges at C. Then verify that these 6-regions covers the whole sphere, but only $\triangle$ABC and the triangle whose vertices are antipodal points of A, B, C are covered 3-times.

(3) Comparing the area of the six crescent shaped regions above and the area of the unit sphere, show that

$$\angle \mathrm{A} + \angle \mathrm{B} + \angle \mathrm{C} = \pi + (\text{the area of } \triangle \mathrm{ABC}).$$

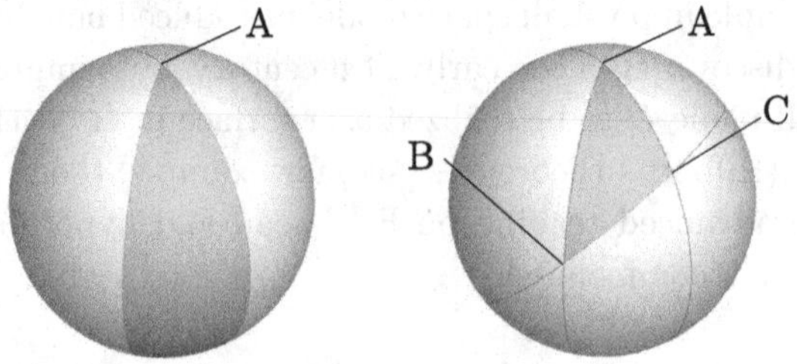

7 Under the situation of Problem **1** in Section 9, show that the absolute value of the geodesic curvature κ_g of γ is $\kappa|\sin\theta|$.

8 Derive the formula (10.16) of the geodesic curvature in general parameter, from the formula (10.13).

9 Show that a curve on the surface whose geodesic curvature identically zero is a geodesic if the parameter is taken as the arc-length.

10 Consider the xy-plane on $\boldsymbol{R}^3$, and choose the unit normal vector as $\nu := (0,0,1)$. Then show that the geodesic curvature κ_g of a curve $\gamma(t) := (x(t), y(t), 0)$ on the xy-plane is equal to the curvature κ of the planar curve $(x(t), y(t))$.

11 For a given θ ($|\theta| < \pi/2$), a space curve

$$\gamma_\theta(t) := (\cos\theta\cos t, \cos\theta\sin t, \sin\theta)$$

is a small circle as a curve on the unit sphere

$$S^2 = \{(x,y,z) \in \boldsymbol{R}^2 \,|\, x^2 + y^2 + z^2 = 1\}.$$

Compute the geodesic curvature of this small circle. Here, the unit normal vector of S^2 is taken as the outward normal.

12 A parametrization $\boldsymbol{p}(u,v)$ of a surface is called *orthogonal* if $F = \boldsymbol{p}_u \cdot \boldsymbol{p}_v$ vanishes identically. When this is the case, verify that (10.10) reduces to the identity

$$K = -\frac{1}{eg}\left[\left(\frac{g_u}{e}\right)_u + \left(\frac{e_v}{g}\right)_v\right],$$

where $e := \sqrt{E} = |p_u|$ and $g := \sqrt{G} = |p_v|$.

13* Let $\gamma(s)$ be a space curve parametrized by the arc-length with positive curvature $\kappa(s)$, and consider S^2. $\boldsymbol{e}(s) := \gamma'(s)$ as a curve on the unit sphere. Show that the geodesic curvature of $\boldsymbol{e}(s)$ is $\tau(s)/\kappa(s)$, where $\tau(s)$ is the torsion of $\gamma(s)$. Here, the unit normal vector of S^2 is taken as the outward normal.

14* Assume a space curve $\gamma(s)$ parametrized by the arc-length lies on the unit sphere S^2. Prove that the absolute value of its geodesic curvature $\kappa_g(s)$ as a curve on S^2 is $\sqrt{\kappa(s)^2 - 1}$, and $\kappa_g'(s) = \kappa(s)^2\tau(s)$. Here, $\kappa(s)$ and $\tau(s)$ are the curvature and the torsion of the space curve $\gamma(s)$, respectively.

11. Proof of the Gauss-Bonnet theorem*

In this section, we give the proof of the Gauss-Bonnet theorem in Section 10. The proof can be applied to 2-dimensional Riemannian manifolds with a slight modification (cf. Problems **3** and **4**). In Section 14, the Gauss-Bonnet theorem for general 2-dimensional Riemannian manifold and an alternative proof are introduced.

The Gauss equation. As a preliminary, we prove that the Gaussian curvature can be expressed by the coefficients of the first fundamental form (equation (10.10) of Section 10).

Proposition 11.1 (Equation (10.7)). *Let* $\boldsymbol{p}(u, v)$ *be a regular surface with whose coefficients of the first and the second fundamental forms are* E, F, G *and* L, M, N, *respectively. Then*

$$\begin{cases} \boldsymbol{p}_{uu} = \Gamma_{11}^1 \boldsymbol{p}_u + \Gamma_{11}^2 \boldsymbol{p}_v + L\nu, \\ \boldsymbol{p}_{uv} = \Gamma_{12}^1 \boldsymbol{p}_u + \Gamma_{12}^2 \boldsymbol{p}_v + M\nu, \\ \boldsymbol{p}_{vv} = \Gamma_{22}^1 \boldsymbol{p}_u + \Gamma_{22}^2 \boldsymbol{p}_v + N\nu \end{cases}$$

hold, where Γ_{ij}^k $(i, j, k = 1, 2)$ *are the Christoffel symbols defined in* (10.6) *in page 104, and* ν *is the unit normal vector of the surface.*

Proof. Since the vectors $\boldsymbol{p}_u(u, v)$, $\boldsymbol{p}_v(u, v)$, $\nu(u, v)$ form a basis of $\boldsymbol{R}^3$ for each (u, v), we can write

$$\begin{cases} \boldsymbol{p}_{uu} = \overline{\Gamma}_{11}^1 \boldsymbol{p}_u + \overline{\Gamma}_{11}^2 \boldsymbol{p}_v + \overline{L}\nu, \\ \boldsymbol{p}_{uv} = \overline{\Gamma}_{12}^1 \boldsymbol{p}_u + \overline{\Gamma}_{12}^2 \boldsymbol{p}_v + \overline{M}\nu, \\ \boldsymbol{p}_{vv} = \overline{\Gamma}_{22}^1 \boldsymbol{p}_u + \overline{\Gamma}_{22}^2 \boldsymbol{p}_v + \overline{N}\nu. \end{cases} \tag{11.1}$$

Here, $\overline{\Gamma}_{ij}^k$, $\overline{L}$, $\overline{M}$ and $\overline{N}$ are functions in (u, v). It is sufficient to show that $\overline{\Gamma}_{ij}^k$'s are equal to the Christoffel symbols, and $\overline{L}$, $\overline{M}$, $\overline{N}$ are equal to the coefficient of the second fundamental form.

Since the unit normal vector ν is perpendicular to both $\boldsymbol{p}_u$ and $\boldsymbol{p}_v$, taking the inner product with ν to (11.1), we have $L = \overline{L}$, $M = \overline{M}$, $N = \overline{N}$ by the definition of the second fundamental form.

Next, we prove $\overline{\Gamma}_{ij}^k = \Gamma_{ij}^k$. Taking the inner product of $\boldsymbol{p}_u$ with the first equation of (11.1) and applying the Leibniz rule (cf. Corollary A.3.10) $(\boldsymbol{p}_u \cdot \boldsymbol{p}_u)_u = 2\boldsymbol{p}_{uu} \cdot \boldsymbol{p}_u$, we have

$$\frac{1}{2}E_u = \boldsymbol{p}_{uu} \cdot \boldsymbol{p}_u = \overline{\Gamma}_{11}^1 E + \overline{\Gamma}_{11}^2 F.$$

Also, since $\boldsymbol{p}_{uu} \cdot \boldsymbol{p}_v = (\boldsymbol{p}_u \cdot \boldsymbol{p}_v)_u - \frac{1}{2}(\boldsymbol{p}_u \cdot \boldsymbol{p}_u)_v$, we have

$$F_u - \frac{1}{2}E_v = \overline{\Gamma}_{11}^1 F + \overline{\Gamma}_{11}^2 G$$

by taking inner product to $\boldsymbol{p}_v$ to the first equation of (11.1). Similarly, taking inner products with $\boldsymbol{p}_u$ and $\boldsymbol{p}_v$ to the second and the third equation of (11.1), we have

$$\frac{1}{2}E_v = \overline{\Gamma}_{12}^1 E + \overline{\Gamma}_{12}^2 F, \qquad \frac{1}{2}G_u = \overline{\Gamma}_{12}^1 F + \overline{\Gamma}_{12}^2 G,$$
$$F_v - \frac{1}{2}G_u = \overline{\Gamma}_{22}^1 E + \overline{\Gamma}_{22}^2 F, \qquad \frac{1}{2}G_v = \overline{\Gamma}_{22}^1 F + \overline{\Gamma}_{22}^2 G.$$

Summing up, we can write the above relations in matrix form as

$$\begin{pmatrix} E & F \\ F & G \end{pmatrix} \begin{pmatrix} \overline{\Gamma}_{11}^1 & \overline{\Gamma}_{12}^1 & \overline{\Gamma}_{22}^1 \\ \overline{\Gamma}_{11}^2 & \overline{\Gamma}_{12}^2 & \overline{\Gamma}_{22}^2 \end{pmatrix} = \frac{1}{2} \begin{pmatrix} E_u & E_v & 2F_v - G_u \\ 2F_u - E_v & G_u & G_v \end{pmatrix}. \tag{11.2}$$

Since the first fundamental matrix is a regular matrix as seen in Section 7, multiplying the inverse matrix of it to (11.2) from the left, we have

$$\widehat{I}^{-1} = \begin{pmatrix} E & F \\ F & G \end{pmatrix}^{-1} = \frac{1}{EG - F^2} \begin{pmatrix} G & -F \\ -F & E \end{pmatrix}.$$

Thus, $\overline{\Gamma}_{ij}^k$ coincides with Γ_{ij}^k. □

Theorem 11.2 (The Gauss equation). *The Gaussian curvature K of a regular surface $p(u, v)$ is expressed in terms of the coefficients E, F, G of the first fundamental form as*

$$\begin{aligned} K = {} & \frac{E(E_v G_v - 2F_u G_v + (G_u)^2)}{4(EG - F^2)^2} \\ & + \frac{F(E_u G_v - E_v G_u - 2E_v F_v - 2F_u G_u + 4F_u F_v)}{4(EG - F^2)^2} \\ & + \frac{G(E_u G_u - 2E_u F_v + (E_v)^2)}{4(EG - F^2)^2} - \frac{E_{vv} - 2F_{uv} + G_{uu}}{2(EG - F^2)}. \end{aligned} \tag{11.3}$$

Proof. By the proof of Proposition 11.1, the Christoffel symbols satisfy

$$\Gamma_{11}^1 E + \Gamma_{11}^2 F = \frac{1}{2} E_u, \quad \Gamma_{11}^1 F + \Gamma_{11}^2 G = F_u - \frac{1}{2} E_v.$$

Here, since $\boldsymbol{p}_{uuv} \cdot \boldsymbol{p}_v = (\boldsymbol{p}_{uu} \cdot \boldsymbol{p}_v)_v - \boldsymbol{p}_{uu} \cdot \boldsymbol{p}_{vv}$, Proposition 11.1 yields

$$\begin{aligned}
\boldsymbol{p}_{uuv} \cdot \boldsymbol{p}_v &= ((\Gamma_{11}^1 \boldsymbol{p}_u + \Gamma_{11}^2 \boldsymbol{p}_v + L\nu) \cdot \boldsymbol{p}_v)_v \\
&\quad - (\Gamma_{11}^1 \boldsymbol{p}_u + \Gamma_{11}^2 \boldsymbol{p}_v + L\nu) \cdot (\Gamma_{22}^1 \boldsymbol{p}_u + \Gamma_{22}^2 \boldsymbol{p}_v + N\nu) \\
&= (\Gamma_{11}^1 F + \Gamma_{11}^2 G)_v \\
&\quad - (\Gamma_{11}^1 E + \Gamma_{11}^2 F)\, \Gamma_{22}^1 - (\Gamma_{11}^1 F + \Gamma_{11}^2 G)\, \Gamma_{22}^2 - LN \\
&= \left(F_u - \frac{1}{2} E_v\right)_v - \frac{1}{2} E_u \Gamma_{22}^1 - \left(F_u - \frac{1}{2} E_v\right) \Gamma_{22}^2 - LN.
\end{aligned}$$

Substituting (10.6) into these equations, we can deduce

(11.4)

$$\begin{aligned}
\boldsymbol{p}_{uuv} \cdot \boldsymbol{p}_v &= F_{uv} - \frac{1}{2} E_{vv} \\
&\quad + \frac{E(E_v G_v - 2F_u G_v)}{4(EG - F^2)} + \frac{G(E_u G_u - 2E_u F_v)}{4(EG - F^2)} \\
&\quad + \frac{F(E_u G_v + E_v G_u - 2F_u G_u - 2E_v F_v + 4F_u F_v)}{4(EG - F^2)} - LN.
\end{aligned}$$

Similarly, since $\boldsymbol{p}_{uvu} \cdot \boldsymbol{p}_v = (\boldsymbol{p}_{uv} \cdot \boldsymbol{p}_v)_u - \boldsymbol{p}_{uv} \cdot \boldsymbol{p}_{uv}$,

$$\boldsymbol{p}_{uvu} \cdot \boldsymbol{p}_v = \frac{1}{2} G_{uu} - \frac{EG_u^2 - 2FE_v G_u + GE_v^2}{4(EG - F^2)} - M^2 \tag{11.5}$$

holds. Since $\boldsymbol{p}_{uuv} = \boldsymbol{p}_{uvu}$, the right-hand side of (11.4) is equal to that of (11.5). Hence, we can write $LN - M^2$ in terms of E, F and G. Then by Equation (8.7), the Gaussian curvature is $K = (LN - M^2)/(EG - F^2)$, so taking the form for $LN - M^2$ that we have just found and dividing it by $EG - F^2$, we arrive at the result. □

Proof of the Gauss-Bonnet theorem. To prove the Gauss-Bonnet theorem, we do need to write the Gaussian curvature in terms of the entries of the first fundamental form as in (11.3), but this equation is very unwieldy. In order to simplify this expression for the Gaussian curvature, we will consider a special choice of coordinates for the surface.

The polar coordinate system of the plane $\boldsymbol{R}^2$ is the correspondence of a pair (r, θ) $(r > 0)$ of two numbers to the point on the plane where the distance from the origin is r and the angle measured from the positive direction of the x-axis is θ. Similarly, for a fixed point P on the surface, take two unit tangent vectors $\boldsymbol{e}_1$, $\boldsymbol{e}_2$ of the surface at P mutually orthogonal. Then the *geodesic polar coordinate system* of the surface around P is the correspondence of a pair (r, θ) $(r > 0)$ to the unique point Q on the surface satisfying the following (Fig. 11.1):

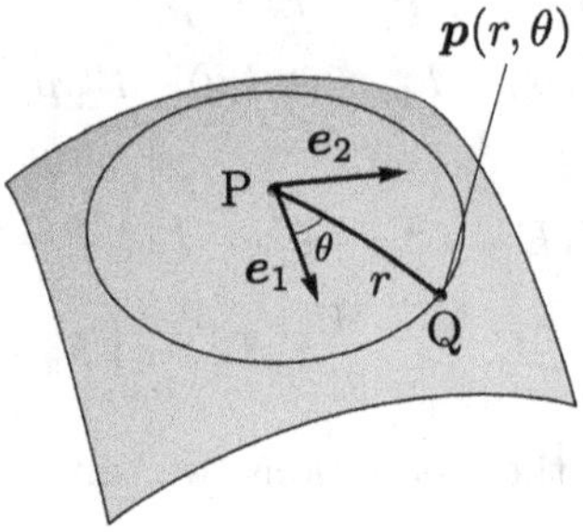

Fig. 11.1 The geodesic polar coordinate system.

- there exists a geodesic from P to Q with length r,
- and the angle between the velocity of the geodesic at P and $\boldsymbol{e}_1$ is θ.

Such a coordinate system exists for a sufficiently small range of r. The proof of this fact will be given at the end of this section.

Parametrize a surface as $\boldsymbol{p} = \boldsymbol{p}(r,\theta)$ by the geodesic polar coordinate system at P. Then for a fixed θ, $\boldsymbol{p}(s,\theta)$ is a geodesic passing through P at $s = 0$, and s is the arc-length parameter.

Lemma 11.3. *For a regular surface* $\boldsymbol{p}(r,\theta)$ *parametrized by geodesic polar coordinates, we have*

$$\boldsymbol{p}_r \cdot \boldsymbol{p}_r = 1, \quad \boldsymbol{p}_r \cdot \boldsymbol{p}_\theta = 0, \quad \lim_{r\to 0+} \frac{\boldsymbol{p}_\theta \cdot \boldsymbol{p}_\theta}{r^2} = 1 \qquad \left(\boldsymbol{p}_r := \frac{\partial \boldsymbol{p}}{\partial r},\ \boldsymbol{p}_\theta := \frac{\partial \boldsymbol{p}}{\partial \theta}\right),$$

where $r \to 0^+$ *means* r *tends to* 0 *from above.*

Proof. For a fixed angle θ, the map $r \mapsto \boldsymbol{p}(r,\theta)$ is a geodesic where r is the arc-length parameter. Hence the first equation follows.

Since $\boldsymbol{p}(0,\theta) = \mathrm{P}$ (constant), and $\boldsymbol{p}_r(0,\theta)$ is the tangent vector of the geodesic $r \mapsto \boldsymbol{p}(r,\theta)$ at $r = 0$, it holds that

$$\boldsymbol{p}_\theta(0,\theta) = 0, \quad \boldsymbol{p}_r(0,\theta) = (\cos\theta)\,\boldsymbol{e}_1 + (\sin\theta)\,\boldsymbol{e}_2.$$

Here, $\boldsymbol{e}_1$ and $\boldsymbol{e}_2$ are the unit vectors which are used in the definition of the geodesic polar coordinate system. We shall prove the second and third equations. Since $\boldsymbol{p}_r \cdot \boldsymbol{p}_r = 1$, we have

$$(\boldsymbol{p}_r \cdot \boldsymbol{p}_\theta)_r = \boldsymbol{p}_{rr} \cdot \boldsymbol{p}_\theta + \boldsymbol{p}_r \cdot \boldsymbol{p}_{\theta r} = \boldsymbol{p}_{rr} \cdot \boldsymbol{p}_\theta + \frac{1}{2}(\boldsymbol{p}_r \cdot \boldsymbol{p}_r)_\theta = \boldsymbol{p}_{rr} \cdot \boldsymbol{p}_\theta.$$

Here, since the map $r \mapsto \boldsymbol{p}(r,\theta)$ is a geodesic, $\boldsymbol{p}_{rr}$ is proportional to the normal vector of the surface, that is $\boldsymbol{p}_{rr} \cdot \boldsymbol{p}_\theta = 0$. Then for a fixed angle θ, $\boldsymbol{p}_r \cdot \boldsymbol{p}_\theta$ does not depend on r. Hence we have the second equation:

$$\boldsymbol{p}_r(r,\theta) \cdot \boldsymbol{p}_\theta(r,\theta) = \boldsymbol{p}_r(0,\theta) \cdot \boldsymbol{p}_\theta(0,\theta) = 0.$$

Moreover, since $\boldsymbol{p}(r,\theta)$ is smooth (i.e., of class C^∞) with respect to (r,θ), using L'Hôpital's rule (Theorem A.1.3 in Appendix A.1) twice, we obtain the third equation:

$$\begin{aligned}
\lim_{r\to 0+} \frac{\boldsymbol{p}_\theta \cdot \boldsymbol{p}_\theta}{r^2} &= \lim_{r\to 0+} \frac{\boldsymbol{p}_{r\theta} \cdot \boldsymbol{p}_\theta}{r} \\
&= \lim_{r\to 0+} (\boldsymbol{p}_{rr\theta} \cdot \boldsymbol{p}_\theta + \boldsymbol{p}_{r\theta} \cdot \boldsymbol{p}_{r\theta}) = \lim_{r\to 0+} \boldsymbol{p}_{r\theta} \cdot \boldsymbol{p}_{r\theta} \\
&= (-\sin\theta\, \boldsymbol{e}_1 + \cos\theta\, \boldsymbol{e}_2) \cdot (-\sin\theta\, \boldsymbol{e}_1 + \cos\theta\, \boldsymbol{e}_2) = 1.
\end{aligned}$$

Here, we used the identity $\lim_{r\to 0+} \boldsymbol{p}_\theta(r,\theta) = \boldsymbol{0}$ and $\lim_{r\to 0+}$ means the right-hand limit. □

For a surface $\boldsymbol{p}(r,\theta)$ parametrized by a geodesic polar coordinate system (r,θ) we define a function

$$h(r,\theta) := \sqrt{\boldsymbol{p}_\theta(r,\theta) \cdot \boldsymbol{p}_\theta(r,\theta)}. \tag{11.6}$$

Then by Lemma 11.3, h satisfies

$$\lim_{r\to 0+} \frac{h}{r} = 1 \qquad \text{and then} \qquad \lim_{r\to 0+} h_r = 1. \tag{11.7}$$

Using this h, the coefficients of the first fundamental form of $\boldsymbol{p}(r,\theta)$ are expressed as

$$E = \boldsymbol{p}_r \cdot \boldsymbol{p}_r = 1, \qquad F = \boldsymbol{p}_r \cdot \boldsymbol{p}_\theta = 0, \qquad G = \boldsymbol{p}_\theta \cdot \boldsymbol{p}_\theta = h^2. \tag{11.8}$$

Thus the first fundamental form can be written as

$$ds^2 = dr^2 + h^2\, d\theta^2. \tag{11.9}$$

Substituting (11.8) into (11.3), the Gaussian curvature is written as

$$K = -\frac{h_{rr}}{h}, \tag{11.10}$$

and the area element dA as in (7.14) s in the form

$$dA = h\,dr\,d\theta. \tag{11.11}$$

Substituting (10.6) and (10.6) into (10.8), we have that the necessary and sufficient condition for a curve $\gamma(s) = \boldsymbol{p}(r(s), \theta(s))$ on the surface $\boldsymbol{p}(r, \theta)$ to be a geodesic is

$$\begin{cases} r'' - h\,h_r(\theta')^2 = 0, \\ \theta'' + 2\dfrac{h_r}{h}r'\theta' + \dfrac{h_\theta}{h}(\theta')^2 = 0 \end{cases} \qquad \left({}' = \frac{d}{ds}\right). \tag{11.12}$$

We show the following lemma using these equations.

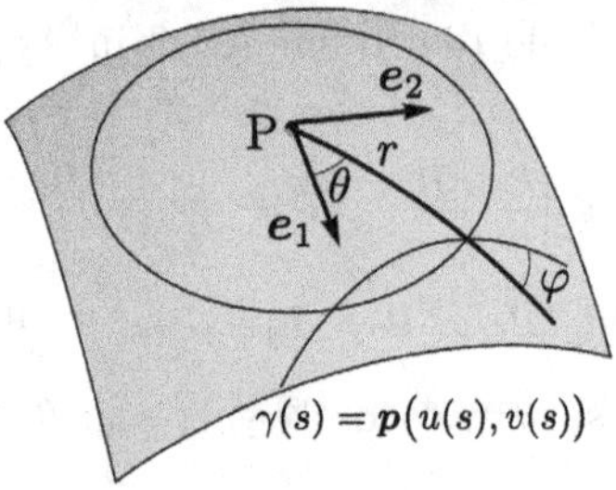

Fig. 11.2 The angle φ in Lemma 11.4.

Lemma 11.4. *For a regular surface $\boldsymbol{p}(r, \theta)$ with geodesic polar coordinates, let $\gamma(s)$ be a unit-speed geodesic (not necessarily passing through the origin) with velocity vector $\gamma'(s)$ making angle $\varphi(s)$ with the vector $\boldsymbol{p}_r$ in $\boldsymbol{R}^3$ (see Fig. 11.2). Then $\varphi' = -\theta' h_r$ holds, where h is the function defined in* (11.9).

Proof. Since $F = 0$ by (11.8), $\boldsymbol{p}_r$ and $\boldsymbol{p}_\theta/h$ form an orthonormal basis on the tangent plane. Then an arbitrary unit vector tangent to the surface is expressed as

$$(\cos\varphi)\boldsymbol{p}_r + \left(\frac{\sin\varphi}{h}\right)\boldsymbol{p}_\theta.$$

The "angle" in the statement of the lemma refers to this angle φ. (Fig. 11.2).

Letting $\gamma(s) = \boldsymbol{p}(r(s), \theta(s))$, we have $\gamma' = r'\boldsymbol{p}_r + \theta'\boldsymbol{p}_\theta$, and then by definition of the angle φ, it holds that

$$r' = \cos\varphi, \qquad \theta' = \frac{1}{h}\sin\varphi. \tag{11.13}$$

Then the first equation of (11.12) implies that

$$0 = (\cos\varphi)' - hh_r\frac{1}{h^2}\sin^2\varphi = -(\varphi' + h_r\theta')\sin\varphi.$$

Hence the conclusion follows if $\sin\varphi \neq 0$. When $\sin\varphi(s) = 0$ at $s = s_0$, if s_0 is a limit of a sequence $\{s_n\}$ such that $\sin\varphi(s_n) \neq 0$, the continuity of φ implies the conclusion. On the other hand, if $\sin\varphi(s) = 0$ on an open interval including s_0, $\varphi' = 0$, and the conclusion follows because $\theta' = 0$ by (11.13). □

Proof of the Gauss-Bonnet theorem (Theorem 10.6). Under the preliminaries above, we give a proof of the Gauss-Bonnet theorem (Theorem 10.6) for geodesic triangles. Dividing geodesic triangle into smaller geodesic triangles, if necessary, each triangle is contained in a geodesic polar coordinate system centered at a vertex of the triangle. (The proof of this fact is given in Section 19.) If the theorem is proved for such a small triangle, the Gauss-Bonnet theorem for general geodesic triangles by summing up the result for the small triangles (see Fig. 11.3).

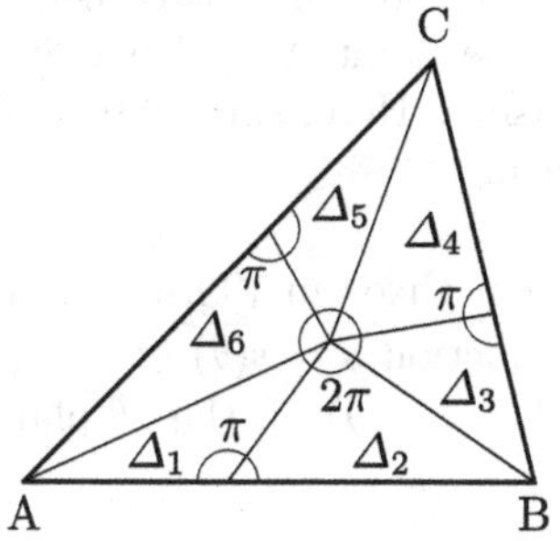

Fig. 11.3

For example, if Theorem 10.6 holds for the triangles Δ_j $(j = 1, \dots, 6)$ in Fig. 11.3, then we have

$$\begin{aligned}\iint_{\triangle\mathrm{ABC}} K\,dA &= \sum_{j=1}^{6}\iint_{\Delta_j} K\,dA = \angle\mathrm{A} + \angle\mathrm{B} + \angle\mathrm{C} + 3\pi + 2\pi - 6\pi \\ &= \angle\mathrm{A} + \angle\mathrm{B} + \angle\mathrm{C} - \pi,\end{aligned}$$

that is, the theorem holds for the triangle $\triangle$ABC.

Thus, we assume from now on that the surface is parametrized by the geodesic polar coordinate system (r, θ) centered at A as $\boldsymbol{p}(r, \theta)$, and the geodesic triangle $\triangle$ABC is contained in the image of a region determined by this parametrization.

Since the interior angle of the vertex A is $\angle$A, the coordinates of B and C may be assumed as $(r_{\mathrm{B}}, 0)$ and $(r_{\mathrm{C}}, \angle\mathrm{A})$ respectively, without loss of generality, where r_{B} and r_{C} are the lengths of the edges AB and AC, respectively. By the definition of the geodesic polar coordinate system, $s \mapsto \boldsymbol{p}(s,\theta)$ is a geodesic passing through A for each fixed θ. In particular, $s \mapsto \boldsymbol{p}(s,0)$ represents the edge AB and $s \mapsto \boldsymbol{p}(s,\angle\mathrm{A})$ represents the edge AC.

We parametrize the geodesic BC as

$$\gamma(s) = \gamma_{\mathrm{BC}}(s) = \boldsymbol{p}(r(s), \theta(s)) \qquad (0 \leq s \leq l),$$

where s is the arc-length parameter. Here, l is the length of BC, $\gamma_{\mathrm{BC}}(0) =$ B, and $\gamma_{\mathrm{BC}}(l) =$ C. Then by the reason below, $\theta'(s) \neq 0$ holds for each s, and θ is an increasing function.

In fact, if there exists s_0 such that $\theta'(s_0) = 0$, the edge BC tangents to the geodesic $s \mapsto \boldsymbol{p}(s, \theta(s_0))$ passing through A. Then by the uniqueness of geodesics (cf. page 106), these two geodesics must coincide. In particular γ_{BC} passes the point A, which is impossible. Hence $\theta(s)$ is increasing or decreasing. Here, since $\theta(0) = 0$ and $\theta(l) = \angle\mathrm{A}$, the function $\theta(s)$ is increasing.

By the inverse function theorem (Theorem A.1.5 in Appendix A.1), there exists the inverse function $s = s(\theta)$ of $\theta = \theta(s)$. Then the edge BC is represented as a graph $r = r(\theta)$ on the $r\theta$-plane. Hence $\triangle$ABC is the image of the set

$$\{(r,\theta) \mid 0 \leq r \leq r(\theta), 0 \leq \theta \leq \angle\mathrm{A}\}$$

on the $r\theta$-plane. Using this, (11.10), (11.11) and (11.7), we have

$$\begin{aligned}
\iint_{\triangle\mathrm{ABC}} K\, dA &= -\iint_{\triangle\mathrm{ABC}} h_{rr}\, dr\, d\theta = -\int_0^{\angle\mathrm{A}} d\theta \int_0^{r(\theta)} h_{rr}\, dr \\
&= -\int_0^{\angle\mathrm{A}} [h_r]_{r=0}^{r=r(\theta)}\, d\theta = -\int_0^{\angle\mathrm{A}} \{h_r(r(\theta),\theta) - h_r(0,\theta)\}\, d\theta \\
&= -\int_0^{\angle\mathrm{A}} \{h_r(r(\theta),\theta) - 1\}\, d\theta = \angle\mathrm{A} - \int_0^{\angle\mathrm{A}} h_r(r(\theta),\theta)\, d\theta.
\end{aligned}$$

Regarding the geodesic $\gamma = \gamma_{\mathrm{BC}}$, take the angle φ as in Lemma 11.4, then we have $h_r = -\varphi'/\theta'$. By the parameter change $s = s(\theta)$, φ is considered as a function of θ.

Thus, we have

$$\iint_{\triangle ABC} K\,dA = \angle A + \int_0^{\angle A} \frac{d\varphi}{d\theta}\,d\theta = \angle A + \varphi(\angle A) - \varphi(0).$$

Here, the interior angle $\angle B$ at the vertex B is equal to $\pi - \varphi(0)$, the interior angle $\angle C$ at the vertex C is equal to $\varphi(\angle A)$ (Fig. 11.4).

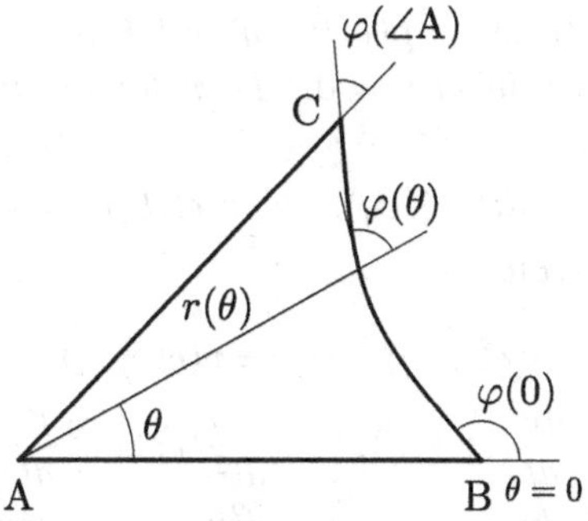

Fig. 11.4

Summing up, we have

$$\iint_{\triangle ABC} K\,dA = \angle A + \varphi(\angle A) - \varphi(0)$$
$$= \angle A + \angle C - (\pi - \angle B) = \angle A + \angle B + \angle C - \pi,$$

which is the conclusion. □

Existence of the geodesic polar coordinate system. We now prove the existence of the geodesic polar coordinate system described in page 120. Fix a point P on the surface. Applying a rotation and a translation if necessary, we may assume that the point P is the origin, and the tangent plane of the surface at P is the xy-plane. Then by the proof of Theorem 8.7 in Section 8, the surface is parametrized as $\boldsymbol{p}(u,v) = (u, v, f(u,v))$, that is, the graph of $z = f(x,y)$. In this case, it holds that

$$E(0,0) = G(0,0) = 1, \qquad F(0,0) = 0,$$

and $\boldsymbol{p}_u(0,0) = (1,0,0)$ and $\boldsymbol{p}_v(0,0) = (0,1,0)$ are mutually orthogonal unit tangent vectors of the surface at P. Then an arbitrary $\boldsymbol{w}$ tangent to the surface at P is expressed uniquely in the form

$$\boldsymbol{w} = (\xi, \eta, 0) = \xi\,\boldsymbol{p}_u(0,0) + \eta\,\boldsymbol{p}_v(0,0) \qquad (\xi, \eta \in \boldsymbol{R}). \tag{11.14}$$

Denote by

$$\gamma(t;\boldsymbol{w}) = \boldsymbol{p}(u(t;\xi,\eta), v(t;\xi,\eta))$$

the geodesic passing through P with velocity $\boldsymbol{w}$ at $t = 0$. Then $u(t) = u(t;\xi,\eta)$, $v(t) = v(t;\xi,\eta)$ are the solution of the differential equation (10.9) for geodesics satisfying the initial conditions

$$(u(0), v(0), \tilde{u}(0), \tilde{v}(0)) = (0, 0, \xi, \eta). \tag{11.15}$$

Lemma 11.5. *Let* $\gamma(t;\boldsymbol{w}) = \boldsymbol{p}(u(t;\xi,\eta), v(t;\xi,\eta))$ *be the geodesic passing through* P *with velocity* $\boldsymbol{w}$ *at* $t = 0$. *Then for an arbitrary constant* c, *it holds that*

$$u(ct;\xi,\eta) = u(t;c\xi,c\eta), \qquad v(ct;\xi,\eta) = v(t;c\xi,c\eta),$$

that is, $\gamma(ct;\boldsymbol{w}) = \gamma(t;c\boldsymbol{w})$.

Proof. Setting $\hat{u}(t) := u(ct;\xi,\eta)$, $\hat{v}(t) := v(ct;\xi,\eta)$,

$$\frac{d\hat{u}}{dt}(t) = c\frac{du}{dt}(ct;\xi,\eta), \qquad \frac{d^2\hat{u}}{dt^2}(t) = c^2\frac{d^2u}{dt^2}(ct;\xi,\eta),$$
$$\frac{d\hat{v}}{dt}(t) = c\frac{dv}{dt}(ct;\xi,\eta), \qquad \frac{d^2\hat{v}}{dt^2}(t) = c^2\frac{d^2v}{dt^2}(ct;\xi,\eta)$$

hold, and then $\hat{u}(t)$ and $\hat{v}(t)$ are the solution of equation (10.8). On the other hand, $u(t) := u(t;c\xi,c\eta)$, $v(t) := v(t;c\xi,c\eta)$ are the solution of (10.8) with the initial condition $(0,0,c\xi,c\eta)$. Then by the uniqueness of the solution to ordinary differential equations (Theorem A.2.1 of Appendix A.2), we have $\hat{u}(t) = u(t;c\xi,c\eta)$, $\hat{v}(t) = v(t;c\xi,c\eta)$. □

Lemma 11.6. *There exists a positive number* ε *such that for an arbitrary* $(\xi,\eta) \in \boldsymbol{R}^2$ *satisfying* $\xi^2 + \eta^2 < \varepsilon^2$, *the solutions* $u(t;\xi,\eta)$, $v(t;\xi,\eta)$ *of the differential equation* (10.9) *with the initial condition* (11.15) *is defined on the interval of* t *containing the interval* $[0,1]$.

Proof. By the fundamental theorem for ordinary differential equations (Theorem A.2.1 in Appendix A.2), there exist sufficiently small positive numbers $\tilde{\varepsilon}$ and δ, such that if $\xi^2 + \eta^2 < \tilde{\varepsilon}^2$, $u(t;\xi,\eta)$, $v(t;\xi,\eta)$ are defined on the interval $-\delta < t < \delta$. Here, we set $\tilde{\delta} := \delta/2$ and $\varepsilon := \tilde{\varepsilon}\tilde{\delta}$. Then for (ξ,η) satisfying $\xi^2 + \eta^2 < \varepsilon^2$, $u(t;\xi/\tilde{\delta},\eta/\tilde{\delta})$ and $v(t;\xi/\tilde{\delta},\eta/\tilde{\delta})$ are defined on $-\delta < t < \delta$, in particular on $0 \le t \le \tilde{\delta}$, because $(\xi/\tilde{\delta})^2 + (\eta/\tilde{\delta})^2 < \tilde{\varepsilon}^2$. On the other hand, by Lemma 11.5, we have

$$u(t;\xi,\eta) = u\left(\tilde{\delta}t;\frac{\xi}{\tilde{\delta}},\frac{\eta}{\tilde{\delta}}\right), \qquad v(t;\xi,\eta) = v\left(\tilde{\delta}t;\frac{\xi}{\tilde{\delta}},\frac{\eta}{\tilde{\delta}}\right).$$

The right-hand side of each equation above is defined on $0 \le \tilde{\delta}t \le \tilde{\delta}$, then $u(t;\xi,\eta)$ and $v(t;\xi,\eta)$ are defined on $0 \le t \le 1$. □

By Lemma 11.6, we can define a map φ_{P} defined the region $D_\varepsilon := \{(\xi,\eta) \in \boldsymbol{R}^2 \mid \xi^2+\eta^2 < \varepsilon^2\}$ on the $\xi\eta$-plane into the uv-plane as

$$\varphi_{\mathrm{P}} \colon (\xi,\eta) \longmapsto \varphi_{\mathrm{P}}(\xi,\eta) = (u,v) = (u(1;\xi,\eta), v(1;\xi,\eta)). \tag{11.16}$$

This φ_{P} maps the point (ξ,η) to the point on the geodesic with the initial velocity $\xi\,\boldsymbol{p}_u(0,0)+\eta\,\boldsymbol{p}_v(0,0)$ at $t=1$, where t is the parameter of the geodesic[1].

Proposition 11.7. *The map φ_{P} as in (11.16) is a diffeomorphism of a small neighborhood of the origin of the $\xi\eta$-plane to a neighborhood of the origin in the uv-plane.*

Proof. By Lemma 11.5 and (11.15), we have

$$\begin{aligned}
\frac{\partial u}{\partial \xi}(0,0) &= \left.\frac{d}{dt}\right|_{t=0} u(1;t,0) = \left.\frac{d}{dt}\right|_{t=0} u(t;1,0) = 1,\\
\frac{\partial u}{\partial \eta}(0,0) &= \left.\frac{d}{dt}\right|_{t=0} u(1;0,t) = \left.\frac{d}{dt}\right|_{t=0} u(t;0,1) = 0.
\end{aligned}$$

Similarly $v_\xi(0,0)=0$, $v_\eta(0,0)=1$ hold, and hence the Jacobi matrix of φ_{P} is the identity matrix. Then by the inverse function theorem (Theorem A.1.5 of Appendix A.1), we have the conclusion. □

By the proof above, (ξ,η) gives a coordinate system of the surface around P. Such a coordinate system (ξ,η) is called a *normal coordinate system*.

Let (r,θ) be the polar coordinate system on the $\xi\eta$-plane:

$$\xi = r\cos\theta, \qquad \eta = r\sin\theta.$$

Then by (11.16),

$$\varphi_{\mathrm{P}}(r\cos\theta, r\sin\theta) = (u(r;\cos\theta,\sin\theta), v(r;\cos\theta,\sin\theta)).$$

Hence the curve $r \mapsto \varphi_{\mathrm{P}}(r\cos\theta, r\sin\theta)$ on the uv-plane is a geodesic, whose velocity vector at P makes angle θ with $\boldsymbol{p}_u(0,0)$. In this way, (r,θ) gives the geodesic polar coordinate system.

[1] The map φ_{P} is called the *exponential map*, often denoted by "$\mathrm{Exp}_{\mathrm{P}}$". See, for example, Chapter 3, §2 of [5].

Geometric meaning of Gaussian curvature. Take a point P on a surface and geodesic polar coordinate system (r, θ) centered at P. For sufficiently small positive values of r, the geodesic circle with center P and radius r is the curve the surface corresponding to (r, θ) $(0 \leq \theta \leq 2\pi)$. A circle of radius r in the plane has length $2\pi r$, and encloses a region of area πr^2, but on a general surface, the following theorem holds.

Theorem 11.8. *Let $\mathcal{L}(r)$ denote the length of a geodesic circle with center* P *and radius r in a regular surface, and let $\mathcal{A}(r)$ denote the area of the region in the surface enclosed by the geodesic circle. Then*

$$K(\mathrm{P}) = \lim_{r\to 0+} \frac{3}{\pi}\left(\frac{2\pi r - \mathcal{L}(r)}{r^3}\right) = \lim_{r\to 0+} \frac{12}{\pi}\left(\frac{\pi r^2 - \mathcal{A}(r)}{r^4}\right)$$

holds, where $K(\mathrm{P})$ denotes the Gaussian curvature of the surface at P.

By this theorem, one can obtain the Gaussian curvature by comparing the length or the area of geodesic circles with that of circles on the plane.

Proof. The first fundamental form with respect to the polar coordinate system can be expressed as (11.9). Then we have

$$\mathcal{L}(r) = \int_0^{2\pi} h(r, \theta)\, d\theta.$$

Here, by (11.7), $\lim_{r\to 0+} h(r, \theta) = 0$. Then by L'Hôpital's rule (Theorem A.1.3 in Appendix A.1), we have

$$\begin{aligned}\lim_{r\to 0+} \frac{2\pi r - \mathcal{L}(r)}{r^3} &= \lim_{r\to 0+} \frac{1}{r^3}\left(2\pi r - \int_0^{2\pi} h(r, \theta)\, d\theta\right) \\ &= \lim_{r\to 0+} \frac{1}{3r^2}\left(2\pi - \int_0^{2\pi} h_r(r, \theta)\, d\theta\right).\end{aligned}$$

Then applying L'Hôpital's rule again, (11.10) implies that

$$\begin{aligned}\lim_{r\to 0+} \frac{2\pi r - \mathcal{L}(r)}{r^3} &= \lim_{r\to 0+} \frac{1}{6r}\left(-\int_0^{2\pi} h_{rr}(r, \theta)\, d\theta\right) \\ &= \lim_{r\to 0+} \frac{1}{6}\left(\int_0^{2\pi} K(r, \theta)\frac{h(r, \theta)}{r}\, d\theta\right).\end{aligned}$$

Thus the first conclusion follows by (11.7). The second conclusion can be obtained applying L'Hôpital's rule three times to

$$\mathcal{A}(r) = \int_0^r dr \int_0^{2\pi} h(r, \theta)\, d\theta. \qquad \square$$

Exercise 11

1 Let $\gamma(s)$ $(a \le s \le b)$ be a space curve parametrized by the arc-length whose curvature $\kappa(s)$ is positive, and let $\boldsymbol{e}(s)$, $\boldsymbol{n}(s)$, $\boldsymbol{b}(s)$ be the unit tangent vector, the principal normal vector, and the binormal vector of γ, respectively (cf. Section 5).

(1) Let $\tau(s)$ be the torsion of γ and set

$$\boldsymbol{d}(s) := \frac{\tau(s)}{\kappa(s)}\boldsymbol{e}(s) + \boldsymbol{b}(s),$$

which is called the *normalized Darboux vector*. Then show that the map

$$\boldsymbol{f}(s,t) = \gamma(s) + t\boldsymbol{d}(s) \qquad (a \le s \le b,\ |t| \le \varepsilon)$$

gives a regular surface for a sufficiently small $\varepsilon > 0$, whose Gaussian curvature is zero. (For a geometric meaning of the ratio τ/κ, see Problem **13** in Section 10.)

(2) Show that $\gamma(s)$ is a geodesic of the surface.

Chapter III

Surfaces from the Viewpoint of Manifolds*

In the previous chapters, we studied curves and surfaces, but looking at surfaces from the viewpoint of manifolds allows us to obtain a broader perspective. In this chapter, for those readers that are familiar with manifolds, we aim at the Hopf theorem (Theorem 17.4) for constant mean curvature surfaces, and continue with further study of surfaces. We start in Section 12 with a review of differential forms, and use them in Sections 13 and 14 to prove the Gauss-Bonnet theorem on 2-dimensional Riemannian manifolds. As an application of that, in Section 15 we prove the index formula for vector fields on compact oriented 2-manifolds and study the index of umbilics on a surface. In Section 16, we show existence of conformal coordinates for surfaces. Then in Section 17, we introduce the Gauss and Codazzi equations, and the fundamental theorem of surface theory, which will be proven in Appendix B.10, and prove the Hopf theorem. In Section 18, we explain the maximal speed descent property (the property of brachistocrones) for cycloids from the viewpoint of Riemannian geometry. Finally in Section 19, we give a proof of existence of geodesic triangulations on surfaces. In this chapter, we assume that the reader is familiar with vector fields on manifolds, differential forms, wedge products and exterior derivatives. (See the references suggested in the text within this chapter.)

12. Differential forms

We now assume the reader is familiar with manifolds, for example, the knowledge in Chapter 5 of Singer-Thorpe [35].

We will refer to the sets of real-valued differential 0-forms, 1-forms,

2-forms on a 2-manifold (i.e. 2-dimensional differentiable manifold) S as $\mathcal{A}^0(S)$, $\mathcal{A}^1(S)$, $\mathcal{A}^2(S)$, respectively. In fact, $\mathcal{A}^0(S)$ is the same as the set $C^\infty(S)$ of smooth functions on S. The *exterior derivative* operators are linear maps between these sets as follows:

$$d = d_0 \colon \mathcal{A}^0(S) \longrightarrow \mathcal{A}^1(S), \qquad d = d_1 \colon \mathcal{A}^1(S) \longrightarrow \mathcal{A}^2(S).$$

In particular, d_0 is the same as the exterior derivative d defined in Section 7.

Let $\mathfrak{X}^\infty(S)$ denote the set of all smooth vector fields on a manifold S. Then a differential 1-form is a linear map $\alpha : \mathfrak{X}^\infty(S) \to C^\infty(S)$ that satisfies

$$\alpha(fX) = f\alpha(X) \qquad (f \in C^\infty(S), X \in \mathfrak{X}^\infty(S)).$$

A differential 2-form β, or 2-form β for short, satisfies

$$\beta(X,Y) = -\beta(Y,X),$$
$$\beta(fX,Y) = \beta(X,fY) = f\beta(X,Y) \qquad (f \in C^\infty(S), X, Y \in \mathfrak{X}^\infty(S))$$

and so is a bilinear map $\beta \colon \mathfrak{X}^\infty(S) \times \mathfrak{X}^\infty(S) \to C^\infty(S)$. Furthermore, when α and β are 1-forms, we can define their *wedge product*, or *exterior product*, "$\alpha \wedge \beta$" as

$$(12.1) \qquad (\alpha \wedge \beta)(X,Y) := \alpha(X)\beta(Y) - \alpha(Y)\beta(X) \quad (X, Y \in \mathfrak{X}^\infty(S)).$$

(In some textbooks, the wedge product of two 1-forms is defined the same way as it is shown here, but divided by two.) With this, the exterior derivative of the differential 1-form is characterized as a linear map $d(= d_1) \colon \mathcal{A}^1(S) \to \mathcal{A}^2(S)$ with the property

$$(12.2) \qquad d\alpha(X,Y) = X\alpha(Y) - Y\alpha(X) - \alpha([X,Y]) \quad (X, Y \in \mathfrak{X}^\infty(S))$$

for $\alpha \in \mathcal{A}^1(S)$. It then holds that

$$(12.3) \qquad d(f\alpha) = df \wedge \alpha + f\, d\alpha \qquad (f \in C^\infty(S))$$

(see Problem **1** at the end of this section). Here $[X,Y]$ is the *Lie bracket* for vector fields X and Y (see Chapter 1 of [40]).

For a 2-manifold S, we say that S is *orientable* if there exists a differential 2-form ω which is non-zero at every point of S. Once fixing one choice of such an ω, we then say that S is *oriented.* A local coordinate neighborhood $(U;(u,v))$ of S is called *positively oriented* if ω can be expressed on U as

$$(12.4) \qquad \omega = \lambda\, du \wedge dv$$

for some function $\lambda > 0$, and when $\lambda < 0$ we say it is *negatively oriented.* Switching the order of the u and v in the local coordinates (u,v) will cause a switch between positively and negatively oriented coordinates. A local change of coordinates from one positively oriented coordinate chart to another will have the positive Jacobian (see Problem **5** at the end of this section).

Now let us consider a 2-manifold S with a Riemannian metric ds^2 defined on it. Then (S, ds^2) is called a *Riemannian* 2-*manifold.* In the case of surfaces in $\boldsymbol{R}^3$, the first fundamental form gives a Riemannian metric. Here it is useful if the readers are familiar with Riemannian manifolds, for example, the material up through Chapter 3 of do Carmo's textbook [5]. Let $(U; (u, v))$ be a connected local coordinate neighborhood of S, and let $\{\boldsymbol{e}_1, \boldsymbol{e}_2\}$ be an *orthonormal frame field*, that is, a pair of vector fields on U such that it forms an orthonormal basis of the tangent space $T_{\mathrm{P}}S$ at each point $\mathrm{P} \in U$ (cf. Problem **3** at the end of this section). Consider another orthonormal frame field $\{\tilde{\boldsymbol{e}}_1, \tilde{\boldsymbol{e}}_2\}$, then there exists four smooth functions a, b, c, d on U such that

$$\tilde{\boldsymbol{e}}_1 = a\boldsymbol{e}_1 + c\boldsymbol{e}_2, \quad \tilde{\boldsymbol{e}}_2 = b\boldsymbol{e}_1 + d\boldsymbol{e}_2.$$

We can rewrite these relations as

$$(\tilde{\boldsymbol{e}}_1, \tilde{\boldsymbol{e}}_2) = (\boldsymbol{e}_1, \boldsymbol{e}_2)A, \qquad A := \begin{pmatrix} a & b \\ c & d \end{pmatrix}.$$

At each point of U, A gives an orthogonal matrix. By the continuity of A with respect to the parameter u, v, $\det A$ is identically equal to 1 or -1 (cf. (A.3.11) in Appendix A.3). The former case (resp. the latter case) we say that $\{\tilde{\boldsymbol{e}}_1, \tilde{\boldsymbol{e}}_2\}$ has the *same* (resp. *opposite*) *orientation* as $\{\boldsymbol{e}_1, \boldsymbol{e}_2\}$.

Now define two 1-forms ω_1, ω_2 on U by the property that $\omega_j(\boldsymbol{e}_k) = \delta_{jk}$ $(j, k = 1, 2)$, where δ_{jk} is the Kronecker delta given in (A.3.10) in Section A.3. In other words, $\{\omega_1, \omega_2\}$ comprises the *dual basis* of $\{\boldsymbol{e}_1, \boldsymbol{e}_2\}$ for the dual vector space $T^*_{\mathrm{P}}S$ (of the tangent space $T_{\mathrm{P}}S$) for each $\mathrm{P} \in U$ (see Problem **4** at the end of this section). We call $\{\omega_1, \omega_2\}$ the *dual frame field of* $\{\boldsymbol{e}_1, \boldsymbol{e}_2\}$. Let $\{\tilde{\boldsymbol{e}}_1, \tilde{\boldsymbol{e}}_2\}$ have the same orientation as $\{\boldsymbol{e}_1, \boldsymbol{e}_2\}$, and denote by $\{\tilde{\omega}_1, \tilde{\omega}_2\}$ the dual frame field of $\{\tilde{\boldsymbol{e}}_1, \tilde{\boldsymbol{e}}_2\}$. Then, at each point there exists a real value $\theta \in [0, 2\pi)$ so that

$$\begin{cases} (\tilde{\boldsymbol{e}}_1, \tilde{\boldsymbol{e}}_2) = (\boldsymbol{e}_1, \boldsymbol{e}_2)\begin{pmatrix} \cos\theta & -\sin\theta \\ \sin\theta & \cos\theta \end{pmatrix}, \\ \begin{pmatrix} \tilde{\omega}_1 \\ \tilde{\omega}_2 \end{pmatrix} = \begin{pmatrix} \cos\theta & \sin\theta \\ -\sin\theta & \cos\theta \end{pmatrix}\begin{pmatrix} \omega_1 \\ \omega_2 \end{pmatrix}. \end{cases} \tag{12.5}$$

The following assertion holds.

Proposition 12.1. *Let* $\{\tilde{\mathbf{e}}_1, \tilde{\mathbf{e}}_2\}$ *be an orthonormal frame field on* U *and* $\{\tilde{\omega}_1, \tilde{\omega}_2\}$ *its dual frame field. Then* $\tilde{\omega}_1 \wedge \tilde{\omega}_2$ *(resp.* $-\tilde{\omega}_1 \wedge \tilde{\omega}_2$*) coincides with* $\omega_1 \wedge \omega_2$ *on* U *if and only if the orientation of* $\{\tilde{\mathbf{e}}_1, \tilde{\mathbf{e}}_2\}$ *is the same as (resp. opposite of) that of* $\{\mathbf{e}_1, \mathbf{e}_2\}$.

Proof. If $\{\tilde{\mathbf{e}}_1, \tilde{\mathbf{e}}_2\}$ has the same orientation, then the assertion follows by (12.5). If $\{\tilde{\mathbf{e}}_1, \tilde{\mathbf{e}}_2\}$ has the opposite orientation, then $\{\tilde{\mathbf{e}}_1, -\tilde{\mathbf{e}}_2\}$ has the same orientation, and $\{\tilde{\omega}_1, -\tilde{\omega}_2\}$ corresponds to its dual, and so $\tilde{\omega}_1 \wedge \tilde{\omega}_2 = -\omega_1 \wedge \omega_2$. □

Let us now consider the case that S is oriented. Take a positively oriented local coordinate neighborhood $(U; (u, v))$ of S. An orthonormal frame field $\{\mathbf{e}_1, \mathbf{e}_2\}$ on U is said to be *compatible* with respect to the orientation of S (or *positively oriented*) if it satisfies

$$du \wedge dv(\mathbf{e}_1, \mathbf{e}_2) > 0 \tag{12.6}$$

on U. (The $\{\mathbf{e}_1, \mathbf{e}_2\}$ in Problem **3** at the end of this section satisfies (12.6).) We denote its dual frame field by $\{\omega_1, \omega_2\}$, and consider a 2-form on U

$$d\hat{A} := \omega_1 \wedge \omega_2, \tag{12.7}$$

which is independent of the choice of positively oriented coordinate neighborhood, by Proposition 12.1. So $d\hat{A}$ can be considered as a 2-form defined on S, called the (oriented) *area element* of the Riemannian manifold (S, ds^2). Since $d\hat{A}$ is nowhere vanishing, it plays the role of a 2-form ω giving an orientation to S (cf. (12.4)).

Using local coordinates (u, v), we can write the Riemannian metric as[1]

$$ds^2 = E\,du^2 + 2F\,du\,dv + G\,dv^2.$$

Then we have (cf. Problem **7** at the end of this section)

$$d\hat{A} = \sqrt{EG - F^2}\,du \wedge dv. \tag{12.8}$$

In Chapter II, at (7.14), we called

$$dA = \sqrt{EG - F^2}\,du\,dv$$

the area element. Here we consider dA to be different from the $d\hat{A}$ for manifolds, and now call dA the *orientation-independent* area element of S, because it can be defined even when S is not orientable.

[1] du^2, dv^2, $du\,dv$ are symmetric products of 1-forms. In general, for differential 1-forms α, $\beta \in \mathcal{A}^1(S)$, the symmetric product is defined as $(\alpha\beta)(X, Y) := (1/2)\{\alpha(X)\beta(Y) + \alpha(Y)\beta(X)\}$ (with $X, Y \in \mathfrak{X}^\infty(S)$).

We now take S to be a compact manifold. Then the *integral*

$$\int_S f\,dA$$

for each function $f \in C^\infty(S)$ is defined. More precisely, for a covering of S by a finite number of charts $\{(U_j;(u_j,v_j))\}_{j=1,\dots,N}$, one can take a partition of unity $\{\rho_j\}_{j=1}^N$ corresponding to the covering (cf. [40, Chapter 1]). That is, each ρ_j is a non-negative smooth function which takes the value 0 outside U_j, and the sum of the ρ_j's satisfies $\sum_{j=1}^N \rho_j = 1$. Then, we define

$$\int_S f\,dA := \sum_{j=1}^N \iint_{U_j} \rho_j f \sqrt{E_j G_j - F_j^2}\,du_j\,dv_j.$$

Here, for each coordinate chart $(U_j,(u_j,v_j))$, the Riemannian metric takes the form $ds^2 = E_j\,du_j^2 + 2F_j\,du_j\,dv_j + G_j\,dv_j^2$. This integral is a well-defined notion even when S is not orientable. If f is identically 1, the integral $\int_S dA$ gives the (total) area of the manifold S, which coincides with the area of the surface given in (6.12) in Section 6, see also (7.14) in Section 7.

When S is orientable, since it has a differential 2-form $d\hat{A}$ that is never 0, any differential 2-form Ω can be written as $\Omega = f\,d\hat{A}$ for some function f. Then we define the integral of Ω over S by

$$\int_S \Omega := \int_S f\,dA. \tag{12.9}$$

Considering a smooth curve $\gamma(t) = (u(t),v(t))$ $(a \le t \le b)$ within a local coordinate neighborhood $(U;(u,v))$ of S, and taking a differential 1-form $\alpha = f\,du + g\,dv$ on U $(f,\ g \in C^\infty(U))$, we can define a *line integral* as (cf. Appendix B.3)

$$\int_\gamma \alpha := \int_a^b \left(f(u(t),v(t))\frac{du}{dt} + g(u(t),v(t))\frac{dv}{dt} \right) dt. \tag{12.10}$$

This definition does not depend on the choice of parameter for the curve, nor on the choice of local coordinates for S. Therefore, even when γ is a piecewise smooth curve (cf. Appendix B.3) on S, we can separate γ into a collection of smooth pieces, and apply the above definition to each piece, and sum up those integrals to define the line integral $\int_\gamma \alpha$. Then the following holds:

Theorem 12.2 (The Stokes theorem). *Consider a region D in an oriented 2-manifold S whose boundary ∂D consists of a finite number of simple*

[1]Stokes, Sir George Gabriel (1819–1903).

closed piecewise smooth curves, and suppose that $\overline{D} := D \cup \partial D$ is compact (see Fig. 12.1). For a differential 1-form α defined on a domain containing $\overline{D}$, taking the integral of α over ∂D using the orientation on each curve in ∂D that puts D to the left side, we have

$$\int_{\partial D} \alpha = \int_{\overline{D}} d\alpha.$$

Here, on the left-hand side of this equation, we are summing over the line integrals of all closed curves that comprise ∂D.

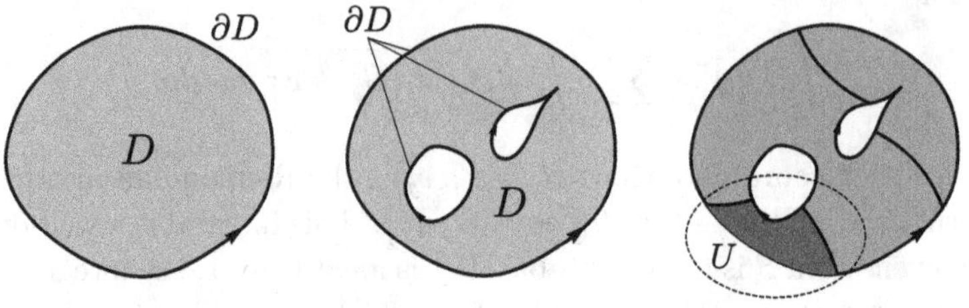

Fig. 12.1 The Stokes theorem.

Proof. We divide the region $\overline{D}$ so that it consists of non-overlapping small subregions whose boundaries are piecewise smooth (cf. [28, Theorem 10.6]). Moreover, we may assume that each subregion is contained in a certain local coordinate neighborhood of the manifold, and the boundary of each small region consists of a single piecewise smooth simple closed regular curve. For those portions of boundary curves that bound two different regions, integration will occur twice, once in each direction for each adjacent region, and those two integrals will cancel. Because it suffices to prove the result for each subregion, without loss of generality, and we may simply regard $\overline{D}$ itself as such a subregion. Then $\overline{D}$ is contained in a certain coordinate neighborhood $(U;(u,v))$ (cf. Fig. 12.1, right), and we can write

$$\alpha = x\,du + y\,dv,$$

where $x = x(u,v)$ and $y = y(u,v)$ are smooth functions on U. It holds that

$$d(x\,du + y\,dv) = dx \wedge du + dy \wedge dv = (-x_v(u,v) + y_u(u,v))\,du \wedge dv.$$

So Green's theorem (cf. Theorem B.3.2) yields that

$$\begin{aligned}\int_{\overline{D}} d\alpha &= \int_{\overline{D}} (-x_v(u,v) + y_u(u,v))\,du \wedge dv \\ &= \int_{\overline{D}} (-x_v(u,v) + y_u(u,v))\,du\,dv = \int_{\partial D} \alpha,\end{aligned}$$

proving the assertion. □

As seen on the left-side figure of Fig. 12.1, a region without punctures is said to be *simply connected* (for more details, see Chapter 3 of [35]).

For a smooth function f on a manifold, $d(df) = 0$ holds (see Problem **2** at the end of this section). The converse assertion for simply connected domains holds as follows.

Theorem 12.3 (Poincaré[2] lemma). *Given a differential 1-form α on a simply connected region D in $\boldsymbol{R}^2$ with $d\alpha = 0$, there exists a smooth function f defined on D which satisfies $\alpha = df$.*

We will provide a proof in Problem **8** at the end of this section, using Theorem 12.2.

Exercises 12

1 Derive (12.3) from equations (12.1) and (12.2).

2 Show that $d(df) = 0$ for a smooth function f on a 2-dimensional manifold.

3 Suppose we have a Riemannian metric $ds^2 = E\,du^2 + 2F\,du\,dv + G\,dv^2$ on a local coordinate neighborhood $(U; (u, v))$. From the pair of vector fields $\{\partial/\partial u, \partial/\partial v\}$, show that the Gram-Schmidt orthogonalization will produce the following orthonormal frame field $\{\boldsymbol{e}_1, \boldsymbol{e}_2\}$ on U:

$$\boldsymbol{e}_1 = \frac{1}{\sqrt{E}}\frac{\partial}{\partial u}, \qquad \boldsymbol{e}_2 = \frac{-1}{\sqrt{EG - F^2}}\left(\frac{F}{\sqrt{E}}\frac{\partial}{\partial u} - \sqrt{E}\frac{\partial}{\partial v}\right).$$

4 Continuing on from Problem **3**, show that

$$\omega_1 = \sqrt{E}\left(du + \frac{F}{E}\,dv\right), \qquad \omega_2 = \sqrt{\frac{EG - F^2}{E}}\,dv$$

comprise the *dual frame* to $\{\boldsymbol{e}_1, \boldsymbol{e}_2\}$, that is, $\omega_j(\boldsymbol{e}_k) = \delta_{jk}$ holds with the Kronecker delta δ_{jk} (cf. (A.3.10)). Furthermore, confirm $ds^2 = {\omega_1}^2 + {\omega_2}^2$ directly.

5 For an oriented 2-manifold, show that coordinate transformations between positively oriented local coordinate neighborhoods are the same as the orientation preserving coordinate transformations described in Section 2.

6 Using (12.5), show that $d\hat{A}$ does not depend on the choice of positively oriented orthonormal frame.

7 Show that $d\hat{A}$ can be expressed as in (12.8) using the components of the Riemannian metric ds^2.

8 Prove the Poincaré lemma (Theorem 12.3) in the following way:

[2]Poincaré, Henri (1854–1912).

(1) Fix a point $\mathrm{P_0}$ in D. Take a curve γ in D connecting $\mathrm{P_0}$ and $\mathrm{P} \in D$, and consider the line integral of α over γ. Show that this line integral depends only on the choice of endpoint P and not on the choice of connecting curve γ. (Hint: Take two curves joining P and $\mathrm{P_0}$. When they bound a domain as in the left-hand figure below, one can apply the Stokes Theorem 12.2 to that domain. In the case of the central figure below, we can take the third path as in the right-hand figure, and can apply the Stokes theorem.)

(2) For the line integral as above, set

$$f(\mathrm{P}) := \int_\gamma \alpha = \int_{\mathrm{P_0}}^{\mathrm{P}} \alpha.$$

Show that f is a smooth function in P, and that $df = \alpha$ holds.

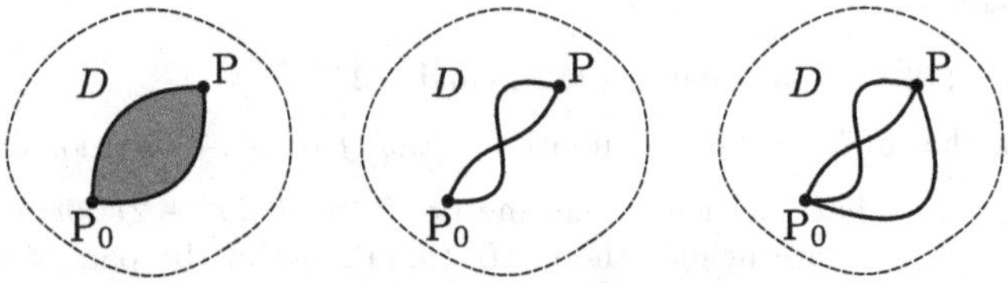

13. Levi-Civita connections

As a preparation for defining Gaussian curvature for Riemannian 2-manifolds, we introduce in this section the notion of Levi-Civita connections.

Connection forms. We fix a Riemannian 2-manifold S with Riemannian metric ds^2. Let $(U; (u, v))$ be a local coordinate neighborhood, and $\{\boldsymbol{e}_1, \boldsymbol{e}_2\}$ an orthonormal frame field. Denoting its dual frame field by $\{\omega_1, \omega_2\}$, the Riemannian metric (first fundamental form) becomes

$$ds^2 = {\omega_1}^2 + {\omega_2}^2, \tag{13.1}$$

where ${\omega_1}^2$, ${\omega_2}^2$ are the squares of ω_1, ω_2 with respect to the symmetric product (see the footnote on page 134 and Problem **4** in the previous Section 12). The exterior derivatives of ω_1 and ω_2 are 2-forms, so there exist $a, b \in C^\infty(U)$ such that

$$d\omega_1 = a(\omega_1 \wedge \omega_2), \qquad d\omega_2 = b(\omega_1 \wedge \omega_2).$$

Using this, we set

$$\mu := -a\,\omega_1 - b\,\omega_2. \tag{13.2}$$

The 1-form μ on U is called the *connection form*[1] with respect to an orthonormal frame field $\{\mathbf{e}_1, \mathbf{e}_2\}$ defined on a coordinate neighborhood of S.

Lemma 13.1. *Let (S, ds^2) be a Riemannian 2-manifold and $\{\mathbf{e}_1, \mathbf{e}_2\}$ an orthonormal frame field on an open subset U on S. Then the connection form μ with respect to it satisfies*

$$d\omega_1 = \omega_2 \wedge \mu, \qquad d\omega_2 = -\omega_1 \wedge \mu. \tag{13.3}$$

Here $\{\omega_1, \omega_2\}$ is the dual frame field with respect to the orthonormal frame field $\{\mathbf{e}_1, \mathbf{e}_2\}$. Conversely, if a 1-form μ defined on U satisfies (13.3), then it coincides with the connection form.

Proof. It is immediate from the definition that the connection form fulfills the above relation. Conversely, if a 1-form μ satisfies (13.3), then μ is a linear combination of ω_1, ω_2, and consequently we have (13.2). □

The definition of the connection form μ depends on the choice of the orthonormal frame field $\{\mathbf{e}_1, \mathbf{e}_2\}$.

Now, choosing another orthonormal frame on U with the same orientation as $\{\mathbf{e}_1, \mathbf{e}_2\}$, and with corresponding dual frame field $\{\tilde{\omega}_1, \tilde{\omega}_2\}$, there exists a function $\theta \in C^\infty(U)$ so that at each point (12.5) in Section 12 holds.

Lemma 13.2. *If $\{\tilde{\mathbf{e}}_1, \tilde{\mathbf{e}}_2\}$ has the same orientation as $\{\mathbf{e}_1, \mathbf{e}_2\}$, then its connection form $\tilde{\mu}$ is related to μ by $\tilde{\mu} = \mu - d\theta$, where θ is as described above.*

Proof. By (12.5) and (13.3), we see that

$$\begin{aligned}
d\tilde{\omega}_1 &= d(\cos\theta\,\omega_1 + \sin\theta\,\omega_2) \\
&= (-\sin\theta\,d\theta \wedge \omega_1 + \cos\theta\,d\omega_1) + (\cos\theta\,d\theta \wedge \omega_2 + \sin\theta\,d\omega_2) \\
&= -\sin\theta(d\theta \wedge \omega_1) + \cos\theta(\omega_2 \wedge \mu) + \cos\theta(d\theta \wedge \omega_2) + \sin\theta(-\omega_1 \wedge \mu) \\
&= d\theta \wedge (-\sin\theta\,\omega_1 + \cos\theta\,\omega_2) + (\cos\theta\,\omega_2 - \sin\theta\,\omega_1) \wedge \mu \\
&= d\theta \wedge \tilde{\omega}_2 + \tilde{\omega}_2 \wedge \mu = -\tilde{\omega}_2 \wedge d\theta + \tilde{\omega}_2 \wedge \mu = \tilde{\omega}_2 \wedge (-d\theta + \mu)
\end{aligned}$$

is satisfied. By a similar computation, we see that $d\tilde{\omega}_2 = -\tilde{\omega}_1 \wedge (-d\theta + \mu)$ also holds, and by Lemma 13.1 we have $\tilde{\mu} = -d\theta + \mu$. □

[1] Generally, the matrix-valued 1-form $\begin{pmatrix} 0 & \mu \\ -\mu & 0 \end{pmatrix}$, taking values in the set of skew symmetric matrices, is called a *connection form*, but here we call μ itself the connection form.

Levi-Civita connections. Using the connection form μ with respect to an orthonormal frame field $\{\mathbf{e}_1, \mathbf{e}_2\}$ on a coordinate neighborhood U, we define

$$\nabla_X \mathbf{e}_1 := -\mu(X)\mathbf{e}_2, \quad \nabla_X \mathbf{e}_2 := \mu(X)\mathbf{e}_1 \qquad (X \in \mathfrak{X}^\infty(U)). \tag{13.4}$$

Then, for a vector field $Y = \eta_1 \mathbf{e}_1 + \eta_2 \mathbf{e}_2$ on U, we have

$$\nabla_X Y := \eta_1 \nabla_X \mathbf{e}_1 + d\eta_1(X)\mathbf{e}_1 + \eta_2 \nabla_X \mathbf{e}_2 + d\eta_2(X)\mathbf{e}_2 \tag{13.5}$$

as the definition of $\nabla_X Y$, which is the *(Levi-Civita) covariant derivative* of Y in the direction of X. By Lemma 13.2, one can show that this is independent of the choice of the orthonormal frame field (see Problem **1** at the end of this section). Thus, for two vector fields X, Y on S, we can define the Levi-Civita covariant derivative $\nabla_X Y$. (This derivation ∇ is called the *Levi-Civita connection* or *Levi-Civita covariant derivative.*)

In particular, for arbitrary vector fields X, Y, $Z \in \mathfrak{X}^\infty(S)$ and function $f \in C^\infty(S)$, we have

$$\nabla_X(Y + Z) = \nabla_X Y + \nabla_X Z, \tag{13.6}$$

$$\nabla_X fY = f\nabla_X Y + df(X)Y, \tag{13.7}$$

$$\nabla_{X+Y} Z = \nabla_X Z + \nabla_Y Z, \tag{13.8}$$

$$\nabla_{fX} Y = f\nabla_X Y. \tag{13.9}$$

In general, a map $\nabla : \mathfrak{X}^\infty(S) \ni (X, Y) \mapsto \nabla_X Y \in \mathfrak{X}^\infty(S)$ having the above properties (13.6)–(13.9) is called a *covariant derivative*, or the *linear connection* on S.

Theorem 13.3. *On the Riemannian manifold (S, ds^2), the Levi-Civita connection ∇ has the following properties:*

$$\nabla_X Y - \nabla_Y X = [X, Y], \tag{13.10}$$

$$X\langle Y, Z\rangle = \langle \nabla_X Y, Z\rangle + \langle Y, \nabla_X Z\rangle. \tag{13.11}$$

Here, $X, Y, Z \in \mathfrak{X}^\infty(S)$, and $\langle\ ,\ \rangle$ is the inner product on the tangent space given by the Riemannian metric ds^2. Conversely, any covariant derivative on S satisfying the above two properties coincides with the Levi-Civita connection ∇.

Proof. Restricting to a local coordinate patch U of S, consider vector fields X, Y, Z on U, and define

$$T(X, Y) := \nabla_X Y - \nabla_Y X - [X, Y],$$
$$Q(X, Y, Z) := Z\langle X, Y\rangle - \langle \nabla_Z X, Y\rangle - \langle X, \nabla_Z Y\rangle.$$

Then for functions f, g, $h \in C^\infty(U)$, we find that

$$T(fX, gY) = fgT(X, Y), \qquad Q(fX, gY, hZ) = fghQ(X, Y, Z)$$

hold, and it suffices to show that the above T and Q are zero in the case that X, Y are chosen from amongst the two vector fields $\boldsymbol{e}_1$, $\boldsymbol{e}_2$.

First of all, by (13.4), $Q(\boldsymbol{e}_j, \boldsymbol{e}_k, Z) = 0$ $(j, k = 1, 2)$ holds, so from $Q = 0$ we conclude (13.11). Next, by (13.4), we have

$$\begin{aligned} T(\boldsymbol{e}_1, \boldsymbol{e}_2) &= \nabla_{\boldsymbol{e}_1}\boldsymbol{e}_2 - \nabla_{\boldsymbol{e}_2}\boldsymbol{e}_1 - [\boldsymbol{e}_1, \boldsymbol{e}_2] \\ &= \mu(\boldsymbol{e}_1)\boldsymbol{e}_1 + \mu(\boldsymbol{e}_2)\boldsymbol{e}_2 - [\boldsymbol{e}_1, \boldsymbol{e}_2], \end{aligned}$$

and then (12.2), (12.1) and Lemma 13.1 give

$$\begin{aligned} \omega_1(T(\boldsymbol{e}_1, \boldsymbol{e}_2)) &= \omega_1(\mu(\boldsymbol{e}_1)\boldsymbol{e}_1 + \mu(\boldsymbol{e}_2)\boldsymbol{e}_2 - [\boldsymbol{e}_1, \boldsymbol{e}_2]) = \mu(\boldsymbol{e}_1) - \omega_1([\boldsymbol{e}_1, \boldsymbol{e}_2]) \\ &= \mu(\boldsymbol{e}_1) + d\omega_1(\boldsymbol{e}_1, \boldsymbol{e}_2) - \boldsymbol{e}_1(\omega_1(\boldsymbol{e}_2)) + \boldsymbol{e}_2(\omega_1(\boldsymbol{e}_1)). \end{aligned}$$

Here, in general, for a general vector field $X \in \mathfrak{X}^\infty(S)$ and a function $f \in C^\infty(S)$, $Xf = df(X)$ is the directional derivative of f in the direction X. Since $\omega_1(\boldsymbol{e}_2)$ and $\omega_1(\boldsymbol{e}_1)$ are constant functions, $\boldsymbol{e}_1(\omega_1(\boldsymbol{e}_2))$ and $\boldsymbol{e}_2(\omega_1(\boldsymbol{e}_1))$ vanish identically. So, we have

$$\begin{aligned} \omega_1(T(\boldsymbol{e}_1, \boldsymbol{e}_2)) &= \mu(\boldsymbol{e}_1) + d\omega_1(\boldsymbol{e}_1, \boldsymbol{e}_2) \\ &= \mu(\boldsymbol{e}_1) + \omega_2 \wedge \mu(\boldsymbol{e}_1, \boldsymbol{e}_2) = \mu(\boldsymbol{e}_1) - \omega_2(\boldsymbol{e}_2)\mu(\boldsymbol{e}_1) = 0. \end{aligned}$$

Similarly, we also know $\omega_2(T(\boldsymbol{e}_1, \boldsymbol{e}_2)) = 0$, and so $T(\boldsymbol{e}_1, \boldsymbol{e}_2) = 0$. Furthermore, because $T(\boldsymbol{e}_2, \boldsymbol{e}_1) = -T(\boldsymbol{e}_1, \boldsymbol{e}_2) = 0$, we have $T = 0$ and (13.10) holds.

Conversely, for another covariant derivative D, if equations (13.10) and (13.11) hold, we can prove that D equals the Levi-Civita covariant derivative ∇. For vector fields X, Y on U, we set

$$A_X Y := D_X Y - \nabla_X Y.$$

As D and ∇ are both covariant derivatives, $A_{fX}(gY) = fgA_X(Y)$ (X, $Y \in \mathfrak{X}^\infty(U)$, f, $g \in C^\infty(U)$). Therefore, if $A_{\boldsymbol{e}_i}\boldsymbol{e}_j = 0$ $(i, j = 1, 2)$, we would know $D = \nabla$. Since D and ∇ both satisfy (13.10),

$$A_X Y = A_Y X \tag{13.12}$$

holds. Furthermore, D and ∇ both satisfy (13.11), so

$$\langle A_X \boldsymbol{e}_1, \boldsymbol{e}_1 \rangle = \langle A_X \boldsymbol{e}_2, \boldsymbol{e}_2 \rangle = 0, \tag{13.13}$$

$$\langle A_X \boldsymbol{e}_1, \boldsymbol{e}_2 \rangle = -\langle \boldsymbol{e}_1, A_X \boldsymbol{e}_2 \rangle \tag{13.14}$$

hold as well. In particular, by (13.13), we have

$$\langle A_{\mathbf{e}_j}\mathbf{e}_1, \mathbf{e}_1\rangle = \langle A_{\mathbf{e}_j}\mathbf{e}_2, \mathbf{e}_2\rangle = 0 \qquad (j = 1, 2).$$

This equality and (13.12) give $\langle A_{\mathbf{e}_2}\mathbf{e}_1, \mathbf{e}_1\rangle = \langle A_{\mathbf{e}_1}\mathbf{e}_2, \mathbf{e}_2\rangle = 0$, and from (13.14), we have

$$\langle A_{\mathbf{e}_1}\mathbf{e}_1, \mathbf{e}_2\rangle = -\langle \mathbf{e}_1, A_{\mathbf{e}_1}\mathbf{e}_2\rangle = -\langle \mathbf{e}_1, A_{\mathbf{e}_2}\mathbf{e}_1\rangle = 0.$$

By the above, $\langle A_{\mathbf{e}_1}\mathbf{e}_i, \mathbf{e}_j\rangle = 0$ $(i, j = 1, 2)$ has been shown. We conclude that $A_{\mathbf{e}_1}\mathbf{e}_i = 0$ $(i = 1, 2)$. Similarly, we know that $A_{\mathbf{e}_2}\mathbf{e}_i = 0$ $(i = 1, 2)$, so $A = 0$ holds, and we have shown that $D = \nabla$. □

Next, we consider the exterior derivative of the connection form μ. We define a function K by

$$d\mu = K\,\omega_1 \wedge \omega_2 \;(= K\,d\hat{A}) \tag{13.15}$$

on the coordinate neighborhood U. Then K does not depend on the choice of orthonormal frame field $\{\mathbf{e}_1, \mathbf{e}_2\}$. (In the case of a frame field with the same orientation, this is easily shown from Lemma 13.2. In the case of an oppositely oriented frame field, it can be shown using part (1) of Problem **1** at the end of this section.) Therefore, at all points P in S, unique values K_{P} for the function K have been determined. By (13.15), we know that K is a smooth function on S. The value K_{P} is called the *Gaussian curvature* at P. In particular, when S is a surface in $\boldsymbol{R}^3$, this is the same as the Gaussian curvature of the surface (see Problem **4** at the end of this section).

Curves on Riemannian manifolds. For a smooth curve $\gamma\colon [a, b] \to S$ on a Riemannian manifold (S, ds^2), the velocity vector $\dot\gamma(t)$ is a tangent vector of S at $\gamma(t)$. Then

$$\mathcal{L}(\gamma) := \int_a^b |\dot\gamma(t)|\,dt, \qquad \left(|\dot\gamma(t)| := \sqrt{ds^2(\dot\gamma(t), \dot\gamma(t))}\right) \tag{13.16}$$

gives the length of the curve of γ. When γ is a piecewise smooth curve, then we divide the curve γ into pieces so that each piece is a smooth curve. Summing up the lengths of all the pieces of the curve, we can define the total length $\mathcal{L}(\gamma)$ of γ. Take two points P, Q on S, and define their distance $d_{ds^2}(\mathrm{P}, \mathrm{Q})$ as the infimum of the lengths of the piecewise smooth curves bounded by P and Q. Then the following fact is well-known (cf. [5, Proposition 2.5]).

Theorem 13.4. *The function $d_{ds^2} : S \times S \to \boldsymbol{R}$ gives a distance whose induced topology on S is compatible with respect to the topology of S.*

For each point P on S, the set

$$\Delta_{\mathrm{P}}(r) := \{\mathrm{Q} \in S \mid d_{ds^2}(\mathrm{P}, \mathrm{Q}) < r\}$$

is called a *geodesic disc* of radius r, and

$$C_{\mathrm{P}}(r) := \{\mathrm{Q} \in S \mid d_{ds^2}(\mathrm{P}, \mathrm{Q}) = r\}$$

is called a *geodesic circle* of radius r (centered at P).

Let $\gamma(t)$ $(a \le t \le b)$ be a smooth curve on S. We suppose that γ lies in a local coordinate neighborhood U, then we can take an orthonormal frame field $\{\boldsymbol{e}_1, \boldsymbol{e}_2\}$ on U. A vector field $X(t)$ along a curve γ is an assignment $t \mapsto X(t)$ such that $X(t) \in T_{\gamma(t)}S$ and there exist smooth functions $f_1(t)$ and $f_2(t)$ on $[a, b]$ such that

$$X(t) = f_1(t)\boldsymbol{e}_1(t) + f_2(t)\boldsymbol{e}_2(t),$$

where $\boldsymbol{e}_j(t)$ $(j = 1, 2)$ is the restriction of $\boldsymbol{e}_j$ on the curve γ. (A typical such an example is $\dot{\gamma}(t)$.)

We now define

$$\nabla_{\dot{\gamma}}X = \dot{f}_1\boldsymbol{e}_1 + \dot{f}_2\boldsymbol{e}_2 + f_1\nabla_{\dot{\gamma}}\boldsymbol{e}_1 + f_2\nabla_{\dot{\gamma}}\boldsymbol{e}_2,$$

where

$$\nabla_{\dot{\gamma}}\boldsymbol{e}_1(t) = -\mu(\dot{\gamma})\boldsymbol{e}_2(t), \qquad \nabla_{\dot{\gamma}}\boldsymbol{e}_2(t) = \mu(\dot{\gamma})\boldsymbol{e}_1(t).$$

This definition does not depend on a choice of orthonormal frame fields. In fact, if $\{\tilde{\boldsymbol{e}}_1, \tilde{\boldsymbol{e}_2}\}$ is another orthonormal frame field. By replacing $\tilde{\boldsymbol{e}}_2$ by $-\tilde{\boldsymbol{e}}_2$ we may assume that $\{\tilde{\boldsymbol{e}}_1, \tilde{\boldsymbol{e}_2}\}$ is expressed as in (12.5). Then we have that

$$\begin{aligned}
\nabla_{\dot{\gamma}}\tilde{\boldsymbol{e}}_1 &= \nabla_{\dot{\gamma}}(\cos\theta\boldsymbol{e}_1 + \sin\theta\boldsymbol{e}_1) \\
&= -\dot{\theta}\sin\theta\boldsymbol{e}_1 + \dot{\theta}\cos\theta\boldsymbol{e}_2 - \cos\theta\mu(\dot{\gamma})\boldsymbol{e}_2 + \sin\theta\mu(\dot{\gamma})\boldsymbol{e}_1 \\
&= -(\mu(\dot{\gamma}) - \dot{\theta})(-\sin\theta\boldsymbol{e}_1 + \cos\theta\boldsymbol{e}_2) = -\tilde{\mu}(\dot{\gamma})\tilde{\boldsymbol{e}}_2.
\end{aligned}$$

Here we applied Lemma 13.2 in the final equality. Similarly, $\nabla_{\dot{\gamma}}\tilde{\boldsymbol{e}}_2$ is equal to $\tilde{\mu}(\dot{\gamma})\tilde{\boldsymbol{e}}_1$, and we get the consistency of the definition of $\nabla_{\dot{\gamma}}X$.

We consider the covariant derivative $\nabla_{\dot{\gamma}}\dot{\gamma}$ as the acceleration vector of the curve γ. Taking an orthonormal frame field $\{\boldsymbol{e}_1, \boldsymbol{e}_2\}$ and writing $\dot{\gamma} = \xi_1\boldsymbol{e}_1 + \xi_2\boldsymbol{e}_2$, (13.4) and (13.5) imply

$$\begin{aligned}
\nabla_{\dot{\gamma}}\dot{\gamma} &= \dot{\xi}_1\boldsymbol{e}_1 + \dot{\xi}_2\boldsymbol{e}_2 + \xi_1\nabla_{\dot{\gamma}}\boldsymbol{e}_1 + \xi_2\nabla_{\dot{\gamma}}\boldsymbol{e}_2 \\
&= (\dot{\xi}_1 + \mu(\dot{\gamma})\xi_2)\boldsymbol{e}_1 + (\dot{\xi}_2 - \mu(\dot{\gamma})\xi_1)\boldsymbol{e}_2.
\end{aligned}$$

When the curve $\gamma(t)$ has acceleration vector $\nabla_{\dot\gamma}\dot\gamma$ that is identically zero, it is called a *geodesic* of the Riemannian manifold (S, ds^2). In this case, since

$$(13.17)\qquad \begin{aligned}\langle \nabla_{\dot\gamma}\dot\gamma, \dot\gamma\rangle &= \xi_1(\xi_1 + \mu(\dot\gamma)\xi_2) + \xi_2(\xi_2 - \mu(\dot\gamma)\xi_1)\\ &= \frac{1}{2}\frac{d}{dt}({\xi_1}^2 + {\xi_2}^2) = \frac{1}{2}\frac{d}{dt}\langle\dot\gamma,\dot\gamma\rangle = \frac{1}{2}\frac{d}{dt}|\dot\gamma|^2,\end{aligned}$$

the length $|\dot\gamma| := \sqrt{\langle\dot\gamma,\dot\gamma\rangle}$ of the tangent vector $\dot\gamma$ is constant (see Proposition 10.1).

Geodesics in Riemannian manifolds are the analogous notion to straight lines in planes. Straight lines in planes can be described as the "curves with curvature 0". Analogously, in the case of manifolds, after giving a notion of the "geodesic curvature" (cf. (14.5)) for curves in the manifold, the geodesics are the curves with vanishing geodesic curvature with constant speed. Let $\gamma : [a, b] \to S$ be a geodesic, then for each $c \in [a, b]$, there exists a positive number ε such that

$$d_{ds^2}(\gamma(c), \gamma(t)) \qquad (|c - t| < \varepsilon)$$

coincides with the length of the subarc $\gamma([c, t])$ (cf. [5, Chapter 3]). In this sense, the geodesics can be considered as locally length minimizing curves of constant speed.

The covariant derivative of a surface in $\boldsymbol{R}^3$. In general, a smooth map $\boldsymbol{p}$ from a manifold S to another manifold M is said to be an *immersion* if, for each $\mathrm{P} \in S$, the differential map $(d\boldsymbol{p})_{\mathrm{P}} : T_{\mathrm{P}}S \to T_{\boldsymbol{p}(\mathrm{P})}M$ at P is of rank $\dim S$ as a linear map from the tangent space $T_{\mathrm{P}}S$ at P of S to the tangent space $T_{\boldsymbol{p}(\mathrm{P})}M$ at $\boldsymbol{p}(\mathrm{P})$ of M. (For the tangent space of a manifold and differential maps, see Chapter 1 of [40].) In particular, when $M = \boldsymbol{R}^3$ and S is a 2-manifold, taking a local coordinate neighborhood $(U; (u, v))$ of S, the condition for $\boldsymbol{p} : U \to \boldsymbol{R}^3$ to be an immersion is equivalent to $\boldsymbol{p}$ being a regular surface (i.e. $\boldsymbol{p}_u \times \boldsymbol{p}_v \neq \boldsymbol{0}$). Thus, by composing with the immersion $\boldsymbol{p}$, each local coordinate system of S can be considered as a parametrization of the surface. For an immersion $\boldsymbol{p} : S \to \boldsymbol{R}^3$, we assume that there is a smoothly-defined unit normal vector ν on the entirety of the surface. Then

$$d\hat{A} := \det(\boldsymbol{p}_u, \boldsymbol{p}_v, \nu)\, du \wedge dv$$

is a smooth 2-form on S without zeros; in fact, it is the area form (see (12.7)), and S is orientable (see Section 12). Conversely, when S is orientable, for a positively oriented coordinate neighborhood $(U; (u, v))$, we define

$$(13.18)\qquad \nu := \frac{\boldsymbol{p}_u \times \boldsymbol{p}_v}{|\boldsymbol{p}_u \times \boldsymbol{p}_v|},$$

and ν does not depend on the choice of positively oriented coordinate system (see Problem **4** in Section 6, and Problem **5** in Section 12). Thus we have produced a global smooth unit normal vector on S.

Now, for an immersion $\boldsymbol{p} = (x, y, z) \colon S \to \boldsymbol{R}^3$ from a 2-dimensional manifold S to $\boldsymbol{R}^3$, and for a tangent vector V at a point P of S, we define the vector $\hat{V}$ in $\boldsymbol{R}^3$ by

$$(13.19)\qquad \hat{V} := d\boldsymbol{p}(V) = (dx(V), dy(V), dz(V)).$$

This is a tangent vector to $\boldsymbol{p}(S)$ at the point $\boldsymbol{p}(\mathrm{P})$. Since $\boldsymbol{p}$ is an immersion, the map $V \mapsto \hat{V}$ from the set of all vectors tangent to S at P to the set of all vectors tangent to the surface $\boldsymbol{p}(S)$ at $\boldsymbol{p}(\mathrm{P})$ is a one-to-one correspondence. This correspondence is essentially the same as that of (7.15) in Section 7. Then, taking X to be a vector field on the manifold S, we have the vector-valued function

$$(13.20)\qquad \hat{X} : S \ni \mathrm{P} \longmapsto \hat{X}_{\mathrm{P}} \in \boldsymbol{R}^3.$$

Writing $\hat{X} = (\alpha, \beta, \gamma)$ as a collection of three coordinate functions, and taking the exterior derivatives of those three functions, we have $d\hat{X} = (d\alpha, d\beta, d\gamma)$.

We can pull back the metric of $\boldsymbol{R}^3$ via an immersion $\boldsymbol{p} : S \to \boldsymbol{R}^3$ to a Riemannian metric ds^2 on S (see [5, Chapter 6]). We call this the *first fundamental form* of $\boldsymbol{p}$, and it is the same as the first fundamental form described in Section 7. In fact, by definition, arbitrarily choosing tangent vectors V, W at a given point, we have

$$(13.21)\qquad ds^2(V, W) = \langle V, W \rangle = \hat{V} \cdot \hat{W} = d\boldsymbol{p}(V) \cdot d\boldsymbol{p}(W).$$

Here " $\cdot$ " is the inner product for $\boldsymbol{R}^3$, and $\langle\ ,\ \rangle$ is the inner product on the tangent space of S determined by the Riemannian metric ds^2. In other words, $ds^2 = d\boldsymbol{p} \cdot d\boldsymbol{p}$, so this is equivalent to (7.11) in Section 7. Using the identification $V \leftrightarrow \hat{V}$ of tangent vectors on the surface, we can obtain the meaning for the covariant derivative given here:

Theorem 13.5. *Let $\boldsymbol{p} : S \to \boldsymbol{R}^3$ be an immersion of a 2-manifold S, and define $\nabla_X Y \in \mathfrak{X}^\infty(S)$ by*

$$\begin{aligned}\nabla_X Y &:= [d\hat{Y}(X)]^{\mathrm{H}} \\ &= (\textit{the component of } d\hat{Y}(X) \textit{ tangent to the surface } \boldsymbol{p}(S)) \\ &= d\hat{Y}(X) - (d\hat{Y}(X) \cdot \nu)\nu \qquad (\nu \textit{ is a unit normal vector})\end{aligned}$$

for X, $Y \in \mathfrak{X}^\infty(S)$. Then ∇ is the Levi-Civita covariant derivative (13.4) *determined by the first fundamental form.*

Proof. In fact, the vector $d\hat{Y}(X) - (d\hat{Y}(X)\cdot\nu)\nu$ in $\boldsymbol{R}^3$ can be identified with the tangent vector $\nabla_X Y$ of the manifold S using (13.19). Although there exists a $\pm$-ambiguity for the choice of the unit vector ν, the definition of $\nabla_X Y$ does not depend on the choice of the sign, since ν appears twice.

Firstly, for $X, Y \in \mathfrak{X}^\infty(S)$ and $f \in C^\infty(S)$, noting that $[\hat{X}]^{\mathrm{H}} = \hat{X}$, $[\hat{Y}]^{\mathrm{H}} = \hat{Y}$, we have

$$\begin{aligned}\nabla_{fX}Y &= [d\hat{Y}(fX)]^{\mathrm{H}} = [f d\hat{Y}(X)]^{\mathrm{H}} = f[d\hat{Y}(X)]^{\mathrm{H}} = f\nabla_X Y,\\ \nabla_X fY &= [d(\widehat{fY})(X)]^{\mathrm{H}} = [d(f\hat{Y})(X)]^{\mathrm{H}} = [df(X)\hat{Y} + f d\hat{Y}(X)]^{\mathrm{H}}\\ &= df(X)\hat{Y} + f[d\hat{Y}(X)]^{\mathrm{H}} = df(X)Y + f\nabla_X Y,\end{aligned}$$

and ∇ satisfies (13.7) and (13.9). Furthermore, one can easily check that (13.6) and (13.8) also hold, so ∇ is a covariant derivative. In addition, from the definition (13.21) of the first fundamental form,

$$\begin{aligned}X\langle Y, Z\rangle &= X(\hat{Y}\cdot\hat{Z}) = d(\hat{Y}\cdot\hat{Z})(X) = d\hat{Y}(X)\cdot\hat{Z} + \hat{Y}\cdot d\hat{Z}(X)\\ &= [d\hat{Y}(X)]^{\mathrm{H}}\cdot\hat{Z} + \hat{Y}\cdot[d\hat{Z}(X)]^{\mathrm{H}} = \langle\nabla_X Y, Z\rangle + \langle Y, \nabla_X Z\rangle\end{aligned}$$

holds. Here we are using that $\hat{Y}$ and $\hat{Z}$ are tangent vectors to the surface. Therefore, ∇ satisfies (13.11).

From the above discussion, if we show that ∇ satisfies (13.10), then by Theorem 13.3, we know that it is the Levi-Civita covariant derivative. In fact, on a local coordinate neighborhood $(U; (u, v))$ of S, we have

$$\nabla_{\frac{\partial}{\partial u}}\frac{\partial}{\partial v} = \left[d\left(\widehat{\frac{\partial}{\partial v}}\right)\left(\frac{\partial}{\partial u}\right)\right]^{\mathrm{H}} = \left[d\left(\frac{\partial \boldsymbol{p}}{\partial v}\right)\left(\frac{\partial}{\partial u}\right)\right]^{\mathrm{H}} = \left[\frac{\partial^2 \boldsymbol{p}}{\partial u\partial v}\right]^{\mathrm{H}},$$

and so the T defined in the proof of Theorem 13.3 satisfies

$$T\left(\frac{\partial}{\partial u}, \frac{\partial}{\partial v}\right) = \left[\frac{\partial^2 \boldsymbol{p}}{\partial u\partial v}\right]^{\mathrm{H}} - \left[\frac{\partial^2 \boldsymbol{p}}{\partial v\partial u}\right]^{\mathrm{H}} - \widehat{\left[\frac{\partial}{\partial u}, \frac{\partial}{\partial v}\right]} = \mathbf{0}.$$

Here, we used the relations $\boldsymbol{p}_{uv} = \boldsymbol{p}_{vu}$ and $[\partial/\partial u, \partial/\partial v] = \mathbf{0}$. We also have $T(fX, Y) = T(X, fY) = fT(X, Y)$, and in particular $T(Y, X) = -T(X, Y)$ holds, and so T vanishes identically and (13.10) holds. □

Exercises 13

1 By establishing the following (1)–(3), prove that the definition of the Levi-Civita covariant derivative $\nabla_X Y$ does not depend on the choice of orthonormal frame field:

(1) Taking the connection form μ for an orthonormal frame field $\{\boldsymbol{e}_1, \boldsymbol{e}_2\}$, show that $-\mu$ is the connection form given by choosing $\{\boldsymbol{e}_1, -\boldsymbol{e}_2\}$.
(2) Take the connection form $\tilde{\mu}$ for an orthonormal frame field $\{\tilde{\boldsymbol{e}}_1, \tilde{\boldsymbol{e}}_2\}$ that has the same orientation as $\{\boldsymbol{e}_1, \boldsymbol{e}_2\}$. Noting that $\nabla_X Y$ is defined by μ via (13.4), (13.5), show that

$$(*) \qquad \nabla_X \tilde{\boldsymbol{e}}_1 = -\tilde{\mu}(X)\tilde{\boldsymbol{e}}_2, \qquad \nabla_X \tilde{\boldsymbol{e}}_2 = \tilde{\mu}(X)\tilde{\boldsymbol{e}}_1.$$

(3) Show also that, when the orthonormal frame field $\{\tilde{\boldsymbol{e}}_1, \tilde{\boldsymbol{e}}_2\}$ has opposite orientation to $\{\boldsymbol{e}_1, \boldsymbol{e}_2\}$, $(*)$ holds, using (1) and (2).

2 For the first fundamental form ds^2 of an immersion of a surface S into $\boldsymbol{R}^3$, and for an orthonormal frame field $\{\boldsymbol{e}_1, \boldsymbol{e}_2\}$ defined on a local coordinate neighborhood U, show that the corresponding connection μ satisfies

$$\mu(X) = d\hat{\boldsymbol{e}}_2(X) \cdot \hat{\boldsymbol{e}}_1 = -d\hat{\boldsymbol{e}}_1(X) \cdot \hat{\boldsymbol{e}}_2 \qquad (X \in \mathfrak{X}^\infty(U)),$$

where $\hat{\boldsymbol{e}}_j$ is a function valued in $\boldsymbol{R}^3$ obtained by (13.19) from $\boldsymbol{e}_j$, for each $j = 1, 2$.

3 For a Riemannian 2-manifold S with Riemannian metric ds^2, using a local coordinate neighborhood $(U; (u, v))$, we can write the metric as

$$ds^2 = E\,du^2 + 2F\,du\,dv + G\,dv^2.$$

For this E, F and G, according to equation (10.6) in Section 10, we have $\{\Gamma_{ij}^k\}_{i,j,k=1,2}$ defined, and these are called the *Christoffel symbols.* Show that the Levi-Civita covariant derivative ∇ of S satisfies

$$\nabla_{\frac{\partial}{\partial u}} \frac{\partial}{\partial u} = \Gamma_{11}^1 \frac{\partial}{\partial u} + \Gamma_{11}^2 \frac{\partial}{\partial v}, \qquad \nabla_{\frac{\partial}{\partial u}} \frac{\partial}{\partial v} = \Gamma_{12}^1 \frac{\partial}{\partial u} + \Gamma_{12}^2 \frac{\partial}{\partial v},$$
$$\nabla_{\frac{\partial}{\partial v}} \frac{\partial}{\partial u} = \Gamma_{21}^1 \frac{\partial}{\partial u} + \Gamma_{21}^2 \frac{\partial}{\partial v}, \qquad \nabla_{\frac{\partial}{\partial v}} \frac{\partial}{\partial v} = \Gamma_{22}^1 \frac{\partial}{\partial u} + \Gamma_{22}^2 \frac{\partial}{\partial v}.$$

4 Taking geodesic polar coordinates (r, θ) on a regular surface S in $\boldsymbol{R}^3$, then the first fundamental form of the surface can be written as (see (11.9) on page 121)

$$ds^2 = dr^2 + h^2\,d\theta^2 \qquad (h = h(r, \theta)).$$

Here $h(r, \theta)$ is a smooth function that takes positive values. Using (13.15) for the definition of Gaussian curvature, show that $K = -h_{rr}/h$ holds. In particular, K is equal to the Gaussian curvature defined in Section 8.

14. The Gauss-Bonnet formula for 2-manifolds

In the study of curves, parametrizations by arc-length play an important role. For surfaces rather, if the Gaussian curvature is not identically zero, then it is not possible to find local coordinates that preserve length, and instead, choosing an orthonormal frame field, the prospect of fruitful discussion of surfaces remains. Taking this stance in this section, we give an alternative proof of the Gauss-Bonnet theorem (Theorem 10.6).

Now, assuming S is oriented, we take a positively oriented orthonormal frame field $\{\boldsymbol{e}_1, \boldsymbol{e}_2\}$ on a local coordinate neighborhood. Consider a smooth regular curve $\gamma\colon [a,b] \to U$, that is, $d\gamma/ds$ does not vanish. When the curve $\gamma(s)$ satisfies $|d\gamma/ds| = 1$, the parameter s is said to be *arc-length.* One can prove existence of the arc-length parameter in the same way as in Section 2. We now assume s is an arc-length parameter. Let $\varphi(s)$ denote the (counterclockwise) angle between $\gamma'(s)$ and $\boldsymbol{e}_1(s) = \boldsymbol{e}_1(\gamma(s))$ $('= d/ds)$. With this, we can write

$$\gamma'(s) = \xi_1(s)\boldsymbol{e}_1(s) + \xi_2(s)\boldsymbol{e}_2(s) \tag{14.1}$$
$$(\boldsymbol{e}_2(s) = \boldsymbol{e}_2(\gamma(s)),\ \xi_1(s) = \cos\varphi(s),\ \xi_2(s) = \sin\varphi(s)).$$

In particular,

$$\xi_1\xi_2{}' - \xi_2\xi_1{}' = \cos\varphi(\sin\varphi)' - \sin\varphi(\cos\varphi)' = \varphi' \tag{14.2}$$

holds. We can write the unit normal vector $\boldsymbol{n}_g(s)$ to the left-hand side of this curve γ as

$$\boldsymbol{n}_g(s) = -\xi_2(s)\boldsymbol{e}_1(s) + \xi_1(s)\boldsymbol{e}_2(s). \tag{14.3}$$

On the other hand, we call the vector

$$\boldsymbol{\kappa}_g := \nabla_{\gamma'}\gamma' = (\xi_1{}' + \mu(\gamma')\xi_2)\boldsymbol{e}_1 + (\xi_2{}' - \mu(\gamma')b\xi_1)\boldsymbol{e}_2 \tag{14.4}$$

the *geodesic curvature vector* of γ. By (13.17), $\boldsymbol{\kappa}_g$ is perpendicular to γ', and so can be written as a scalar multiple of $\boldsymbol{n}_g$, i.e., we can write $\boldsymbol{\kappa}_g(s) = \kappa_g(s)\boldsymbol{n}_g(s)$. This coefficient $\kappa_g(s)$ is called the *geodesic curvature* of $\gamma(s)$, and we have

$$\kappa_g = \langle \boldsymbol{\kappa}_g, \boldsymbol{n}_g \rangle = \langle \gamma'', \boldsymbol{n}_g \rangle, \tag{14.5}$$

where $\gamma'' := \nabla_{\gamma'}\gamma'$. The geodesic curvature for a curve $\gamma(t)$ with non-arclength parametrization is given as follows (cf. (10.16)):

$$\kappa_g(t) = \frac{\langle \ddot{\gamma}(t), \boldsymbol{n}_g(t) \rangle}{\langle \dot{\gamma}(t), \dot{\gamma}(t) \rangle} \qquad (\ddot{\gamma}(t) := \nabla_{\dot{\gamma}(t)}\dot{\gamma}(t)) \,. \tag{14.6}$$

With this definition, the geodesic curvature $\kappa_g(t)$ at the point $\gamma(t)$ on the curve γ is a notion that does not depend on the choice of parameter. Furthermore, when the manifold S is realized as a regular surface in $\boldsymbol{R}^3$, the κ_g defined in Section 10 coincides with the geodesic curvature (10.13) (see Problem **1** at the end of this section). Similarly to the case of surfaces in $\boldsymbol{R}^3$, the geodesic curvature of geodesics is 0. Conversely, a curve with geodesic curvature 0, when changing the parametrization so that it is arc-length, is a geodesic (see Problem **9** in Section 10). By (14.2)–(14.5),

$$\kappa_g\,ds = \xi_1 d\xi_2 - \xi_2 d\xi_1 - \mu = d\varphi - \mu \tag{14.7}$$

holds on γ. Here $\mu = \mu(\gamma')\,ds$ is a 1-form defined along the curve γ. Using this, we will now prove the Gauss-Bonnet theorem for (not necessarily geodesic) triangles:

Proposition 14.1. *Let (S, ds^2) be an oriented Riemannian 2-manifold, and take a local coordinate neighborhood $(U; (u, v))$. Consider a closed simply-connected triangular region $\overline{D}$ in U surrounded by three smooth regular curves γ_j $(j = 1, 2, 3)$, with internal angles φ_{12}, φ_{23} and φ_{31} (see Fig. 14.1, left side). Then*

$$\int_{\partial\overline{D}} \kappa_g(s)\,ds + \int_{\overline{D}} K\,dA = -\pi + (\varphi_{12} + \varphi_{23} + \varphi_{31}) \tag{14.8}$$

holds. Here, s is an arc-length parameter along the edges of the triangle. The geodesic curvature in the first integral on the left-hand side is computed with respect to the orientation that places the region $\overline{D}$ to the left of the edges.

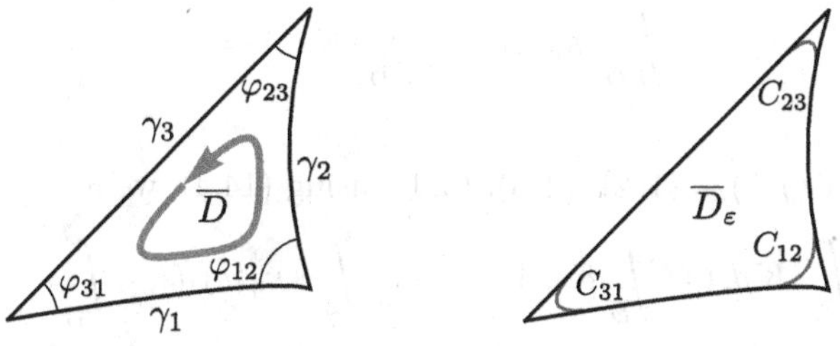

Fig. 14.1

Proof. Let us round off the three corners of $\overline{D}$ to give regions $\overline{D}_\varepsilon$ with C^∞ boundary that converge to $\overline{D}$ when $\varepsilon \to 0$ (for the way to round off corners, see Proposition B.5.5 in Appendix B.5). We denote by $\gamma_{1,\varepsilon}$, $\gamma_{2,\varepsilon}$ and $\gamma_{3,\varepsilon}$ the intersections of $\partial\overline{D}_\varepsilon$ with the original edges γ_1, γ_2 and γ_3, respectively, and by C_{12}, C_{23}, C_{31} the curves on $\partial\overline{D}_\varepsilon$ which correspond to the rounded corners (see the right-hand side of Fig. 14.1).

By the Stokes theorem (cf. Theorem 12.2), (13.15), (14.7) and the definition of the integral of 2-forms in (12.9), we have

$$\begin{aligned}\int_{\partial\overline{D}_\varepsilon} \kappa_g(s)\,ds + \int_{\overline{D}_\varepsilon} K\,dA &= \int_{\partial\overline{D}_\varepsilon} \kappa_g(s)\,ds + \int_{\overline{D}_\varepsilon} K\,\omega_1\wedge\omega_2 \\ &= \int_{\partial\overline{D}_\varepsilon} \kappa_g(s)\,ds + \int_{\overline{D}_\varepsilon} d\mu \\ &= \int_{\partial\overline{D}_\varepsilon} \kappa_g(s)\,ds + \int_{\partial\overline{D}_\varepsilon} \mu = \int_{\partial\overline{D}_\varepsilon} d\varphi.\end{aligned}$$

Because $\partial\overline{D}_\varepsilon$ is a closed regular curve, which we orient so that it contains the region $\overline{D}$ on its left side, $\int_{\partial\overline{D}_\varepsilon} d\varphi$ equals an integer multiple of 2π. Since we have taking a local coordinate system $(U;(u,v))$, we can consider the following variation

$$ds_t^2 := (1-t)ds^2 + t(du^2+dv^2) \qquad (0 \le t \le 1)$$

of the Riemannian metric on U, and the integral $\int_{\partial\overline{D}_\varepsilon} d\varphi$, and because it takes values in integer multiples of 2π, it will not change value when t changes. Because ds_1^2 is the standard metric du^2+dv^2 on $\boldsymbol{R}^2$, the integral is nothing but 2π times the winding number of the closed curve, when $t=1$. In particular, because $\partial\overline{D}_\varepsilon$ winds around exactly once with positive orientation (see Theorem 3.2 in Section 3), we know that $\int_{\partial\overline{D}_\varepsilon} d\varphi = 2\pi$, and

$$\int_{\partial\overline{D}_\varepsilon} \kappa_g(s)\,ds + \int_{\overline{D}_\varepsilon} K\,dA = 2\pi$$

is proven.

Now, for $(j,k) = (1,2)$, $(2,3)$, $(3,1)$, using (14.4), we have

$$\lim_{\varepsilon\to 0}\int_{\overline{D}_\varepsilon} K\,dA = \int_{\overline{D}} K\,dA, \qquad \lim_{\varepsilon\to 0}\int_{\gamma_{j,\varepsilon}} \kappa_g(s)\,ds = \int_{\gamma_j} \kappa_g(s)\,ds,$$

$$\lim_{\varepsilon\to 0}\int_{C_{jk}} \kappa_g(s)\,ds = \lim_{\varepsilon\to 0}\int_{C_{jk}} d\varphi - \lim_{\varepsilon\to 0}\int_{C_{jk}} \mu = \lim_{\varepsilon\to 0}\int_{C_{jk}} d\varphi = \pi - \varphi_{jk},$$

giving us the result. As $\varepsilon \to 0$, each C_{jk} approaches a single point, and the integral of μ also approaches 0. So the desired formula (14.8) is obtained by taking the limit as $\varepsilon \to 0$. □

Using this result, we can prove the following global Gauss-Bonnet theorem as a generalization of Theorem 10.7.

Theorem 14.2 (The global Gauss-Bonnet theorem). *Let (S, ds^2) be an oriented Riemannian 2-manifold with a bounded closed region $\overline{D}$ whose boundary $\partial\overline{D}$ is either empty or is a number of smooth closed curves oriented so that $\overline{D}$ is enclosed on the left-hand side. Let κ_g be the geodesic curvature of $\partial\overline{D}$. Then*

$$\int_{\partial\overline{D}} \kappa_g(s)\, ds + \int_{\overline{D}} K dA = 2\pi\chi(\overline{D})$$

holds. Here, $\chi(\overline{D})$ is the Euler characteristic of $\overline{D}$. In the case that $\overline{D} = S$ is compact and without boundary, we have

$$\int_S K\, dA = 2\pi\chi(S). \tag{14.9}$$

When S is a regular surface in $\boldsymbol{R}^3$, the formula (14.9) is the same as Theorem 10.7 in Section 10.

Remark 14.3. Even when S is compact and non-orientable, (14.9) holds. In fact, there exists a compact orientable manifold $\hat{S}$ and a 2-to-1 immersion $p\colon \hat{S} \to S$ (cf. [5, Exercise 12 of Chapter 0]). Then the Riemann metric of S can be lifted to $\hat{S}$ by p, and the Gaussian curvature $\hat{K}$ of the lifted metric satisfies

$$4\pi\chi(S) = 2\pi\chi(\hat{S}) = \int_{\hat{S}} \hat{K}\, dA = 2\int_S K\, dA.$$

Proof of Theorem 14.2. Triangulating $\overline{D}$ into m triangles with smooth edges, take n_1 to be the total number of triangle vertices in $\partial\overline{D}$, and take n_2 to be the total number of triangle vertices in the interior of $\overline{D}$ (the existence of such a triangulation is shown in Chapter 1 of [25] or Theorem 10.6 of [28]). Also, take l_1 to be the total number of triangle edges in $\partial\overline{D}$, and l_2 the number of triangle vertices in the interior of $\overline{D}$. Summing up (over all triangles) the result of Proposition 14.1 about each triangle, the integrals of the geodesic curvature over the edges in the interior of $\overline{D}$ will cancel out (because each edge is being integrated on twice with opposite orientation, due to the two adjacent triangles), and κ_g will be integrated precisely on the boundary $\partial\overline{D}$. Furthermore, at each triangle vertex in $\partial\overline{D}$ the interior triangle angles will sum to π, and at each triangle vertex in the interior of $\partial\overline{D}$ the interior triangle angles will sum to 2π, and we have

$$\int_{\partial\overline{D}} \kappa_g(s)\, ds + \int_{\overline{D}} K\, dA = \pi n_1 + 2\pi n_2 - \pi m.$$

On the other hand, each triangle has three edges, and the triangle edges in the interior of $\overline{D}$ each have two adjacent triangles, so $l_1 + 2l_2 = 3m$, and also $n_1 = l_1$, thus

$$\begin{aligned}\pi n_1 + 2\pi n_2 - \pi m &= 2\pi(n_1 + n_2) - \pi n_1 - 3\pi m + 2\pi m \\ &= 2\pi\{(n_1 + n_2) - (l_1 + l_2) + m\} = 2\pi\chi(\overline{D})\end{aligned}$$

holds, proving the result. □

Exercise 14

1 Consider a regular space curve $\gamma(t)$ lying in an oriented surface S in $\boldsymbol{R}^3$, and consider the unit normal vector ν to the surfaces S that corresponds to the given orientation (see (13.18)). Confirm that the geodesic curvature of the curve $\gamma(s)$ on the surface as defined in Section 10 is equivalent to the definition of geodesic curvature as given in (14.5) and (14.6).

15. Poincaré-Hopf index theorem

As one application of the Gauss-Bonnet theorem (Proposition 14.1 and Theorem 14.2), we shall give the index formula for vector fields on a compact oriented 2-manifold S.

It is well-known that there exists a Riemannian metric ds^2 on a manifold S (see [35, Theorem 13], for example). In this section, we fix one such choice for ds^2.

Let X be a smooth vector field on S that is defined away from a finite number of points $\mathrm{P}_1, \ldots, \mathrm{P}_n$ in S, and is never zero away from those points. This field X could be zero at a point P_j ($j = 1, 2, \ldots, n$), or it could be undefined at that point.

When X is zero at P_j, we say that P_j is an *isolated zero* of X. On the other hand, when X is not defined at P_j, we say that P_j is an *isolated singularity* of X. Taking one point P_j from amongst P_k ($k = 1, \ldots, n$), there exists a sufficiently small positively oriented local coordinate neighborhood $(U_j; (u_j, v_j))$ so that none of the other $n-1$ points $\{\mathrm{P}_k \in S, | \, k \neq j\}$ lie in U_j. We can take U_j to be simply connected, that is, we can take it to be diffeomorphic to a disc in $\boldsymbol{R}^2$. Then $\boldsymbol{e}_1 := X/|X|$ ($|X|$ denotes the norm of X) is well-defined on $U_j \setminus \{\mathrm{P}_j\}$. Because $\boldsymbol{e}_1$ is a unit vector field, there exists another unit vector field $\boldsymbol{e}_2$ (rotation of $\boldsymbol{e}_1$ by 90°) so that $\{\boldsymbol{e}_1, \boldsymbol{e}_2\}$ is a positively oriented orthonormal frame field on $U_j \setminus \{\mathrm{P}_j\}$.

Now, consider a simple closed regular curve $\gamma(s)$, $(0 \le s \le l)$, inside $(U_j; (u_j, v_j))$ that wraps counterclockwise once about P_j. We take s to be an arc-length parameter, so $\gamma'(s)$ is always a unit vector. Let $\psi_j(s)$ be the angle between $\gamma'(s)$ and $\boldsymbol{e}_1$. (The angle ψ_j is the opposite of the angle φ defined in (14.1). Just like for the definition of φ, also $\psi_j(s)$ is defined so that it is a continuous real-valued function.) Then, since $\gamma'(0) = \gamma'(l)$, $\psi_j(0)$ and $\psi_j(l)$ differ only by an integer multiple of 2π. Thus, this difference will not change when the curve γ is continuously deformed. We define the *index* of the vector field X about the point P_j to be the integer

$$\text{(15.1)} \qquad \operatorname{ind}_{\mathrm{P}_j} X := 1 + \frac{1}{2\pi}(\psi_j(l) - \psi_j(0)).$$

This counts the number of times that the unit vector field $X/|X|$ wraps counterclockwise when traveling once along the path γ around P_j. The number 1 in the definition (15.1) offsets the rotation of $\gamma(s)$ itself, which travels about P_j exactly once. The following proposition holds.

Proposition 15.1. *The definition of the index of a vector field at a given point does not depend on the Riemannian metric on the manifold.*

Proof. Let ds_i^2 $(i = 0, 1)$ be two Riemannian metrics on S. Then

$$d\tilde{s}_t^2 := (1-t)ds_0^2 + t\,ds_1^2 \qquad (0 \le t \le 1)$$

is a continuous deformation between the Riemannian metrics ds_0^2 and ds_1^2. Because the index is an integer, its value cannot change under a continuous deformation of the metric, giving us the desired conclusion. □

For example, consider vector fields on $\boldsymbol{R}^2$ given by

$$X_1 = u\frac{\partial}{\partial u} + v\frac{\partial}{\partial v}, \qquad X_2 = u\frac{\partial}{\partial u} - v\frac{\partial}{\partial v},$$
$$X_3 = -v\frac{\partial}{\partial u} + u\frac{\partial}{\partial v}, \qquad X_4 = (u^2 - v^2)\frac{\partial}{\partial u} + 2uv\frac{\partial}{\partial v}.$$

These all have an isolated zero at the origin, and their respective indices are 1, −1, 1 and 2 (see Fig. 15.1). We have the following formula regarding the sum of the indices of a vector field:

Theorem 15.2 (The Poincaré-Hopf[1] index theorem). *Let S be a compact oriented 2-manifold. Let X be a vector field defined on S with a finite number of points $\mathrm{P}_1, \ldots, \mathrm{P}_n$ removed, so that X is never zero away from*

[1]Hopf, Heinz (1894–1971). See page 137 for Poincaré.

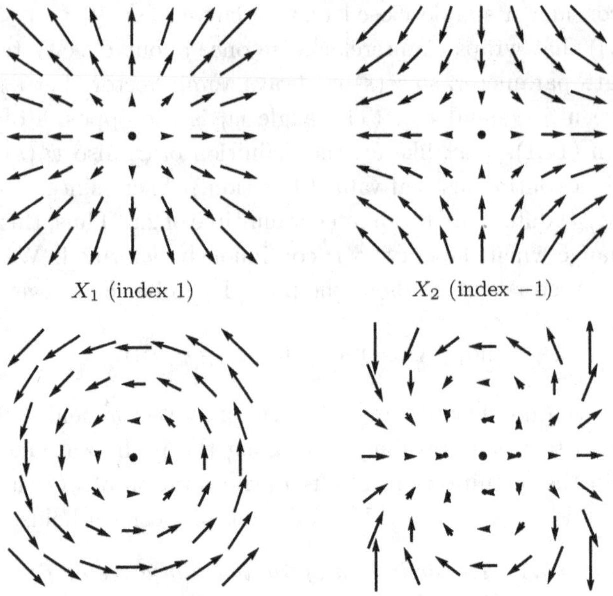

Fig. 15.1 Examples of indices of vector fields.

those n points, that is, only those n points can be isolated zeroes or isolated singularities of X. Then

$$\sum_{j=1}^{n} \mathrm{ind}_{\mathrm{P}_j} X = \chi(S),$$

where $\chi(S)$ is the Euler number of S (see (10.11) *in Section 10).*

Proof. Take a local coordinate chart neighborhood $(U_j; (u_j, v_j))$ around P_j compatible to the orientation of S, for $j = 1, \ldots, n$. Taking one choice of Riemannian metric ds^2 on S and fixing a sufficiently small positive number ε, we consider each geodesic circle $C_j(\varepsilon)$ of radius ε traveling counterclockwise once around each point P_j in U_j. Removing the discs in S surrounded by the $C_j(\varepsilon)$ $(j = 1, \ldots, n)$, we call the remaining region $\overline{D}_\varepsilon$. Let ε_0 be one fixed choice of such an ε. Let us choose a local coordinate neighborhood $(U_j; (u_j, v_j))$ for each P_j $(j = 1, \ldots, n)$ so that P_j corresponds to the origin (see page 127), and so $C_j(\varepsilon) \subset U_j$. We define a new Riemannian metric $d\tilde{s}^2$ on S, at each point Q in S, as follows:

$$d\tilde{s}^2_{\mathrm{Q}} := \begin{cases} ds^2 & (\text{for } \mathrm{Q} \in \overline{D}_{\varepsilon_0}), \\ \rho(|\mathrm{Q}|)ds^2 + (1 - \rho(|\mathrm{Q}|))(du_j^2 + dv_j^2) & (\text{for } \mathrm{Q} \in U_j), \end{cases}$$

where $|Q|$ ($:= d(Q, P_j)$) means the distance from Q to P_j with respect to ds^2, for $Q \in U_j$ ($j = 1, \ldots, n$), and $\rho : \boldsymbol{R} \to [0,1]$ is a C^∞-function, which takes value 1 on the interval $[-\varepsilon_0/3, \varepsilon_0/3]$, and 0 on the set $\boldsymbol{R} \setminus [-2\varepsilon_0/3, 2\varepsilon_0/3]$.

The formula we want to prove does not depend on the choice of Riemannian metric on S (see Proposition 15.1), so in this proof we may now use $d\tilde{s}^2$ instead of ds^2.

We choose ε to be a positive number smaller than ε_0. About each P_j, we can take the angular function ψ_j on $C_j(\varepsilon)$ as explained in the definition of the index, and then, for the curve $C_j(\varepsilon)$, the φ defined in (14.1) is the same as $-\psi_j$. Rotating counterclockwise 90° from the unit vector field $\boldsymbol{e}_1 := X/|X|$ on $\overline{D}_\varepsilon$, we can define the vector field $\boldsymbol{e}_2$ so that $\{\boldsymbol{e}_1, \boldsymbol{e}_2\}$ forms an orthonormal frame field on $\overline{D}_\varepsilon$. Let μ be the connection form on $\overline{D}_\varepsilon$ with respect to the above mentioned basis. By (14.7), the geodesic curvature κ_g on the curve $C_j(\varepsilon)$ satisfies $\kappa_g\, ds = -d\psi_j - \mu$. Let $-C_j(\varepsilon)$ denote the same geodesic circle $C_j(\varepsilon)$, but with the opposite orientation, and then the outer side of $-C_j(\varepsilon)$ will be on the left-hand side. By Theorem 14.2 and the Stokes theorem (Theorem 12.2), we have

$$\begin{aligned}
2\pi\chi(S) &= \int_S K\,dA = \lim_{\varepsilon\to 0}\int_{\overline{D}_\varepsilon} K\,dA = \lim_{\varepsilon\to 0}\int_{\overline{D}_\varepsilon} d\mu = \lim_{\varepsilon\to 0}\int_{\partial\overline{D}_\varepsilon} \mu \\
&= \lim_{\varepsilon\to 0}\sum_{j=1}^{n}\int_{-C_j(\varepsilon)} \mu = \lim_{\varepsilon\to 0}\sum_{j=1}^{n}\int_{C_j(\varepsilon)} (-\mu) \\
&= \lim_{\varepsilon\to 0}\sum_{j=1}^{n}\int_{C_j(\varepsilon)} (\kappa_g\,ds + d\psi_j) = 2\pi n + \lim_{\varepsilon\to 0}\sum_{j=1}^{n}\int_{C_j(\varepsilon)} d\psi_j.
\end{aligned}$$

In fact, for small ε, $d\tilde{s}^2$ coincides with the Euclidean metric on $S \setminus D_\varepsilon$, and by Theorem 3.2 in Section 3, we have $\int_{C_j(\varepsilon)} \kappa_g\, ds = 2\pi$, which is used in the process of computing the lines above.

Since each integral on the right-hand side is the variation of the angle ψ_j traveling on $C_j(\varepsilon)$ leftward, we have $2\pi(\operatorname{ind}_{P_j} X - 1)$. This concludes the proof. □

A vector field give rise to a "flow along oriented curves in the 2-dimensional manifold[2]". The formula (15.2) gives the value of the sum of the indices of such a flow. But furthermore, in the case that the curve flow in not necessarily oriented, the indices might take non-integer values

[2]These are called *integral curves* of a vector field. For a detailed explanation, see Chapter 1 of [40].

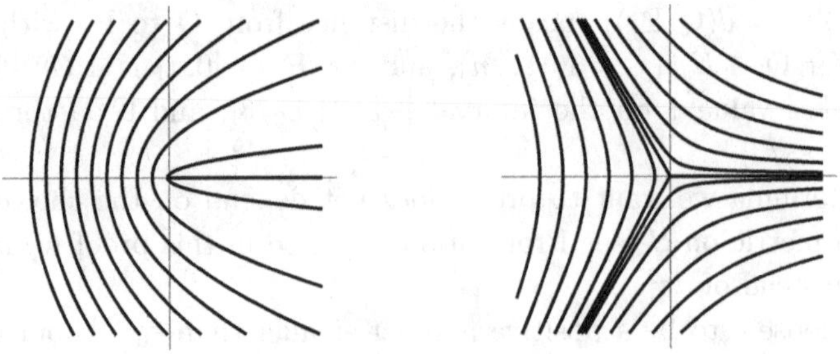

η: index $1/2$ $\qquad$ ζ: index $-1/2$

Fig. 15.2 Flows corresponding to non-orientable families of curves (cf. Example 15.3).

like $\pm 1/2$, $\pm 3/2$, ..., and so half-integer-valued indices are also defined, and we obtain a generalization of the formula we gave above for indices.

For explicit examples, see the families of curves in Fig. 15.2. For these two flows, one cannot choose the direction of the flow on the whole figure. In fact, for the family of curves wrapping once about the origin in the direction of the arrows, we switch to the opposite orientation. These are explicit examples of "directional fields", as defined in the following:

At a point P on a manifold S, consider two non-zero tangent vectors V, $W \in T_{\mathrm{P}}S$ at P. We say that V and W are *projectively equivalent* if there exists a real number c so that $W = cV$. We write this equivalence relation as $V \sim W$, and the equivalence class represented by V is written as $[V]$. In particular, the statements $V \sim W$ and $[V] = [W]$ have the same meaning. Also, the equivalence class $[V]$ is a 1-dimensional subspace of the tangent space $T_{\mathrm{P}}S$.

Now a "directional field" on S is defined as follows, as a map for which each point P corresponds to a 1-dimensional subspace of $T_{\mathrm{P}}S$: Suppose that the manifold S is 2-dimensional, and that U is an open set in S. We denote by $\mathfrak{X}^\infty(U)$ the collection of all smooth vector fields on U, and let X, $Y \in \mathfrak{X}^\infty(U)$ be two vector fields without zeros on U. We say that X and Y are *projectively equivalent* on U if there exists a smooth function f without zeros on U such that $Y = fX$ holds. This equivalence relation is written as $X \underset{U}{\sim} Y$. When $X \underset{U}{\sim} Y$ holds, we have, in particular, that $X_{\mathrm{P}} \sim Y_{\mathrm{P}}$ at each point in U (projective equivalence as tangent vectors).

Given an open covering $\{U_j\}_{j=1,2,\dots,N}$ of S and vector fields $X_j \in \mathfrak{X}^\infty(U_j)$ defined on the respective U_j, we consider the set of pairs $\xi := \{(U_j, X_j)\}_{j=1,\dots,N}$. We call ξ a *directional field* or a *projective vector field*, if

$$X_j \underset{U_j \cap U_k}{\sim} X_k \tag{15.2}$$

holds for any choice of two numbers j, k ($1 \leq j, k \leq N$). In this case, at each point P of S, ξ determines a 1-dimensional subspace $\xi_{\mathrm{P}} = [X_j(\mathrm{P})]$ of the vector space $T_{\mathrm{P}}S$. Here, $\mathrm{P} \in U_j$ and $X_j(\mathrm{P})$ is the (vector) value of the vector field X_j at P. In fact, if $\mathrm{P} \in U_k$, from (15.2) we have $[X_j(\mathrm{P})] = [X_k(\mathrm{P})]$, and so ξ is a well-defined map, that is, independent of the choice of j at each point.

For a smooth vector field X on S without zeros, in the case that $N = 1$ and $U_1 = S$, the vector field X then determines a directional field $\xi = \{(S, X)\}$. At each point P, ξ determines a 1-dimensional subspace $[X(\mathrm{P})]$ of $T_{\mathrm{P}}S$, so ξ can be written as $[X]$. In this way, a general vector field X naturally determines one directional field $[X]$. Below we introduce a less trivial example of a directional field.

Example 15.3. Consider $\boldsymbol{R}^2 \setminus \{(0,0)\}$ as a union of two open sets

$$U_1 := \boldsymbol{R}^2 \setminus \{(u,0) \in \boldsymbol{R}^2 \mid u \leq 0\}, \quad U_2 := \boldsymbol{R}^2 \setminus \{(u,0) \in \boldsymbol{R}^2 \mid u \geq 0\}.$$

We give the following two vector fields on U_1:

$$\begin{aligned} Y_1 &:= \left(u + \sqrt{u^2+v^2}\right)\frac{\partial}{\partial u} + v\frac{\partial}{\partial v}, \\ Z_1 &:= \left(u + \sqrt{u^2+v^2}\right)\frac{\partial}{\partial u} - v\frac{\partial}{\partial v}. \end{aligned} \tag{15.3}$$

Also, on U_2 we define the two vector fields

$$\begin{aligned} Y_2 &:= -v\frac{\partial}{\partial u} + \left(u - \sqrt{u^2+v^2}\right)\frac{\partial}{\partial v}, \\ Z_2 &:= v\frac{\partial}{\partial u} + \left(u - \sqrt{u^2+v^2}\right)\frac{\partial}{\partial v}. \end{aligned} \tag{15.4}$$

Since

$$\begin{aligned} Y_2 &= -v\frac{\partial}{\partial u} + \left(u - \sqrt{u^2+v^2}\right)\frac{\partial}{\partial v} \\ &= -v\frac{\partial}{\partial u} - \frac{v^2}{u+\sqrt{u^2+v^2}}\frac{\partial}{\partial v} = \frac{-v}{u+\sqrt{u^2+v^2}}Y_1, \end{aligned}$$

Y_2 is proportional to Y_1 on $U_1 \cap U_2$. Similarly, Z_2 is proportional to Z_1 on $U_1 \cap U_2$. So $\eta := \{(U_j, Y_j)\}_{j=1,2}$ and $\zeta := \{(U_j, Z_j)\}_{j=1,2}$ determine a directional field on $\boldsymbol{R}^2 \setminus \{(0,0)\}$. The left-hand side of Fig. 15.2 shows the flow generated by η, and the right-hand side of the figure shows the flow generated by ζ.

Let ξ be a directional field defined on a manifold with a finite number of points $\mathrm{P}_1, \dots, \mathrm{P}_n$ removed. Then we can define the index $\mathrm{ind}_{\mathrm{P}_j} \xi$ at P_j $(j = 1, \dots, n)$ just like we do for ordinary vector fields. In fact, for each $j = 1, \dots, n$, taking a local coordinate neighborhood $(U_j; (u_j, v_j))$ around P_j and simple closed regular curve $\gamma_j : [0, l] \to S$ that wraps once counterclockwise about P_j in U_j, we can correspondingly define a smooth vector field $W_j(t)$ $(0 \le t \le l)$ on γ_i by $[W_j(t)] = \xi_{\gamma_j(t)}$, as a map defined on the interval $[0, l]$. As t varies from $t = 0$ to $t = l$, we define the number $\mathrm{ind}_{\mathrm{P}_j} \xi$ of $W_j(t)$, like in (15.1) for general vector fields. We also define $\psi_j(t)$ to be the angle between $\dot\gamma(t)$ and $W_j(t)$, and then define

$$\mathrm{ind}_{\mathrm{P}_j} \xi := 1 + \frac{1}{2\pi}(\psi_j(l) - \psi_j(0)) \qquad (j = 1, 2, \dots, n).$$

If the directional field ξ satisfies $\xi = [X]$ for some ordinary vector field X, then $W_j(0) = W_j(l)$ holds and the index of ξ coincides with that of X. However, in general it is possible that $W_j(0) = -W_j(l)$, and then the index of the directional field takes a value in the set of half integers.

For example, in Example 15.3 we had two directional fields η, ζ, and on the unit circle $\gamma(t) = (\cos t, \sin t)$ $(0 \le t \le 2\pi)$, we can take smooth vector fields defined on γ as follows:

$$W := \cos\frac{t}{2}\frac{\partial}{\partial u} + \sin\frac{t}{2}\frac{\partial}{\partial v}, \qquad \widetilde{W} := -\cos\frac{t}{2}\frac{\partial}{\partial u} + \sin\frac{t}{2}\frac{\partial}{\partial v} \qquad (0 \le t \le 2\pi).$$

Then we have $[W(t)] = \eta_{\gamma(t)}$ and $[\widetilde{W}(t)] = \zeta_{\gamma(t)}$. The field $W(t)$ turns counterclockwise about half a circle as t traverses from $t = 0$ to $t = 2\pi$, and so η has index $1/2$ about the origin (see the left-hand side of Fig. 15.2). On the other hand, $\widetilde{W}(t)$ turns clockwise about half a circle as t traverses from $t = 0$ to $t = 2\pi$, and so ζ has index $-1/2$ about the origin (see the right-hand side of Fig. 15.2).

We now introduce the eigendirections of symmetric matrices (that is, the directions that the eigenvectors point in) as a way to determine directional fields. Below, we will consider a map

$$A(u, v) := \begin{pmatrix} a(u, v) & b(u, v) \\ b(u, v) & c(u, v) \end{pmatrix} \tag{15.5}$$

from a domain D to the set of 2×2 symmetric matrices. Set

$$d := \sqrt{(a(u, v) - c(u, v))^2 + 4b(u, v)^2}. \tag{15.6}$$

Then the eigenvalues of A are

$$\lambda_1 := \frac{a + c + d}{2}, \qquad \lambda_2 := \frac{a + c - d}{2}.$$

Assume that, away from the origin, A has two distinct eigenvalues, which is equivalent to saying that $d > 0$ is satisfied on $D \setminus \{(0,0)\}$. We are interested in the case that $d(0,0) = 0$ at the origin $(0,0)$. In this case, $A(0,0)$ is a scalar multiple of the identity matrix and we are unable to determine two eigendirections in a canonical way. We then define vector fields X_k $(k = 1,2)$ by

$$X_1 := 2b\frac{\partial}{\partial u} + (c - a + d)\frac{\partial}{\partial v} \qquad X_2 := 2b\frac{\partial}{\partial u} + (c - a - d)\frac{\partial}{\partial v}.$$

Each X_k $(k = 1,2)$ determines an eigendirection of A for λ_k whenever it is non-zero. Because two eigenvectors of a symmetric matrix with two different eigenvalues are perpendicular, the inner product of X_1 and X_2 is 0. Let $X_1^{\perp}$, $X_2^{\perp}$ denote the 90° rotations to the left of X_1, X_2, respectively, in other words,

$$X_1^{\perp} := (a - c - d)\frac{\partial}{\partial u} + 2b\frac{\partial}{\partial v} \qquad X_2^{\perp} := (a - c + d)\frac{\partial}{\partial u} + 2b\frac{\partial}{\partial v}.$$

Since X_1 is perpendicular to X_2, it is obvious that X_1 and $X_2^{\perp}$ (also X_2 and $X_1^{\perp}$) are proportional to each other.

Now, we take two open sets in $D \setminus \{(0,0)\}$ to be

$$U_1 := \{\mathrm{P} \in D \setminus \{(0,0)\} \mid X_1(\mathrm{P}) \neq \mathbf{0}\},$$
$$U_2 := \{\mathrm{P} \in D \setminus \{(0,0)\} \mid X_2(\mathrm{P}) \neq \mathbf{0}\},$$

and since the two vector fields X_1 and X_2 will not both become zero, we have $D \setminus \{(0,0)\} = U_1 \cup U_2$. Because of this, $\{U_1, U_2\}$ is an open covering of $D \setminus \{(0,0)\}$. Then, on $U_1 \cap U_2$, X_1 is proportional to $X_2^{\perp}$, and likewise $X_1^{\perp}$ is proportional to X_2, and so

$$\xi_1 := \{(U_1, X_1),\ (U_2, X_2^{\perp})\}, \qquad \xi_2 := \{(U_1, X_1^{\perp}),\ (U_2, X_2)\} \tag{15.7}$$

determine two directional fields ξ_1, ξ_2 on $D \setminus \{(0,0)\}$. The eigenvalue for A with respect to the eigenspace ξ_1 is λ_1, and the eigenvalue for A with respect to the eigenspace ξ_2 is λ_2. Now we define the vector field

$$V_A(u,v) := (a(u,v) - c(u,v))\frac{\partial}{\partial u} + 2b(u,v)\frac{\partial}{\partial v}, \tag{15.8}$$

using the coefficients of the symmetric matrix $A(u,v)$. Then the following proposition holds.

Proposition 15.4. *Assume the symmetric matrix $A(u,v)$ in equation (15.5) has distinct eigenvalues at all points away from the origin. The eigenvector directional fields ξ_1, ξ_2 of $A(u,v)$ defined in (15.7) satisfy*

$$\operatorname{ind}_{\mathrm{O}} \xi_1 = \operatorname{ind}_{\mathrm{O}} \xi_2 = \frac{1}{2} \operatorname{ind}_{\mathrm{O}} V_A.$$

In particular, ξ_1 and ξ_2 have the same index at the origin.

By this proposition, investigating only the coefficients of the matrix, we can compute the indices of the directional fields in the eigendirections. In fact, we will apply this proposition and Theorem 15.6 stated later in this section to the proof of the Hopf theorem in Section 17.

Proof of Proposition 15.4. Associating the vector $(\alpha, \beta)^T$ with the complex number $\alpha + i\beta$, we can write

$$X_1 = 2b + i(c - a + d), \qquad X_2^{\perp} = i(2b + i(c - a - d)),$$

and then direct computation shows that the squares of the complex numbers X_1 and $X_2^{\perp}$ satisfy

$$(X_1)^2 = 2(c - a + d)V_A, \qquad (X_2^{\perp})^2 = 2(a - c + d)V_A. \tag{15.9}$$

In other words, $(X_1)^2$ and $(X_2^{\perp})^2$ are both real scalar multiples of V_A. Also, because

$$(c - a + d)(a - c + d) = -(c - a)^2 + d^2 = 4b^2 \geq 0,$$

$(X_1)^2$ and $(X_2^{\perp})^2$ differ only by a positive scalar multiple. Then there exists $\varepsilon \in \{1, -1\}$ such that X_1 and $\varepsilon X_2^{\perp}$ differ only by a positive scalar multiple, that is, the arguments of these two complex numbers $X_1, \varepsilon X_2^{\perp}$ coincide, unless one of them vanishes. We fix a sufficiently small number r (> 0). Then we can define the common (continuous) argument function α_θ $(\theta \in \boldsymbol{R})$ uniquely such that $-\pi < \alpha_0 \leq \pi$ and

$$\frac{X_1(r\cos\theta, r\sin\theta)}{e^{i\alpha_\theta}}, \quad \frac{\varepsilon X_2(r\cos\theta, r\sin\theta)^{\perp}}{e^{i\alpha_\theta}} \in [0, \infty),$$

since X_1 and $X_2^{\perp}$ do not vanish at the same time. By (15.9), two times $\alpha_{2\pi} - \alpha_0$ is equal to $\operatorname{ind}_{\mathbf{o}} V_A$. So we have

$$\operatorname{ind}_{\mathbf{o}} V_A = 2 \operatorname{ind}_{\mathbf{o}} \xi_1.$$

Similarly, we can show $\operatorname{ind}_{\mathbf{o}} V_A = 2 \operatorname{ind}_{\mathbf{o}} \xi_2$. □

Example 15.5. As an explicit example, consider these two symmetric-matrix-valued functions:

$$A_1(u, v) = \begin{pmatrix} u & v \\ v & -u \end{pmatrix}, \qquad A_2(u, v) = \begin{pmatrix} u & -v \\ -v & -u \end{pmatrix}. \tag{15.10}$$

Take the Y_1, Y_2 (resp. Z_1, Z_2) from Example 15.3, which are the eigenvectors for A_1 (resp. A_2) at the point (u, v). Then η and ζ are the directional fields of the symmetric matrices A_1 and A_2, respectively. In Example 15.3, using a computation from the definition, we showed that the indices of η

and ζ about the origin are $1/2$ and $-1/2$, respectively, and here, once more, we will find the indices via an application of Proposition 15.4. The vectors in (15.8) are

$$V_{A_1} = 2\begin{pmatrix} u \\ v \end{pmatrix}, \qquad V_{A_2} = 2\begin{pmatrix} u \\ -v \end{pmatrix}.$$

Since the map $(u, v) \mapsto (u, v)$ is the identity map, the unit circle is mapped to the unit circle with the same orientation, and the map $(u, v) \mapsto (u, -v)$ takes the unit circle to the unit circle with the opposite orientation. Thus $\operatorname{ind}_{\mathrm{O}} V_{A_1} = 1$ and $\operatorname{ind}_{\mathrm{O}} V_{A_2} = -1$ hold. By Proposition 15.4, η and ζ have indices at half the value, $1/2$ and $-1/2$, respectively.

Also, for directional fields, following through the proof of Theorem 15.2, we see that the next theorem holds.

Theorem 15.6 (The Poincaré-Hopf index theorem for directional fields). *Let ξ be a directional field defined on a compact orientable 2-manifold S with a finite number of points $\mathrm{P}_1, \ldots, \mathrm{P}_n$ removed. Then it holds that*

$$\sum_{j=1}^{n} \operatorname{ind}_{\mathrm{P}_j} \xi = \chi(S).$$

Proof. This can be proven in an essentially parallel way to that of Theorem 15.2. Choosing local coordinates $(U_j; (u_j, v_j))$ about each point P_j, we let $\overline{D}_\varepsilon$ be a domain obtained by S removing discs of radii ε in U_j centered at P_j for $j = 1, \ldots, n$. The boundary of $\overline{D}_\varepsilon$ is the circles $C_j(\varepsilon)$ of radius ε centered about each P_j. We orient those circles so that P_j is to the left-hand side of each circle. From the definition of directional fields, there exists a unit vector field $\boldsymbol{e}_1$ on any local neighborhood U of any given point in $\overline{D}_\varepsilon$ so that $[\boldsymbol{e}_1] = \xi$ holds. Then let $\boldsymbol{e}_2$ be the vector field obtained by rotating $\boldsymbol{e}_1$ counterclockwise $90°$, and let μ be the connection form with respect to $\{\boldsymbol{e}_1, \boldsymbol{e}_2\}$. If we switch $\boldsymbol{e}_1$ to $-\boldsymbol{e}_1$, $\boldsymbol{e}_2$ also switches sign, and this determines the same connection form μ on U, which depends only on ξ. Hence μ determines a differential 1-form on $\overline{D}_\varepsilon$. With this, by the same argument as in the proof of Theorem 15.2,

$$\begin{aligned} 2\pi\chi(S) &= \int_S K\,dA = -\lim_{\varepsilon\to 0} \sum_{j=1}^{n} \int_{C_j(\varepsilon)} \mu \\ &= \lim_{\varepsilon\to 0} \sum_{j=1}^{n} \int_{C_j(\varepsilon)} (\kappa_g\,ds + d\psi_j) = 2\pi n + \lim_{\varepsilon\to 0} \sum_{j=1}^{n} \int_{C_j(\varepsilon)} d\psi_j. \end{aligned}$$

By the definition of ψ_j,

$$\int_{C_j(\varepsilon)} d\psi_j = 2\pi(-1 + \mathrm{ind}_{\mathrm{P}_j}\,\xi)$$

holds, and the result is shown. □

As an application of the Poincaré-Hopf index theorem, we compute the index at the umbilics of an ellipsoid in $\boldsymbol{R}^3$.

Recall that a *regular surface* means an immersion $\boldsymbol{p}$ of a 2-manifold into $\boldsymbol{R}^3$. We can define directional fields on the surface with the umbilic points removed, by considering the principal curvature directions (see Section 9) for each non-umbilic point. In fact, we can regard the principal curvature directions as the eigendirections of the Weingarten matrix $A = \widehat{I}^{-1}\,\widehat{II}$, and a pair of directional fields is induced. In particular, when the umbilics are isolated, we will call the index of these directional fields about an isolated umbilic the *index of the umbilic* itself.[3] Recall also that an immersion $\boldsymbol{p} : S \to \boldsymbol{R}^3$ (a surface) is said to be *closed* if S is a compact manifold without boundary. Theorem 15.6 implies the following corollary.

Corollary 15.7. *If all of the umbilics of a closed surface S in $\boldsymbol{R}^3$ are isolated, then the sum of the indices at the umbilics is equal to the Euler number of S.*

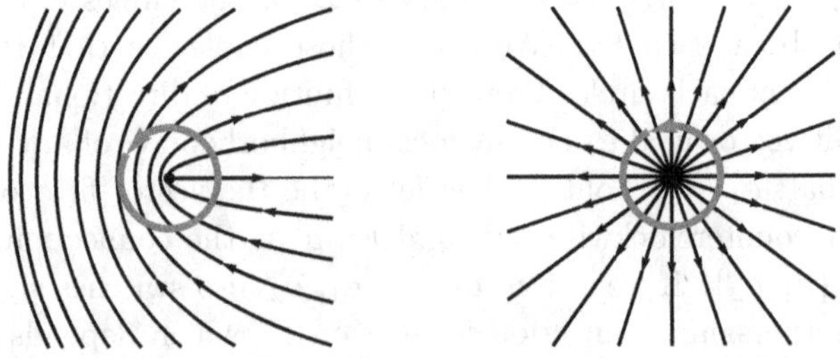

The case $a > b > c$ (index 1/2) The case $a = b > c$ (index 1)

Fig. 15.3 Fields for the principal directions of the ellipsoid $\frac{x^2}{a^2} + \frac{y^2}{b^2} + \frac{z^2}{c^2} = 1$.

[3]At an umbilic point, the indices of the two principal directional fields coincide. In fact, on the isothermal coordinate system (cf. Section 16), the Weingarten matrix A is symmetric because $\widehat{I}$ is a scalar multiple of the identity matrix. Hence two eigendirections of A are mutually orthogonal as vectors of $\boldsymbol{R}^2$, and then they have the same index.

Example 15.8. For example, for an ellipsoid whose lengths along the three axes all differ, the vector fields for the principal curvature directions about an umbilic are as in the left-hand side of Fig. 15.3, and the indices of the umbilics are 1/2. Let us confirm this for the ellipsoid

$$\mathcal{E} := \left\{ (x, y, z) \in \boldsymbol{R}^3 \;\middle|\; \frac{x^2}{a^2} + \frac{y^2}{b^2} + \frac{z^2}{c^2} = 1 \right\} \qquad (a \geq b \geq c > 0).$$

When $a = b = c$, $\mathcal{E}$ becomes a sphere, and in this case all points are umbilics, which is a case that is excluded.

The ellipsoid has symmetry across the xy-plane, so we consider only the region above the xy-plane, on which the ellipsoid is the graph of the function

$$z = f(x, y) := c\sqrt{1 - \left(\frac{x}{a}\right)^2 - \left(\frac{y}{b}\right)^2}.$$

Computing the first fundamental form (Problem **5** in Section 7), and the second fundamental form (Problem **3** in Section 8) of this graph, the Weingarten matrix A as in (8.4) becomes

$$A = \frac{c^4}{a^4 b^4 \delta(x, y)^3 f(x, y)^3} B,$$

where $\delta := \sqrt{1 + f_x^2 + f_y^2}$ and

$$B := \begin{pmatrix} -a^2 b^4 - a^2(c^2 - b^2)y^2 & a^2(c^2 - b^2)xy \\ b^2(c^2 - a^2)xy & -a^4 b^2 - b^2(c^2 - a^2)x^2 \end{pmatrix}.$$

By Proposition 9.6 in Section 9, a necessary and sufficient condition for $(x, y, f(x, y))$ to be an umbilic is that B is a scalar multiple of the identity matrix. Noting that the case $a = b = c$ is excluded, this is equivalent to

$$xy = 0 \qquad \text{and} \qquad a^2 b^4 + a^2(c^2 - b^2)y^2 = a^4 b^2 + b^2(c^2 - a^2)x^2.$$

Therefore, noticing $a \geq b \geq c$, the umbilics in the ellipsoid $\mathcal{E}$ that do not lie in the xy-plane are as follows:

- when $a > b > c$, $(x, y) = \left(\pm a\sqrt{\frac{a^2 - b^2}{a^2 - c^2}}, 0\right)$,
- when $a = b > c$, $(x, y) = (0, 0)$.

Similarly, for graphs $x = g(y, z)$, $y = h(z, x)$, umbilics exist on the xy-plane only when $a > b = c$, and they are $(\pm a, 0, 0)$.

Collecting the above facts, the umbilics of the ellipsoid $\mathcal{E}$ are as follows:

- when $a > b > c$, the four points $\left(\pm a\sqrt{\frac{a^2 - b^2}{a^2 - c^2}}, 0, \pm c\sqrt{\frac{b^2 - c^2}{a^2 - c^2}}\right)$,

- when $a = b > c$, the two points $(0, 0, \pm c)$,
- when $a > b = c$, the two points $(\pm a, 0, 0)$.

Because the ellipsoid is topologically equivalent to a sphere, its genus g is 0, and its Euler number is 2 (see (10.12) in Section 10).

By Corollary 15.7, the sum of the indices of the umbilics is 2. From the symmetries of the surface, the indices of the umbilics must all be the same, so

- when $a > b > c$, the umbilics have index $1/2$,
- when $a = b > c$ or $a > b = c$, the umbilics have index 1.

The indices of the umbilics of ellipsoids here are positive, but we have an example where they become negative, in Problem **3** of Section 17. Additionally, the following are proven in this book:

- Under inversion of the space, and under transformation to parallel surfaces, curvature lines are preserved. In particular, the indices of umbilics are also preserved (see Problem **1** in Appendix B.5, or an alternative proof of Theorem B.7.2 in Appendix B.7).
- The index of an umbilic on a constant mean curvature surfaces is always negative (see Lemma 17.10 in Section 17).

A closed surface with non-negative Gaussian curvature is called a *convex surface*. The still unsolved conjecture that "a convex surface will have at least two umbilics" is called the *Caratheodory conjecture*.[4] Connected to this, for a given regular surface with isolated umbilic P, we have the *Loewner conjecture*[5] that "those umbilics have index at most 1". Because the Euler number of convex surfaces is 2, the Caratheodory conjecture would be a consequence of the Loewner conjecture. At this time, these two conjectures are still unsolved.

Exercises 15

1 The complex number $z = u + iv \in \boldsymbol{C}$ and vector $(u, v) \in \boldsymbol{R}^2$ in the plane can be identified with each other. Let k be a positive integer.

(1) Consider the map $\boldsymbol{C} \ni z \mapsto z^k \in \boldsymbol{C}$ as a vector field which corresponds the vector $z^k \in \boldsymbol{R}^2$ to each (u, v). Show that the vector field has an isolated zero of index k at the origin.

[4] Carathéodory, Constantin (1873–1950).
[5] Loewner, Charles (1893–1968).

(2) Similarly, regarding the map $z \mapsto \bar{z}^k$ as a map on vector fields, show that at the origin we have an index $-k$ isolated zero. Here, $\bar{z} = u - iv$ is the conjugate of $z = u + iv$.

2 Let S be an elliptic paraboloid given by the graph of the function $z = \frac{x^2}{a^2} + \frac{y^2}{b^2}$ $(a \geq b > 0)$.

(1) Show that the umbilics of S are given as follows:

$$\left(0, \pm\frac{b\sqrt{a^2 - b^2}}{2}, \frac{a^2 - b^2}{4}\right).$$

In particular, the number of umbilics is one if $a = b$.

(2) Regarding the fact that S is diffeomorphic to a punctured sphere and applying the Poincaré-Hopf index formula (Theorem 15.6), prove that the index of each umbilic on S is equal to $1/2$ if $a \neq b$ and is equal to 1 if $a = b$.

16. The Laplacian and isothermal coordinates

The Laplacian. Here we consider a compact oriented Riemannian 2-manifold (S, ds^2), and let $\{e_1, e_2\}$ be a local orthonormal frame field defined on a coordinate neighborhood of S. Let $\{\omega_1, \omega_2\}$ be the dual frame field of $\{e_1, e_2\}$ (see Problem **4** in Section 12). Define a linear operator $*\colon \mathcal{A}^1(S) \to \mathcal{A}^1(S)$ of the space of differential forms $\mathcal{A}^1(S)$ (as described in Section 12) so that

$$*\omega_1 = \omega_2, \quad *\omega_2 = -\omega_1$$

and $*(f\omega) = f*\omega$ $(\omega \in \mathcal{A}^1(S),\ f \in C^\infty(S))$ hold. By the formula (12.5) for the change of bases, this definition is independent of the choice of basis. This map is called the $*$-*operator* (or Hodge[1] star operator). In particular, $* \circ * = -1$ holds.

For a function $f \in C^\infty(S)$, we can consider the differential 2-form $d{*}df$, which is proportional to the area form $d\hat{A}$ defined in (12.7) of Section 12:

$$d * df = (\Delta_{ds^2} f)\, d\hat{A}, \tag{16.1}$$

where the function $\Delta_{ds^2} f$ is defined by this equation. We call this linear map $\Delta_{ds^2}\colon C^\infty(S) \to C^\infty(S)$ the *Laplacian* of f. (In some textbooks, the Laplacian is defined rather as $-\Delta_{ds^2}$.)

[1]Hodge, Sir William Vallance Douglas (1903–1975).

Theorem 16.1. *For a function $f \in C^\infty(S)$ defined on a compact connected oriented Riemannian manifold (S, ds^2), if*

$$\int_S f\, dA = 0$$

holds, then there exists some $g \in C^\infty(S)$ so that $f = \Delta_{ds^2} g$.

The Hodge-de Rham[2] theorem is used to prove this important result. For example, see Chapter 6 of [40].

Isothermal Coordinates. If a Riemannian 2-manifold (S, ds^2), with local coordinates $(U; (u, v))$ has Riemannian metric of the form

$$ds^2 = e^{2\sigma}(du^2 + dv^2) \qquad (\sigma := \sigma(u, v) \in C^\infty(U)),$$

we say that (u, v) is an *isothermal coordinate system*, or *conformal coordinate system.*

For a Riemannian manifold (S, ds^2) and an interval I in $\boldsymbol{R}$, we can consider the following differential equation for a function f defined on the direct product $I \times S$:

$$\frac{\partial f}{\partial t} = \Delta_{ds^2} f,$$

where Δ_{ds^2} is the Laplacian for the Riemannian metric on S. This is called the *heat equation.* When heat propagates on the manifold, we can regard $f(t, \mathrm{P})$ as the temperature at the point P at time t. This is an equation that informs us of the heat conduction. In particular, the function f that satisfies $\Delta_{ds^2} f = 0$ is called a *harmonic function,* and expresses the steady state of the temperature.

The coordinate functions u, v of isothermal coordinates are both harmonic, and so, regarding u, v as functions that represent temperature, for curves on which either u or v is fixed, we have level curves for the temperature distribution.

At an arbitrary point of a Riemannian 2-manifold, there exists an isothermal coordinate chart about that point (Theorem 16.4). First we establish two lemmas to prove this fact.

Lemma 16.2. *If the Gaussian curvature K on a Riemannian 2-manifold (S, ds^2) (we do not restrict to orientable S) is identically zero, then at each point of S there exists a local coordinate neighborhood $(U; (u, v))$ containing that point so that the Riemannian metric becomes*

$$ds^2 = du^2 + dv^2.$$

[2] de Rham, Georges (1903–1990).

The meaning of this lemma is that a flat (i.e. zero Gaussian curvature) 2-dimensional Riemannian manifold is locally realized as a subset of $\boldsymbol{R}^2$, preserving distance.

Proof. Let $\{\boldsymbol{e}_1, \boldsymbol{e}_2\}$ be an orthonormal frame field defined on a simply connected coordinate neighborhood U of S, with dual $\{\omega_1, \omega_2\}$. Then the corresponding connection form μ satisfies $d\mu = 0$, since the Gaussian curvature is 0 (cf. (13.15) in Section 13). Then, as U is simply connected, by the Poincaré lemma (cf. Theorem 12.3), there exists a C^∞-function θ on U so that $\mu = d\theta$ holds. With

$$(\tilde{\boldsymbol{e}}_1, \tilde{\boldsymbol{e}}_2) = (\boldsymbol{e}_1, \boldsymbol{e}_2) \begin{pmatrix} \cos\theta & -\sin\theta \\ \sin\theta & \cos\theta \end{pmatrix},$$

and letting $\tilde{\mu}$ denote the connection form for $\{\tilde{\boldsymbol{e}}_1, \tilde{\boldsymbol{e}}_2\}$, by Lemma 13.2, we have that $\tilde{\mu} = \mu - d\theta = 0$. For the field of dual basis $\{\tilde{\omega}_1, \tilde{\omega}_2\}$ of $\{\tilde{\boldsymbol{e}}_1, \tilde{\boldsymbol{e}_2}\}$, (13.3) implies that $d\tilde{\omega}_1 = d\tilde{\omega}_2 = 0$. Therefore, again by the Poincaré lemma, there exist C^∞-functions u, v on U so that

$$\tilde{\omega}_1 = du, \qquad \tilde{\omega}_2 = dv$$

hold. Then by (13.1), we have

$$ds^2 = \tilde{\omega}_1{}^2 + \tilde{\omega}_2{}^2 = du^2 + dv^2,$$

and (u, v) are the desired coordinate functions. □

Lemma 16.3. *Let (S, ds^2) be an oriented Riemannian 2-manifold with Gaussian curvature K. For a function $\sigma \in C^\infty(S)$, defining a new Riemannian metric on S by*

$$d\tilde{s}^2 := e^{2\sigma}\, ds^2,$$

the Gaussian curvature $\widetilde{K}$ with respect to $d\tilde{s}^2$ is

$$e^{2\sigma}\widetilde{K} = K - \Delta_{ds^2}\sigma.$$

Here Δ_{ds^2} is the Laplacian with respect to ds^2.

Proof. We take a positively oriented orthonormal frame field $\{\boldsymbol{e}_1, \boldsymbol{e}_2\}$ on a coordinate neighborhood U on S, with dual frame field $\{\omega_1, \omega_2\}$, and associated connection form μ. Setting $\tilde{\boldsymbol{e}}_j := e^{-\sigma}\boldsymbol{e}_j$ $(j = 1, 2)$, $\{\tilde{\boldsymbol{e}}_1, \tilde{\boldsymbol{e}}_2\}$ is the orthonormal frame field for $d\tilde{s}^2$, and its dual frame $\{\tilde{\omega}_1, \tilde{\omega}_2\}$ is expressed as

$$\tilde{\omega}_1 = e^\sigma \omega_1, \qquad \tilde{\omega}_2 = e^\sigma \omega_2.$$

Noting that in general

$$\alpha = \alpha(\mathbf{e}_1)\omega_1 + \alpha(\mathbf{e}_2)\omega_2 \tag{16.2}$$

for a differential 1-form α, by (12.3) and (13.3) we have

$$\begin{aligned}
d\tilde{\omega}_1 &= d(e^\sigma) \wedge \omega_1 + e^\sigma d\omega_1 \\
&= (de^\sigma(\mathbf{e}_1)\omega_1 + de^\sigma(\mathbf{e}_2)\omega_2) \wedge \omega_1 + e^\sigma(\omega_2 \wedge \mu) \\
&= de^\sigma(\mathbf{e}_2)(\omega_2 \wedge \omega_1) + e^\sigma(\omega_2 \wedge \mu) \\
&= \frac{de^\sigma(\mathbf{e}_2)}{e^\sigma}(\tilde{\omega}_2 \wedge \omega_1) + \tilde{\omega}_2 \wedge \mu = \frac{e^\sigma d\sigma(\mathbf{e}_2)}{e^\sigma}(\tilde{\omega}_2 \wedge \omega_1) + \tilde{\omega}_2 \wedge \mu \\
&= d\sigma(\mathbf{e}_2)\tilde{\omega}_2 \wedge \omega_1 + \tilde{\omega}_2 \wedge \mu = \tilde{\omega}_2 \wedge (d\sigma(\mathbf{e}_2)\omega_1 + \mu).
\end{aligned}$$

Similarly,

$$d\tilde{\omega}_2 = -\tilde{\omega}_1 \wedge (-d\sigma(\mathbf{e}_1)\omega_2 + \mu)$$

can be shown. Let $\tilde{\mu}$ be the connection form with respect to the orthonormal frame field $\{\tilde{\mathbf{e}}_1, \tilde{\mathbf{e}}_2\}$ of the metric $d\tilde{s}^2$. Then by (16.2),

$$\tilde{\mu} = d\sigma(\mathbf{e}_2)\omega_1 - d\sigma(\mathbf{e}_1)\omega_2 + \mu = -(*d\sigma) + \mu$$

holds. Here, $*$ is the $*$-operator for the metric ds^2. From this,

$$\begin{aligned}
e^{2\sigma}\widetilde{K}\omega_1 \wedge \omega_2 = \widetilde{K}\tilde{\omega}_1 \wedge \tilde{\omega}_2 &= d\tilde{\mu} = -d * d\sigma + d\mu \\
&= (-\Delta_{ds^2}\sigma + K)\omega_1 \wedge \omega_2
\end{aligned}$$

follows, and we obtain the result. The final equality is obtained from (16.1) and (13.15). □

Theorem 16.4 (The existence of isothermal coordinates). *For each point of a Riemannian 2-manifold (S, ds^2) (not necessarily orientable), there exists an isothermal coordinate in a neighborhood of that point.*

Proof. We prove the theorem according to the method given in [19]. First, we fix an arbitrary point P on the manifold S, and let $(U; (u, v))$ be a local coordinate neighborhood at P. (See the left-hand side of Fig. 16.1.) Choose U so that it contains the disc $\overline{\Delta_0(2)} \subset \boldsymbol{R}^2$ of radius 2 centered at the origin with respect to the canonical metric $du^2 + dv^2$ of the uv-plane, and so that the origin is mapped to P (see the central picture in Fig. 16.1). Let $\rho : \Delta_0(2) \to [0, 1]$ be a C^∞-function which is identically 1 on the disc $\Delta_0(1/2)$ of radius $1/2$ centered at the origin, and is identically zero outside of the disc $\Delta_0(1)$ of radius 1 centered at the origin. Then, defining the new Riemannian metric

$$d\hat{s}^2 = \rho\, ds^2 + (1 - \rho)(du^2 + dv^2)$$

on U, $d\hat{s}^2$ is the same as the original metric ds^2 on $\Delta_0(1/2)$, and is equal to the standard Euclidean metric $du^2 + dv^2$ of $\boldsymbol{R}^2$ outside of $\Delta_0(1)$.

Now consider a square in $\Delta_0(2)$ centered at the origin such that the length of one side is $\sqrt{2}$. Then the square lies in $\Delta_0(2)$ and contains the region $\Delta_0(1)$. This can be thought of as a 2-dimensional torus T^2 by identifying opposite edges, and then the metric $d\hat{s}^2$ becomes a well-defined metric on T^2 (see the right-hand side of Fig. 16.1).

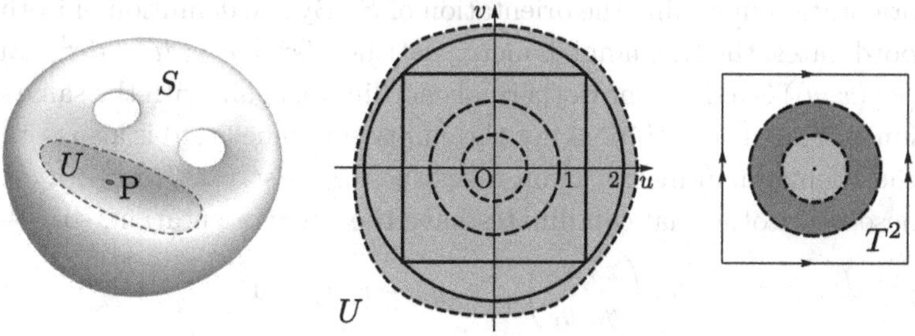

Fig. 16.1

Here, letting $K_{d\hat{s}^2}$ be the Gaussian curvature on T^2 with respect to $d\hat{s}^2$, since T^2 is compact and orientable and with Euler number 0 (see (10.11) in Section 10,), the Gauss-Bonnet theorem (Theorem 14.2) yields

$$\int_{T^2} K_{d\hat{s}^2}\, dA_{d\hat{s}^2} = \chi(T^2) = 0,$$

where $dA_{d\hat{s}^2}$ is the area form for $d\hat{s}^2$. Then by Theorem 16.1, there exists a smooth function σ on T^2 so that $-K_{d\hat{s}^2} = \Delta_{d\hat{s}^2}\sigma$, for the Laplacian $\Delta_{d\hat{s}^2}$ with respect to the metric $d\hat{s}^2$.

Define a new metric on the torus T^2 by $d\tau^2 = e^{-2\sigma} d\hat{s}^2$. Then the orientability of the torus and Lemma 16.3 implies that the Gaussian curvature for $d\tau^2$ is $K_{d\tau^2} = e^{2\sigma}(K_{d\hat{s}^2} + \Delta_{d\hat{s}^2}\sigma) = 0$. Therefore, there exist coordinates $(\tilde{u}, \tilde{v})$ on a neighborhood V of the origin so that $d\tau^2 = d\tilde{u}^2 + d\tilde{v}^2$ holds. Then $d\hat{s}^2 = e^{2\sigma}(d\tilde{u}^2 + d\tilde{v}^2)$, and in particular, because $d\hat{s}^2 = ds^2$ on the intersection of V and $B_0(1/2)$,

$$ds^2 = d\hat{s}^2 = e^{2\sigma}(d\tilde{u}^2 + d\tilde{v}^2)$$

holds, and we have existence of the isothermal coordinates $(\tilde{u}, \tilde{v})$. □

When S is orientable, the existence of isothermal coordinates means that S has the structure of a complex manifold of dimension 1. In other words, orientable Riemannian 2-manifolds have the structure of 1-dimensional complex manifolds (i.e. Riemann surfaces) in a natural way.

Theorem 16.5. *About each point of an oriented 2-dimensional Riemannian manifold S, we can take isothermal coordinates $(U;(u,v))$ with the same orientation, and then $z = u + iv$ ($i = \sqrt{-1}$) becomes a complex structure for the complex 1-manifold S.*

Proof. By Theorem 16.4, in a neighborhood of each point of (S, ds^2) we can take isothermal coordinates $(U;(u,v))$. When necessary, we can change (u,v) to $(u,-v)$ so that the coordinates (u,v) about all points all initially have orientation matching the orientation of S. By the definition of isothermal coordinates, the Riemannian metric satisfies $ds^2 = e^{2\sigma}(du^2+dv^2)$. Suppose $(V;(x,y))$ is a different isothermal coordinate, again with the same orientation as S, and that $U \cap V$ is not the empty set. Sine (x,y) is also isothermal, the Riemannian metric again takes the form $ds^2 = e^{2\tilde{\sigma}}(dx^2+dy^2)$.

These two isothermal coordinates have the same orientation, so

$$\text{(16.3)} \qquad \det\begin{pmatrix} x_u & x_v \\ y_u & y_v \end{pmatrix} = x_u y_v - x_v y_u > 0$$

must hold. Also, since the Riemannian metric is independent of choice of coordinates, on $U \cap V$ we have

$$\begin{aligned} e^{2\sigma}(du^2+dv^2) = ds^2 &= e^{2\tilde{\sigma}}(dx^2+dy^2) \\ &= e^{2\tilde{\sigma}}((x_u\,du + x_v\,dv)^2 + (y_u\,du + y_v\,dv)^2) \\ &= e^{2\tilde{\sigma}}((x_u^2+y_u^2)\,du^2 + 2(x_u x_v + y_u y_v)\,du\,dv \\ &\quad + (x_v^2+y_v^2)\,dv^2), \end{aligned}$$

so, comparing the two sides of the above equation, we have

$$\text{(16.4)} \qquad x_u^2 + y_u^2 = x_v^2 + y_v^2,$$

$$\text{(16.5)} \qquad x_u x_v + y_u y_v = 0.$$

By (16.5), the vectors (x_u, y_u) and (x_v, y_v) in $\boldsymbol{R}^2$ are perpendicular and cannot be $\mathbf{0}$; thus there exists a function φ so that

$$\begin{pmatrix} x_u \\ y_u \end{pmatrix} = \varphi \begin{pmatrix} y_v \\ -x_v \end{pmatrix}.$$

Inserting this into (16.4), we find that $\varphi = \pm 1$, and in particular, by (16.3), $\varphi = 1$ holds. Therefore, we obtain the Cauchy-Riemann equations

$$\text{(16.6)} \qquad x_u = y_v, \qquad x_v = -y_u.$$

Thus we have shown that the transformation $u + iv \mapsto x + iy$ is then a holomorphic function. From this, we have succeeded in giving S the structure of a complex manifold of dimension 1. □

Exercises 16

1 For the metric $ds^2 = du^2 + dv^2$ on $\boldsymbol{R}^2$, show that the Laplacian is

$$\Delta_{ds^2} = \frac{\partial^2}{\partial u^2} + \frac{\partial^2}{\partial v^2}.$$

2 For a Riemannian 2-manifold (S, ds^2) with isothermal coordinate system (u, v) and Riemannian metric $ds^2 = e^{2\sigma}(du^2 + dv^2)$, show that the Laplacian becomes

$$\Delta_{ds^2} = e^{-2\sigma}\left(\frac{\partial^2}{\partial u^2} + \frac{\partial^2}{\partial v^2}\right).$$

In particular, applying Lemma 16.3 show that the Gaussian curvature K becomes

$$K = -e^{-2\sigma}(\sigma_{uu} + \sigma_{vv}).$$

3 Consider the Riemannian metric

$$ds_{\mathrm{H}}^2 := \frac{1}{v^2}(du^2 + dv^2)$$

on the upper half-plane $D := \{(u, v) \in \boldsymbol{R}^2 \mid v > 0\}$ (see page 113, Section 10).

(1) Show that the Gaussian curvature of ds_{H}^2 is identically -1.
(2) Show that any line parallel to the v-axis in D, when parametrized by arc-length, is a geodesic.
(3) Regarding $\boldsymbol{R}^2$ as the complex plane, and setting $z = u + iv$, show that the pull-back of the metric ds_{H}^2 via the map

$$f\colon D \ni z \longmapsto \frac{az + b}{cz + d} \in D \qquad (a, b, c, d \in \boldsymbol{R},\ ad - bc = 1)$$

is just ds_{H}^2 itself (and therefore this map is an isometry of (D, ds_{H}^2)).
(4) In particular, when $cd \neq 0$, show that under the above transformation the lines parallel to the v-axis are mapped to circles that intersect the u-axis perpendicularly. (In particular, the circles that intersect the u-axis perpendicularly are geodesics with respect to the metric ds_{H}^2.)
(5) Show that $d_{\mathrm{H}}(f(z), f(w)) = d_{\mathrm{H}}(z, w)$ holds, for the distance given in page 114 of Section 10:

$$d_{\mathrm{H}}(z, w) = \log\frac{|z - \bar{w}| + |z - w|}{|z - \bar{w}| - |z - w|},$$

where we identify $\boldsymbol{R}^2$ with the complex plane.
(6) Connecting two points z, w in D by a geodesic segment with respect to the metric ds_{H}^2, show that the length of that segment is $d_{\mathrm{H}}(z, w)$.

4 If we change the v^2 in the metric in Problem **3** to v, the Riemannian metric on the upper half-plane $D = \{(u, v) \in \boldsymbol{R}^2 \mid v > 0\}$ becomes

$$ds^2 := \frac{1}{v}(du^2 + dv^2).$$

(1) Show that any geodesic for the Riemannian metric ds^2 on the uv-plane is either a cycloid with u-axis as its base line (see Problem **4** in Section 1), or a half-line perpendicular to the u-axis. In particular, a geodesic with respect to ds^2 is perpendicular to the boundary (the u-axis).

(2) Take a point P on the boundary ∂D of D and a point Q in the interior of D, and suppose that the line PQ is not parallel to the v-axis. In this situation, show there is only one cycloid that connects P and Q and is perpendicular to the u-axis.

(3) Taking two points P and Q in the interior of D, show there is only one geodesic with respect to ds^2 that connects P and Q.

Remark. A curve $\gamma(t) = (u(t), v(t))$ $(a \leq t \leq b)$ in the upper half-plane has length

$$\int_a^b \sqrt{\frac{\dot{u}^2 + \dot{v}^2}{v}}\, dt$$

with respect to this metric. Considering part (1) of Problem **4** in Appendix B.1 (and noting that we have an upward-pointing curve rather than a downward-pointing one), for the initial point P and final point Q for the curve γ (assume Q lies above P), if P happens to lie on the u-axis, then seeking the shortest connecting path PQ instead becomes seeking a *brachistochrone*, i.e., the steepest descent curve. Unlike ds^2_{H}, this metric on D is not complete, so the notion of "shortest path" is not so immediately clear, but as shown in Section 18, the cycloid does in fact give the shortest path between two points, as well as the steepest descent path (the brachistochrone).

17. The Gauss and Codazzi equations

The fundamental theorem for surface theory. In this section, the word *surface* means an immersion $\boldsymbol{p}\colon S \to \boldsymbol{R}^3$ from a 2-manifold S to $\boldsymbol{R}^3$. This allows for the possibility that the image of $\boldsymbol{p}$ in $\boldsymbol{R}^3$ is a surface with self-intersections. A Riemannian metric (the first fundamental form) ds^2 is induced on S, by (13.21). Let ν be a unit normal vector to the surface.

For a local chart $(U; (u, v))$ on S, we write the first and second fundamental forms as

(17.1) $ds^2 = E\,du^2 + 2F\,du\,dv + G\,dv^2, \quad I\!I = L\,du^2 + 2M\,du\,dv + N\,dv^2,$

and the Weingarten matrix A as in (8.4) is

$$A = \begin{pmatrix} a_{11} & a_{12} \\ a_{21} & a_{22} \end{pmatrix} = \begin{pmatrix} E & F \\ F & G \end{pmatrix}^{-1} \begin{pmatrix} L & M \\ M & N \end{pmatrix}. \tag{17.2}$$

Furthermore, we have the Christoffel symbols Γ_{ij}^k $(i, j, k = 1, 2)$ as in (10.6).

Theorem 17.1. *In the above setting, we regard* $\boldsymbol{p}_u = \partial \boldsymbol{p}/\partial u$, $\boldsymbol{p}_v = \partial \boldsymbol{p}/\partial v$ *and the unit normal vector* ν *as column vectors, which we put together to form a* 3×3 *matrix* $\mathcal{F} := (\boldsymbol{p}_u, \boldsymbol{p}_v, \nu)$. *Then* $\mathcal{F}(u, v)$ *is a regular matrix for each* (u, v), *and satisfies*

$$\frac{\partial}{\partial u}\mathcal{F} = \mathcal{F}\Omega, \qquad \frac{\partial}{\partial v}\mathcal{F} = \mathcal{F}\Lambda, \tag{17.3}$$

where

$$\Omega = \begin{pmatrix} \Gamma_{11}^1 & \Gamma_{12}^1 & -a_{11} \\ \Gamma_{11}^2 & \Gamma_{12}^2 & -a_{21} \\ L & M & 0 \end{pmatrix}, \qquad \Lambda = \begin{pmatrix} \Gamma_{12}^1 & \Gamma_{22}^1 & -a_{12} \\ \Gamma_{12}^2 & \Gamma_{22}^2 & -a_{22} \\ M & N & 0 \end{pmatrix}. \tag{17.4}$$

Proof. We are considering a regular surface, so the vectors $\{\boldsymbol{p}_u, \boldsymbol{p}_v, \nu\}$ are linearly independent. Thus $\mathcal{F}$ is a regular matrix.

By Proposition 11.1 of Section 11, we have

$$\begin{cases} \boldsymbol{p}_{uu} = \Gamma_{11}^1 \boldsymbol{p}_u + \Gamma_{11}^2 \boldsymbol{p}_v + L\nu, \\ \boldsymbol{p}_{uv} = \Gamma_{12}^1 \boldsymbol{p}_u + \Gamma_{12}^2 \boldsymbol{p}_v + M\nu, \\ \boldsymbol{p}_{vv} = \Gamma_{22}^1 \boldsymbol{p}_u + \Gamma_{22}^2 \boldsymbol{p}_v + N\nu. \end{cases}$$

On the other hand, by Proposition 8.5 (the Weingarten formula), we have

$$\nu_u = -a_{11}\boldsymbol{p}_u - a_{21}\boldsymbol{p}_v, \qquad \nu_v = -a_{12}\boldsymbol{p}_u - a_{22}\boldsymbol{p}_v,$$

so the theorem's assertion is established. □

With further differentiation of formula (17.3), and noting that $\mathcal{F}_{uv} = \mathcal{F}_{vu}$, we have

$$\mathcal{F}_{uv} = (\mathcal{F}\Omega)_v = \mathcal{F}_v\Omega + \mathcal{F}\Omega_v = \mathcal{F}(\Lambda\Omega + \Omega_v),$$
$$\mathcal{F}_{vu} = (\mathcal{F}\Lambda)_u = \mathcal{F}_u\Lambda + \mathcal{F}\Lambda_u = \mathcal{F}(\Omega\Lambda + \Lambda_u),$$

and so we know that

$$\mathcal{F}(\Lambda\Omega + \Omega_v) = \mathcal{F}(\Omega\Lambda + \Lambda_u)$$

holds. Since $\mathcal{F}$ is invertible, we can eliminate $\mathcal{F}$ from both sides by multiplying on the left by $\mathcal{F}$'s inverse to arrive at (cf. Lemma B.10.1 in Appendix B.10)

$$\Lambda\Omega + \Omega_v = \Omega\Lambda + \Lambda_u. \tag{17.5}$$

In fact, this equation is the condition that guarantees the existence of a solution $\mathcal{F}$ to equation (17.3), as we now see.

Theorem 17.2 (Fundamental theorem for surface theory). *Taking smooth functions E, F, G, L, M, N on a simply connection region D of the uv-plane (see page 137 for simple connectedness), suppose that $ds^2 := E\,du^2 + 2F\,du\,dv + G\,dv^2$ is a Riemannian metric on D. From those functions,* (10.6), (17.2) *determine the Γ_{ij}^k $(i, j, k = 1, 2)$ and a_{ij} $(i, j = 1, 2)$. Let Ω and Λ be determined by* (17.4), *and assume they satisfy* (17.5). *Then there exists a surface $\mathbf{p}(u, v)$ defined (uniquely up to orientation preserving isometries of $\boldsymbol{R}^3$) on D whose fundamental forms are as in* (17.1).

This is called the *fundamental theorem for surface theory.* A proof of this theorem will be laid out in Appendix B.10. As an application of this theorem to surfaces of constant Gaussian curvature, we refer to Problems **4**, **5** at the end of this section.

Here, to make equation (17.5) simpler, let us take an isothermal local coordinate neighborhood $(U; (u, v))$ with respect to the first fundamental form (see Theorem 16.4). In this case, we can take some smooth function $\sigma = \sigma(u, v)$ so that

$$ds^2 = e^{2\sigma}(du^2 + dv^2) \tag{17.6}$$

holds. Then (17.4) implies the following simpler expressions:

$$\Omega = \begin{pmatrix} \sigma_u & \sigma_v & -e^{-2\sigma}L \\ -\sigma_v & \sigma_u & -e^{-2\sigma}M \\ L & M & 0 \end{pmatrix}, \quad \Lambda = \begin{pmatrix} \sigma_v & -\sigma_v & -e^{-2\sigma}M \\ \sigma_u & \sigma_u & -e^{-2\sigma}N \\ M & N & 0 \end{pmatrix}. \tag{17.7}$$

We have the following theorem directly from the metric:

Theorem 17.3. *Let $(U; (u, v))$ be an isothermal coordinate system such that the first fundamental form is written as in* (17.6). *Then* (17.5) *is equivalent to the following three equations:*

$$\sigma_{uu} + \sigma_{vv} + e^{-2\sigma}(LN - M^2) = 0, \tag{17.8}$$

$$L_v - M_u = \sigma_v(L + N), \tag{17.9}$$

$$N_u - M_v = \sigma_u(L + N). \tag{17.10}$$

Letting Δ_{ds^2} denote the Laplacian for ds^2, by Problem **2** in Section 16, the Gaussian curvature in Section 13 becomes

$$K = -\Delta_{ds^2}\sigma = -e^{-2\sigma}(\sigma_{uu} + \sigma_{vv}).$$

Using this and (17.8), we can rewrite K as

$$K = e^{-4\sigma}(LN - M^2). \tag{17.11}$$

Equation (17.8) is called the *Gauss equation*, and it is equivalent to the equation introduced in Theorem 11.2 of Section 11. Equations (17.9), (17.10) are called the *Codazzi*[1] *equations.* (This equation is equivalent to

[1] Codazzi, Delfino (1824–1873).

the covariant derivative ∇II of the second fundamental form being a symmetric tensor. See, for example, Chapter 6 of [5].) The Gauss equation has already appeared when we discussed the Gauss-Bonnet theorem in Section 10. Next we will introduce an application of the Codazzi equations to constant mean curvature surfaces.

Characterizing the round sphere using mean curvature. As seen in Example 8.4 in Section 8, the sphere in $\boldsymbol{R}^3$ is a closed surface of constant mean curvature. We now prove a theorem giving a characterization of the sphere:

Theorem 17.4 (The Hopf[2] theorem [12]). *If S is a 2-dimensional manifold homeomorphic to a sphere, and an immersion $p\colon S \to \boldsymbol{R}^3$ has constant mean curvature, then $\mathbf{p}(S)$ is congruent to a round sphere.*

We now make some preparations in order to prove this theorem.

We consider an immersion $\boldsymbol{p}\colon S \to \boldsymbol{R}^3$ from an oriented 2-dimensional manifold S to $\boldsymbol{R}^3$. By Theorem 16.5, taking coordinates (u, v) compatible with the orientation and isothermal with respect to the first fundamental form ds^2, we can give S the structure of a complex manifold. The first fundamental form is as in (17.6), and the coefficients L, M, N of the second fundamental form can be used to define the *Hopf differential* Q of $\boldsymbol{p}$ as

$$(17.12) \qquad Q := \frac{1}{4}((L - N) - 2iM)\,dz^2 \qquad (dz := du + i\,dv),$$

for the complex coordinate $z = u + iv$.

Lemma 17.5. *The Hopf differential Q is independent of choice of complex coordinate.*

Proof. Taking another complex coordinate $w = \xi + i\eta$, let the entries for the second fundamental form with respect to this coordinate be $\widetilde{L}$, $\widetilde{M}$, $\widetilde{N}$, and then by (8.3) in Section 8 we have

$$(17.13) \qquad \begin{pmatrix} \widetilde{L} & \widetilde{M} \\ \widetilde{M} & \widetilde{N} \end{pmatrix} = \begin{pmatrix} u_\xi & u_\eta \\ v_\xi & v_\eta \end{pmatrix}^T \begin{pmatrix} L & M \\ M & N \end{pmatrix} \begin{pmatrix} u_\xi & u_\eta \\ v_\xi & v_\eta \end{pmatrix}.$$

Using the Cauchy-Riemann equations $u_\xi = v_\eta$ and $u_\eta = -v_\xi$, it suffices to show

$$(\widetilde{L} - \widetilde{N} - 2i\widetilde{M})\,dw^2 = (L - N - 2iM)\,dz^2.$$

[2]Hopf, Heinz (1894–1971).

In fact, on the left-hand side, using (17.13) to replace the $\widetilde{L}$, $\widetilde{M}$, $\widetilde{N}$ with L, M, N, and using

$$dz = d(u+iv) = (u_\xi + iv_\xi)\,d\xi + (u_\eta + iv_\eta)\,d\eta = (u_\xi - iu_\eta)\,dw$$

we can confirm the desired equality. □

From (17.6), the first fundamental form satisfies $E = G = e^{2\sigma}$ and $F = 0$, and so by (8.7) the mean curvature is

$$H = \frac{EN - 2FM + GL}{2(EG - F^2)} = \frac{1}{2}e^{-2\sigma}(L+N). \tag{17.14}$$

By (17.11) and (17.14), $4e^{4\sigma}(H^2 - K)$ equals $(L-N)^2 + 4M^2$, so *the Hopf differential Q being zero at a point z is equivalent to the point in the surface corresponding to z being an umbilic* (see Proposition 9.4 in Section 9). When all points of the surface are umbilics, we say that the surface is *totally umbilic*. As we saw in Proposition 9.7 in Section 9, the only totally umbilic surfaces are parts of a sphere or plane.

Using a complex coordinate and the Hopf differential, the Codazzi equations takes the following simpler form.

Lemma 17.6. *In the above setting, writing the Hopf differential* (17.12) *as $Q = q(z)\,dz^2$, the Codazzi equations* (17.9), (17.10) *are equivalent to*

$$\frac{1}{2}\frac{\partial H}{\partial z} = e^{-2\sigma}\frac{\partial q}{\partial \bar{z}} \qquad \left(q := \frac{1}{4}(L - N - 2iM)\right).$$

Here, H is the mean curvature, and

$$\frac{\partial}{\partial z} := \frac{1}{2}\left(\frac{\partial}{\partial u} - i\frac{\partial}{\partial v}\right), \qquad \frac{\partial}{\partial \bar{z}} := \frac{1}{2}\left(\frac{\partial}{\partial u} + i\frac{\partial}{\partial v}\right). \tag{17.15}$$

Now suppose that an immersion $\boldsymbol{p}\colon S \to \boldsymbol{R}^3$ from a 2-manifold S to $\boldsymbol{R}^3$ has constant mean curvature. If the mean curvature is non-zero, we can define a global unit vector field on the surface, and S is automatically orientable (see Problem **2** at the end of this section). In particular, S has the structure of a complex manifold.

Proposition 17.7. *A non-zero constant mean curvature surface has holomorphic Hopf differential. In other words, when writing the Hopf differential as $Q(z) = q(z)\,dz^2$ for some complex coordinate z, the function $q(z)$ is holomorphic.*

Proof. With the mean curvature H constant, $\partial H/\partial z = 0$ holds, and by Lemma 17.6, we have $\partial q/\partial \bar{z} = 0$. Writing $q = x(u,v) + iy(u,v)$ separated into real and imaginary parts, (17.15) implies

$$0 = \frac{\partial q}{\partial \bar{z}} = \frac{1}{2}\left(\frac{\partial(x+iy)}{\partial u} + i\frac{\partial(x+iy)}{\partial v}\right) = \frac{1}{2}((x_u - y_v) + i(x_v + y_u)),$$

It follows that $x_u - y_v = x_v + y_u = 0$, and these are the Cauchy-Riemann equations (16.6) for q with respect to z. Thus q is holomorphic. □

A surface whose points are all umbilic is said to be *totally umbilic* (cf. Proposition 9.7 in Section 9). One product of Proposition 17.7 is the following assertion.

Corollary 17.8. *A constant mean curvature surface that is not totally umbilic has only isolated umbilic points. In particular, if such a surface is closed, it has only finitely many umbilic points.*

Proof. Since the statement is local, we may assume that S is orientable. Then S has a canonical complex structure, and the Hopf differential $Q = q(z)\,dz^2$ is holomorphic even when the surface is of zero mean curvature. Because the surface is not totally umbilic, Q is not identically zero. By Proposition 17.7, $q(z)$ is a holomorphic function, and so its zeros are isolated (cf. Section 3.2 in Chapter 4 of [1]).

In particular, the umbilics of the surface are isolated. In the case of a closed surface, if there are infinitely many umbilic points, by compactness there must be an accumulation point of those umbilics, contradicting the fact that the umbilics are isolated. □

We now consider an immersion $\boldsymbol{p}\colon S \to \boldsymbol{R}^3$ from a compact 2-manifold S into $\boldsymbol{R}^3$ that has constant mean curvature and is not totally umbilic, and suppose that $\{\mathrm{P}_1, \ldots, \mathrm{P}_n\}$ are the umbilic points of $\boldsymbol{p}$. Away from these n umbilic points, we may assume the two principal curvature functions λ_1, λ_2 are ordered as $\lambda_1 < \lambda_2$. Let the set $S \setminus \{\mathrm{P}_1, \ldots, \mathrm{P}_n\}$ denote S with the points $\{\mathrm{P}_1, \ldots, \mathrm{P}_n\}$ removed, and let P with local neighborhood U having complex coordinate $z = u + iv$ lie in that set. Then the Weingarten matrix is the symmetric matrix (8.4):

$$A := \widehat{I}^{-1}\,\widehat{II} = e^{-2\sigma}\begin{pmatrix} L & M \\ M & N \end{pmatrix}. \tag{17.16}$$

On U, the eigenvalues λ_1, λ_2 of A have corresponding eigendirections (principal curvature directions), and we can take directional fields ξ_1, ξ_2 in those directions. Since the eigendirections do not depend on choice of coordinates, ξ_1, ξ_2 are each a directional field defined on all of $S \setminus \{\mathrm{P}_1, \dots, \mathrm{P}_n\}$. With this, the following proposition holds.

Proposition 17.9. *At each umbilic point* P_j, *we have that*

$$\operatorname{ind}_{\mathrm{P}_j} \xi_1 = \operatorname{ind}_{\mathrm{P}_j} \xi_2 = -\frac{1}{2} \operatorname{ord}_{\mathrm{P}_j} Q \qquad (j = 1, \dots, n) \tag{17.17}$$

hold. Here $\operatorname{ord}_{\mathrm{P}_j} Q$ *is the order of the zero of the holomorphic function* $q(z)$ *in the Hopf differential* $Q = q(z)\,dz^2$ *at* P_j *with respect to the complex coordinate* z.

Proof. We can suppose that P_j is the origin of a local complex coordinate $(U; z)$. We write the Hopf differential as $Q = q(z)dz^2$. The metric ds^2 is as in (17.6), and we can write the Weingarten matrix in the form in (17.16).

Then

$$\begin{aligned} V_A &= e^{-2\sigma} \begin{pmatrix} L - N \\ 2M \end{pmatrix} \\ &= e^{-2\sigma}(L - N + 2iM) = e^{-2\sigma}\overline{(L - N - 2iM)} = e^{-2\sigma}\overline{q(z)} \end{aligned} \tag{17.18}$$

is the vector field determined as in (15.8) by the symmetric matrix A. By Proposition 15.4, the directional fields ξ_1, ξ_2 determined by A satisfy

$$\operatorname{ind}_{\mathrm{P}_j} \xi_1 = \operatorname{ind}_{\mathrm{P}_j} \xi_2 = \frac{1}{2} \operatorname{ind}_{\mathrm{P}_j} V_A \tag{17.19}$$

at each umbilic point P_j. By (15.8), V_A is proportional to the conjugate $\overline{q(z)}$ of $q(z)$. Since the map taking the complex conjugate $z \mapsto \bar{z}$ is a diffeomorphism of $\boldsymbol{R}^2$ that reverses the orientation, the index of this vector field will change sign under an inversion, and we have

$$\operatorname{ind}_{\mathrm{P}_j} V_A = -\operatorname{ind}_{\mathrm{P}_j} q(z).$$

We recall that the meaning of $\operatorname{ind}_{\mathrm{P}_j} q(z)$ is the index of $q(z)$ as a vector field about $z = 0$. When $q(z)$ has an order k zero at $z = 0$, there is a holomorphic function $h(z)$ defined in a neighborhood of the origin so that $q(z) = z^k e^{h(z)}$ holds. Then the map

$$\psi_t : z \longmapsto z^k e^{(1-t)h(z)} \in \boldsymbol{C} = \boldsymbol{R}^2,$$

for each fixed choice of $t \in [0, 1]$, can be regarded as a vector field with an isolated zero at the origin. The index of this vector field is k and is also continuous with respect to t, so the index about the origin does not change with respect to t. Thus the index of $q(z)$ $(= \psi_0)$ as a vector field is the same as the index k of the vector field z^k $(= \psi_1)$ (see Problem **1** in Section 15).

In conclusion, $\operatorname{ind}_{\mathrm{P}_j} q(z)$ is equal to the order of $q(z)$ at its zero point, and (17.19) is equivalent to (17.17). □

Because the order k of the zero point of the holomorphic function $q(z)$ is at least 1, by (17.17) we have

$$\mathrm{ind}_{\mathrm{P}_j}\,\xi_1 = \mathrm{ind}_{\mathrm{P}_j}\,\xi_2 \leq -\frac{1}{2},$$

and we obtain the following assertion.

Lemma 17.10. *The index of an umbilic point on a constant mean curvature surface is always negative.*

With the above preparations, let us now prove the Hopf theorem.

The proof of the Hopf Theorem 17.4. Take a constant mean curvature immersion $\boldsymbol{p}\colon S^2 \to \boldsymbol{R}^3$ from the sphere S^2 to $\boldsymbol{R}^3$. Suppose that $\boldsymbol{p}$ is not totally umbilic. By Corollary 17.8, there are only a finite number of umbilic points, which we list as $\{\mathrm{P}_1, \ldots, \mathrm{P}_n\}$. The index of the umbilic at each point P_j is equal to the index at P_j of either of the directional fields ξ_1 or ξ_2 representing the principle curvature directions, so applying Theorem 15.6 to either of these directional fields, we have

$$\sum_{j=1}^{n} \mathrm{ind}_{\mathrm{P}_j}\,\xi_1 = \chi(S).$$

(Of course, if there are no umbilic points, then $n = 0$, and the left-hand side of this equation is 0.) By the assumption, S is diffeomorphic to a sphere, so the genus is 0 and $\chi(S) = 2$. Also, by Lemma 17.10, each $\mathrm{ind}_{\mathrm{P}_j}\,\xi_1$ is negative, so the left-hand side is less than or equal to 0, which is a contradiction. We conclude that $\boldsymbol{p}$ must be totally umbilic, and by Proposition 9.7 in Section 9, it must then be a round sphere in $\boldsymbol{R}^3$. □

Alexandrov, without any assumption on the genus, showed that if a constant mean curvature closed surface does not intersect itself (i.e. an embedded surface), then it must be a round sphere [2]. The Hopf theorem (Theorem 17.4) does not require any assumption about the surface intersecting itself, so it is not implied by Alexandrov's result. If we do not require the surface to be closed, in addition to the sphere and cylinder, the Delaunay surfaces introduced in Appendix B.7 are another collection of non-trivial examples of constant mean curvature surfaces. We can also ask: if we allow self-intersections, are there other examples of closed constant mean curvature surfaces with positive genus? This question remained unanswered for a long time, and finally in 1986, Wente discovered closed constant mean curvature tori [41], see Fig. 6.4. After that, Kapouleas [15] found such examples of higher genus. Based on these discoveries, research on constant mean curvature surfaces took big steps forward. For more on recent developments in this field, we refer the reader to [16].

Exercises 17

1 Regarding an immersion $\boldsymbol{p}\colon S \to \boldsymbol{R}^3$ on S as a triple of three coordinate functions in $\boldsymbol{R}^3$, show that

$$\Delta_{ds^2}\boldsymbol{p} = 2H\nu.$$

Here Δ_{ds^2} is the Laplacian with respect to the first fundamental form ds^2 on S, H is the mean curvature and ν is the unit normal field.

2 If the mean curvature is never 0 for an immersion $\boldsymbol{p}\colon S \to \boldsymbol{R}^3$, prove that S is orientable.

3 Show that the coordinates (u, v) of the surface

$$\boldsymbol{p}(u,v) := \left(-\frac{u^5}{5} + 2u^3v^2 - uv^4 + u, -u^4v + 2u^2v^3 - \frac{v^5}{5} - v, \frac{2u^3}{3} - 2uv^2\right)$$

are isothermal, that the surface is a minimal immersion, and that the Hopf differential is $-2z\,dz^2$ with $z := u + iv$. Show also that we have an umbilic point at the origin $(u, v) = (0, 0)$ with index $-1/2$.

4 Confirm that the following first and second fundamental matrices

$$\widehat{I} = \begin{pmatrix} 1 & \cos\theta(u,v) \\ \cos\theta(u,v) & 1 \end{pmatrix}, \qquad \widehat{II} = \begin{pmatrix} 0 & \sin\theta(u,v) \\ \sin\theta(u,v) & 0 \end{pmatrix}$$

satisfy (17.5) if the function $\theta(u, v)$ satisfies the equation $\theta_{uv} - \sin\theta = 0$ and also $0 < \theta < \pi$. Furthermore, applying the fundamental theorem for surface theory, show that any surface with these fundamental forms will have Gaussian curvature constantly -1.

5 Confirm that the following first and second fundamental matrices

$$\widehat{I} = \begin{pmatrix} \cosh\theta(u,v) + 1 & 0 \\ 0 & \cosh\theta(u,v) - 1 \end{pmatrix}, \; \widehat{II} = \begin{pmatrix} \sinh\theta(u,v) & 0 \\ 0 & \sinh\theta(u,v) \end{pmatrix}$$

satisfy (17.5) if the function $\theta(u, v)$ satisfies $\theta_{uu} + \theta_{vv} + 4\sinh\theta = 0$ and also $\theta > 0$. Furthermore, applying the fundamental theorem for surface theory, show that any surface with these fundamental forms will have Gaussian curvature constantly 1.

18. Cycloids as brachistochrones

Consider an object in a vertical plane, starting at a point P and following a curve to another point Q, with only gravity creating its motion. If the time required to travel from P to Q is minimized, the curve is called a *brachistochrone* or the path of steepest descent (see Fig. B.1.3 of Appendix B.1 on page 217). It is known that this brachistochrone is a cycloid with horizontal base line passing through P. We prove this in this section. We will refer to the properties of cycloids found in Appendix B.1.

Consider the Riemannian metric

$$ds^2 := \frac{1}{v}(du^2 + dv^2) \tag{18.1}$$

on the upper half-plane[1] $D = \{(u, v) \in \boldsymbol{R}^2 \mid v > 0\}$ (see Problem **4** of Section 16). Then the length of a curve $\gamma(t) = (u(t), v(t))$ $(a \leq t \leq b)$ in D is given by

$$\int_a^b \sqrt{\frac{\dot{u}^2 + \dot{v}^2}{v}}\, dt.$$

Then by Problem **4** (1) in Appendix B.1, we know that finding the brachistochrone is equivalent to finding the shortest curve joining P and Q where P lies on the u-axis and Q lies higher than P (see the remark after Problem **4** of Section 16).

We first assume that P lies on the upper half-plane, and we will show that the shortest path between P and Q is the cycloid passing through both P and Q whose baseline is the u-axis (see Fig. 18.1). As a limit of this, we will show that the cycloid whose baseline is the u-axis and passing through Q is the shortest curve.

We have already shown in Problem **4** in Section 16 that if there exists a shortest path, this must be a cycloid. A Riemannian manifold (S, ds^2) is *complete* if any arbitrarily chosen geodesic $\gamma(s)$ can be extended so that its parameter s is extended to all of $\boldsymbol{R}$ (this definition can be found in [5, Chapter 7]). The existence of the shortest path passing through two distinct arbitrarily given points on S is shown if ds^2 is complete (see [5, Corollary 2.8]). However, our metric given in (18.1) is not complete, and it is not clear that each cycloid is length-minimizing. So the discussion below is necessary:

[1]Comparing this metric with the Poincaré metric $(du^2+dv^2)/v^2$ having constant Gaussian curvature -1 on the upper half-plane, the geodesics for the Poincaré metric are the upper halves of circles that intersect the u-axis vertically, but for the metric ds^2 here, they are the upward cycloids whose *baseline* is the u-axis.

Theorem 18.1. *For points* P, Q $\in D$, *there is a unique shortest path from* P *to* Q *with respect to the metric* ds^2 *in* (18.1), *and this path is the cycloid passing through* P *and* Q *whose baseline is the* u*-axis.*

To prove the theorem, we need a notion of convexity for open neighborhoods of Riemann manifolds.

Definition 18.2. An open neighborhood U of a Riemannian manifold (S, ds^2) is said to be *convex* if any two points on U can be joined by a unique shortest geodesic segment in U.

Proof of Theorem 18.1. Let $d(\mathrm{P}, \mathrm{Q})$ denote the lower bound for the lengths of all piecewise smooth curves from P to Q in D with respect to ds^2. As pointed out in Theorem 13.4, this d defines a distance function on D which is compatible with respect to the topology of S. A shortest length path from P to Q would be one that has length $d(\mathrm{P}, \mathrm{Q})$. If a shortest length curve (one that is piecewise smooth) exists, it will be a geodesic if we give it an arc-length parametrization (see, for example, Corollary 3.9 in Chapter 3 of [5]). Furthermore, as proven in Problem **4** in Section 16, such a shortest length path is unique and is a cycloid whose baseline is the u-axis. Thus it now suffices to show the existence of a shortest length piecewise smooth curve between any two points.

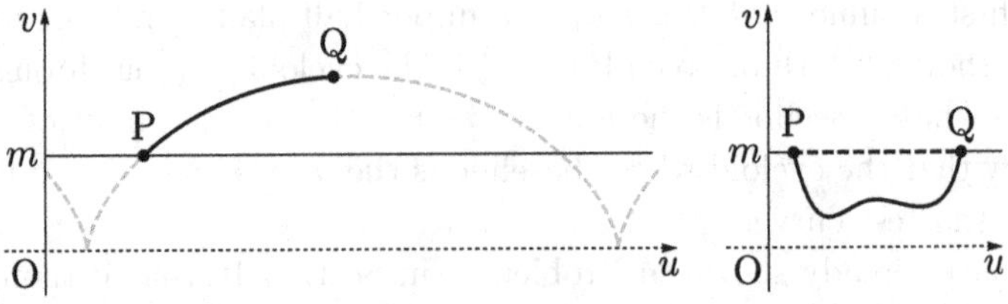

Fig. 18.1 Proof of Theorem 18.3.

Without loss of generality, we may assume the point Q lies above P. Setting $\mathrm{P} = (u_0, v_0)$, consider a line m in the plane through P and parallel to the u-axis (cf. Fig. 18.1, left). If a path connecting P to Q has parts lying below m, then we could replace those parts with portions of m itself, and the length of the path becomes shorter.

In fact, consider a piecewise smooth curve $\gamma(t) = (u(t), v(t))$ $(a \leq t \leq b)$ emanating from P, ending at a point Q on the line m such that γ lies in the closed lower half region with respect to m (cf. Fig. 18.1, right). Since

$$\int_a^b \sqrt{\frac{\dot{u}^2+\dot{v}^2}{v}}\,dt \geq \int_a^b \frac{1}{\sqrt{v_0}}|\dot{u}|\,dt \geq \left|\int_a^b \frac{1}{\sqrt{v_0}}\dot{u}\,dt\right| \geq \frac{1}{\sqrt{v_0}}|u(b)-u(a)|$$

holds and the right-hand side is the length of the line segment PQ, the shortest length path can be assumed to lie above m.

In conclusion, we can consider just the piecewise smooth curves in the closed region $\{v \geq v_0\}$ connecting P and Q, and we can take a sequence of such curves $\{\gamma_n\}_{n=1,2,3,\dots}$ so that the lengths of the γ_n converge to $d(\mathrm{P},\mathrm{Q})$ as $n \to \infty$. We may assume that the lengths are monotonically decreasing with respect to n. Let l_1 denote the length of γ_1. Defining

$$C := \{(u,v) \in \boldsymbol{R}^2 \mid v \geq v_0\} \cap \overline{B(\mathrm{Q},l_1)},$$
$$\text{where } \overline{B(\mathrm{Q},l_1)} := \{\mathrm{R} \in D \mid d(\mathrm{Q},\mathrm{R}) \leq l_1\},$$

all of the γ_n lie in the compact set C. Since C is compact, there exists a constant $\delta_0 > 0$ such that the geodesic disc $\Delta_{\mathrm{P}}(\delta_0)$ is convex (in the sense of Definition 18.2) for each $\mathrm{P} \in C$ (see Theorem 5.14 in Chapter 5 of [3]). Thus any two points on C whose distance is less than δ_0 can be joined by a minimizing geodesic. Without loss of generality, we may assume that all γ_n are defined on the interval $[0,1]$, with parameter proportional to the arc-length parameter.

Choosing a natural number N so that $l_1/N < \delta_0$, we divide the interval $[0,1]$ into N equally spaced intervals using the points

$$0 = t_0 < t_1 < \cdots < t_N = 1$$

on any of the curves two adjacent points have distance smaller than δ_0, and we can replace any curve on that interval by a shortest path with the same endpoints. In this way we can obtain a family of piecewise smooth curves $\{\hat{\gamma}_n\}_{n=1,2,3,\dots}$ consisting of unions of geodesic segments. The length of each $\hat{\gamma}_n$ is less than or equal to the length of γ_n, so $\hat{\gamma}_n$ also converges to $d(\mathrm{P},\mathrm{Q})$. Since C is compact, we can take a subsequence if necessary so that for each $j = 1,\dots,N$, the sequence of points $\{\gamma_n(t_j)\,;\, n = 1,2,\dots\}$ has a limit $\mathrm{A}_j = \lim_{n\to\infty}\gamma_n(t_j)$ for $j = 1,\dots,N$. The length of $\hat{\gamma}_n$ is

$$\sum_{j=1}^{N} d(\gamma_n(t_j),\gamma_n(t_{j-1})).$$

Letting $n \to \infty$ in this equation, this value converges to the length of the piecewise smooth geodesic obtained by joining $\mathrm{P} = \mathrm{A}_0$, A_1, $\dots$, $\mathrm{A}_N = \mathrm{Q}$, and we see that this piecewise smooth curve is the shortest path. □

Theorem 18.3. *For a point* P *in the* u*-axis and a point* $\mathrm{Q} \in D$*, there exists a unique path of shortest length between* P *and* Q*, and it is the cycloid through* Q *whose baseline is the* u*-axis.*

Proof. Let l_0 denote the length of the cycloid, whose baseline is the u-axis, that connects P and Q. Suppose by way of contradiction that there exists a piecewise smooth curve γ from P to Q with length strictly less than l_0. Let $\{\mathrm{P}_n\}$ be a sequence of points in the path γ so that

$$\lim_{n\to\infty} \mathrm{P}_n = \mathrm{P}.$$

By the proof of Theorem 18.1, we may assume that γ intersects the u-axis only at the starting point. Letting σ_n denote the shortest path between P_n and Q, the lengths $\mathcal{L}(\sigma_n)$, $\mathcal{L}(\gamma)$ of σ_n and γ satisfy the inequalities

$$\mathcal{L}(\sigma_n) \leq \mathcal{L}(\gamma) < l_0.$$

As $n \to \infty$, by Theorem 18.1, and noting that σ_n is an arc of a cycloid,

$$l_0 = \lim_{n\to\infty} \mathcal{L}(\sigma_n) \leq \mathcal{L}(\gamma) < l_0$$

holds, and we arrive at a contradiction.

Thus, we have seen that the cycloid connecting P and Q minimizes length. Regarding possible existence of a minimizer other than the cycloid, such a curve would necessarily locally be a cycloid (whose baseline is the u-axis). Therefore, we have shown there is only one length minimizing path. □

19. Geodesic triangulations of compact Riemannian 2-manifolds

Distance functions on a Riemannian manifold. In this section, we prove the existence of a geodesic triangulation of a given compact Riemannian 2-manifold. For this purpose, we prepare a property of distance functions on Riemannian manifolds: Let (S, ds^2) be a Riemannian manifold and d the distance function on S induced by ds^2, which is compatible with the topology of S (cf. Theorem 13.4). The following assertion holds.

Lemma 19.1. *Let* $d\tilde{s}^2$ *be another Riemannian metric on an* n*-manifold* S $(n \geq 2)$*. For any compact subset* K *on* S*, there exist positive constants* m *and* M *such that*

$$md(\mathrm{P},\mathrm{Q}) < \tilde{d}(\mathrm{P},\mathrm{Q}) < Md(\mathrm{P},\mathrm{Q}) \qquad (\mathrm{P},\mathrm{Q} \in K) \tag{19.1}$$

holds, where $\tilde{d}$ *is the distance function induced by* $d\tilde{s}^2$*.*

Proof. There exists a $(1,1)$-tensor field A on S such that $d\tilde{s}^2(X,Y) = ds^2(AX,Y) = d\tilde{s}^2(X,AY)$ holds for any vector fields X, Y on S. Then for each $\mathrm{P} \in S$, A_P is a self-adjoint linear transformation on the tangent space $T_\mathrm{P}S$ at P, and the eigenvalues of A_P are real. We denote by n the dimension of S. By the well-known continuity of the roots of a polynomial with respect to its coefficients, there exist continuous functions $\lambda_j : S \to \boldsymbol{R}$ $(j = 1,\ldots,n)$ satisfying $\lambda_1 \leq \cdots \leq \lambda_n$ on S such that each $\lambda_j(\mathrm{P})$ is an eigenvalue of A_P. We set

$$m := \sqrt{\min_{\mathrm{Q}\in K} \lambda_1(\mathrm{Q})}, \qquad M := \sqrt{\max_{\mathrm{Q}\in K} \lambda_n(\mathrm{Q})}.$$

Fix a point $\mathrm{P} \in S$. Then we can take an orthonormal basis $\{\boldsymbol{e}_1,\ldots,\boldsymbol{e}_n\}$ of $T_\mathrm{P}M$ with respect to the metric ds^2 such that each $\boldsymbol{e}_j$ is the eigenvector of A_P. Take a vector $\boldsymbol{v}$ in $T_\mathrm{P}M$ arbitrarily. Then it can be expressed as $\boldsymbol{v} = \sum_{i=1}^n b_i\boldsymbol{e}_i$ and it holds that

$$d\tilde{s}^2_\mathrm{P}(\boldsymbol{v},\boldsymbol{v}) = ds^2(A_\mathrm{P}\boldsymbol{v},\boldsymbol{v}) = ds^2\left(\sum_{i=1}^n \lambda_i(\mathrm{P})b_i\boldsymbol{e}_i, \sum_{j=1}^n b_j\boldsymbol{e}_j\right) = \sum_{i=1}^n \lambda_i(\mathrm{P})b_i^2.$$

Hence, for any tangent vector $\boldsymbol{v}$ of M^2, we have

$$\begin{aligned} m^2 ds^2(\boldsymbol{v},\boldsymbol{v}) &= m^2\sum_{i=1}^n b_i^2 \\ &\leq \sum_{i=1}^n \lambda_i b_i^2 = d\tilde{s}^2(\boldsymbol{v},\boldsymbol{v}) \\ &\leq M^2\sum_{i=1}^n b_i^2 = M^2 ds^2(\boldsymbol{v},\boldsymbol{v}). \end{aligned}$$

In particular, the formula (13.16) yields that $m\mathcal{L}(\gamma) \leq \widetilde{\mathcal{L}}(\gamma) \leq M\mathcal{L}(\gamma)$, where $\mathcal{L}(\gamma)$ (resp. $\widetilde{\mathcal{L}}(\gamma)$) is the length of a smooth curve γ on S with respect to the metric ds^2 (resp. $d\tilde{s}^2$). Then (19.1) follows immediately. □

The existence of tubular neighborhoods. Let (S, ds^2) be a compact Riemannian 2-manifold, and let $\gamma : [a,b] \to S$ be a smooth regular curve without self-intersection parametrized by the arc-length parameter s. By definition, there exists a positive number ε and a smooth regular curve $\tilde{\gamma} : (a-\varepsilon, b+\varepsilon) \to S$ such that γ is the restriction of $\tilde{\gamma}$ on $[a,b]$. We denote by Exp_P the exponential map (cf. [5, Chapter 3]) of the Riemannian manifold (S, ds^2) at $\mathrm{P} \in S$, and define a map $\psi : (a-\varepsilon, b+\varepsilon) \times \boldsymbol{R} \to S$ by

$$\psi(s,t) := \mathrm{Exp}_{\gamma(s)}(t\boldsymbol{n}(s)),$$

where $\boldsymbol{n}(s) \in T_{\gamma(s)}S$ is a smooth unit normal vector field of $\tilde{\gamma}(s)$. In this situation, we can prove the following lemma.

Lemma 19.2. *There exist positive numbers d', $d''(< \varepsilon)$ satisfying the following properties:*

(1) *The restriction of ψ to $[a - d', b + d'] \times [-d', d']$ gives an injection into S.*
(2) *If $s_1, s_2 \in [a, b]$ satisfy $0 < s_2 - s_1 < d''$, then there exists a continuous function $f : [s_1, s_2] \to (-d', d')$ such that the image of the map*
$$[s_1, s_2] \ni s \longmapsto \psi(s, f(s)) \in S$$
coincides with the image of the geodesic segment bounded by $\gamma(s_1)$ and $\gamma(s_2)$.

The map ψ is called a *tubular neighborhood* of the curve γ, and we call the constant d' the *graphical constant* of the tubular neighborhood.

Proof. It can be easily checked that the differential of ψ at $(s, 0)$ for $a \leq s \leq b$ is of rank two. So there exists a positive number ε' satisfying $0 < \varepsilon' < \varepsilon$ such that the restriction of ψ into $(a - \varepsilon', b + \varepsilon') \times (-\varepsilon', \varepsilon')$ is an immersion. Let n_0 be a positive integer satisfying $1/n_0 < \varepsilon'$. Suppose that for any integer $n \geq n_0$ the restriction of ψ into $[a - 1/n, b + 1/n] \times [-1/n, 1/n]$ is not an injection. Then, for each n, there exist two points
$$(x_n, z_n),\ (x'_n, z'_n) \in \left(a - \frac{1}{n}, b + \frac{1}{n}\right) \times \left(-\frac{1}{n}, \frac{1}{n}\right)$$
such that $\psi(x_n, z_n) = \psi(x'_n, z'_n)$ and $(x_n, z_n) \neq (x'_n, z'_n)$. We can take a strictly monotone increasing subsequence $\{n_j\}_{j=1,2,3,\ldots}$ of $\{n\}_{n \geq n_0}$ such that
$$\lim_{j \to \infty} x_{n_j} = x_\infty, \qquad \lim_{j \to \infty} y_{n_j} = y_\infty,$$
where $x_\infty, y_\infty \in [a, b]$. Then, we have that
$$\begin{aligned} \gamma(x_\infty) = \psi(x_\infty, 0) &= \lim_{j \to \infty} \psi(x_{n_j}, z_{n_j}) \\ &= \lim_{j \to \infty} \psi(x'_{n_j}, z'_{n_j}) = \psi(x'_\infty, 0) = \gamma(x_\infty). \end{aligned}$$
Since γ is injective, we have $(s_0 :=)\ x_\infty = y_\infty$. Here, there exists an open neighborhood W of $(s_0, 0)$ such that the restriction of ψ on W is injective because the differential of ψ is of rank 2 at $(s_0, 0)$. Noticing that (x_{n_j}, z_{n_j}), $(x'_{n_j}, z'_{n_j}) \in W$ for sufficiently large j, we have
$$\psi(x_{n_j}, z_{n_j}) \neq \psi(x'_{n_j}, z'_{n_j}),$$

a contradiction. So for some n, the restriction of ψ to $[a-1/n, b+1/n] \times [-1/n, 1/n]$ is an injection. Setting $d' = 1/n$, we get the first assertion.

We next prove the second assertion. By setting

$$K := [a-1/n, b+1/n] \times [-1/n, 1/n],$$

$\gamma([a,b])$ lies in the interior of $\psi(K)$. So there is a number d'' (> 0) such that $d(\mathrm{P},\mathrm{Q}) \geq d''$ holds for any $\mathrm{P} \in \gamma([a,b])$ and $\mathrm{Q} \in \psi(\partial K)$, where ∂K is the boundary of the rectangular domain K. Since S is compact, there exists a minimizing geodesic segment σ which bounds $\gamma(s_1)$ and $\gamma(s_2)$ (cf. [5, Theorem 2.9]). Let Q be a point on the segment. Since

$$d(\gamma(s_1), \mathrm{Q}) \leq d(\gamma(s_1), \gamma(s_2)) \leq s_2 - s_1 < d'',$$

we can conclude that Q lies in $\psi(K^\circ)$. So σ lies in $\psi(K^\circ)$. We set $\sigma_0 := \psi^{-1}(\sigma)$. Using the inverse function theorem, one can easily see that σ_0 is the image of a regular embedded curve. Suppose that the tangent vector of σ_0 is vertical at some point (s_0, t_0). Then, by the uniqueness of a geodesic with a given initial condition, σ must coincide with the vertical geodesic $t \mapsto \psi(s_0, t)$. However, that contradicts the fact that σ meets the horizontal axis twice in K. Thus, the tangential direction of σ at each point can never be vertical, and so σ meets the image of a vertical line $t \mapsto \psi(s,t)$ exactly once for each $s_1 \leq s \leq s_2$. In particular, σ can be expressed as the graph of a certain function of s. □

Geodesic triangulations. Let AB, BC, CA be three line segments on a plane $\boldsymbol{R}^2$ whose endpoints are three distinct points, and the union AB ∪ BC ∪ CD consists of a triangle that surrounds a bounded closed domain △ABC. We call such a closed domain △ABC a *linear triangular region* in $\boldsymbol{R}^2$.

Definition 19.3. A compact domain T of a Riemannian 2-manifold (S, ds^2) is called a *triangular region* if there is a local coordinate system $\varphi : U \to \boldsymbol{R}^2$ such that $\varphi(T)$ is a linear triangle in $\boldsymbol{R}^2$, and the interior T° of T is mapped to the interior of the linear triangular region.

In this setting, three points (resp. edges) on ∂T corresponding to the vertices (resp. edges) of the linear triangular region $\varphi(T)$ are called the *vertices* (resp. *edges*) of T. By definition, the endpoints of each edge consist of two vertices.

We now fix a Riemannian metric ds^2 on S. If each edge of a triangular region is the image of a geodesic, we call T a *geodesic triangular region.* We now give the definition of a triangulation.

Definition 19.4. Let (S, ds^2) be a Riemannian 2-manifold. A finite union of triangular regions $\{T_i\}_{i=1,2,\dots,N}$ is called a *triangulation* of S if it satisfies the following properties:

(1) $S = T_1 \cup \cdots \cup T_N$,
(2) $T_i^\circ \cap T_j^\circ$ is empty if $i \neq j$, where T_i°, T_j° are the interiors of T_i, T_j, respectively, and
(3) if $T_i \cap T_j$ $(i \neq j)$ is not empty, then it is a common edge or a common vertex of T_i and T_j.

Remark 19.5. Let $\{T_1, \dots, T_N\}$ be a *triangulation* of S. Suppose that S is oriented. By the definition of a triangular region, we may assume that each T_j $(j = 1, \dots, n)$ lies in a coordinate neighborhood U_j of S. Since S is oriented, we may assume that the local coordinate system $\varphi : U \to \boldsymbol{R}^2$ is compatible with respect to the orientation of S. Then we can give an orientation of the boundary ∂T_j so that $\varphi(\partial T_j)$ surrounds the linear triangle $\varphi(T_j)$ counterclockwisely. Then, any two adjacent triangles have opposite orientation along their common edge (see the right-hand side of Fig. 10.5), that is, the triangles $T_1, \dots, T_N$ have compatible boundary orientation.

Here we prove existence of geodesic triangulations, which was used in the proof of the Gauss-Bonnet theorem (Theorem 10.7). More precisely, we shall prove the following assertion.

Theorem 19.6. *Let (S, ds^2) be a compact Riemannian 2-manifold. Then there exists a geodesic triangulation of S.*

Consider two 2-manifolds S_1 and S_2, and from each remove regions D_1, D_2 diffeomorphic to the disc. Then gluing $S_1 \setminus D_1$ and $S_2 \setminus D_2$ along their boundaries, we have a manifold $S_1 \# S_2$ called the *connected sum* of S_1 and S_2. It is known that compact orientable 2-manifolds are diffeomorphic to the connected sum of a sphere and a finite number of tori. The number of tori in this connected sum equals the genus of the manifold. On the other hand, non-orientable compact 2-manifolds are diffeomorphic to the connected sum of a finite number of real projective planes. Using triangulations of tori and projective planes, we can easily find a triangulation of the surface. Thus, an arbitrary compact 2-manifold can have a (not necessarily geodesic) triangulation. (See for example, Chapter 1 of [25] or Theorem 10.6 of [28].)

Proof of Theorem 19.6. Since S is compact, there exists a constant $\delta_0 > 0$ such that the geodesic disc $\Delta_{\mathrm{P}}(r)$ is convex (cf. Definition 18.2 in Section 18)

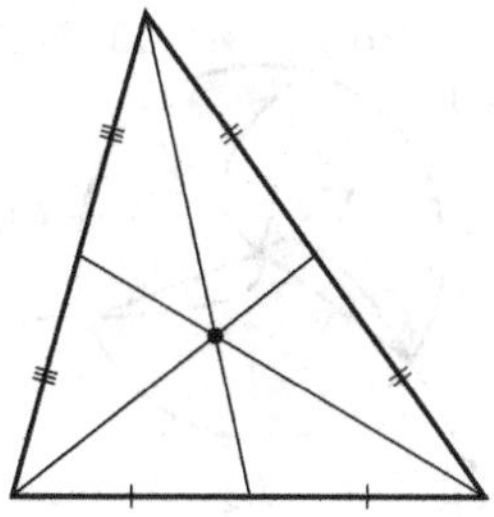

Fig. 19.1 The barycentric subdivision.

if $r < \delta_0$ (see Theorem 5.14 in Chapter 5 of [3]). We call δ_0 the *convex radius* of S.

We fix a triangulation of S. Let T be a triangular region in the triangulation. Then there exists a local coordinate system $\varphi : U \to \boldsymbol{R}^2$ of S such that $T \subset U$ and $\varphi(T)$ is a linear triangular region. Then we can give the (first) barycentric subdivision (cf. Fig. 19.1 and [35, Chapter 4]). Iterating, we get the k'th barycentric subdivision of $\varphi(T)$, which is a union of 6^k triangles. For an arbitrarily given positive number ε, we can choose the number k so that the diameter of each triangle in the k'th barycentric subdivision is smaller than ε (cf. Theorem 4 in Chapter 4 of [35]). This subdivision of $\varphi(T)$ induces a subdivision of T. Then we get the k'th barycentric subdivision of the triangulation of S. Applying Lemma 19.1 by setting $K = \varphi(T)$, and comparing the canonical metric on $\boldsymbol{R}^2$ and the pull-back metric of ds^2 by φ^{-1}, we may assume that our triangulation of S is taken so that each triangular region is contained in a convex domain of S.

We let $\{E_1, \dots, E_m\}$ be the set of edges associated to the triangulation. By Lemma 19.2, each $E_1, \dots, E_m$ has a tubular neighborhood. We denote by d_i'' $(i = 1, 2, \dots, m)$ the graphical constant of the tubular neighborhood of E_i (cf. Lemma 19.2). We fix a vertex V of the triangulation, and suppose that

$$\{\hat{E}_1, \dots, \hat{E}_k\} \ \ (\subset \{E_1, \dots, E_m\})$$

are the set of edges whose terminal point is V.

Let r be a small positive number satisfying

$$r < d_i'' \qquad (i = 1, \dots, m). \tag{19.2}$$

Consider the closed geodesic disc $\overline{\varDelta_{\mathrm{V}}(r)}$ of radius r centered at $\mathrm{V} \in S$. We replace parts of each edge $\hat{E}_j$ $(1 \leq j \leq k)$ by the shortest geodesic segment L_j joining V and the final intersection points of the edge $\hat{E}_j$ for $\overline{\varDelta_{\mathrm{V}}(r)}$. If

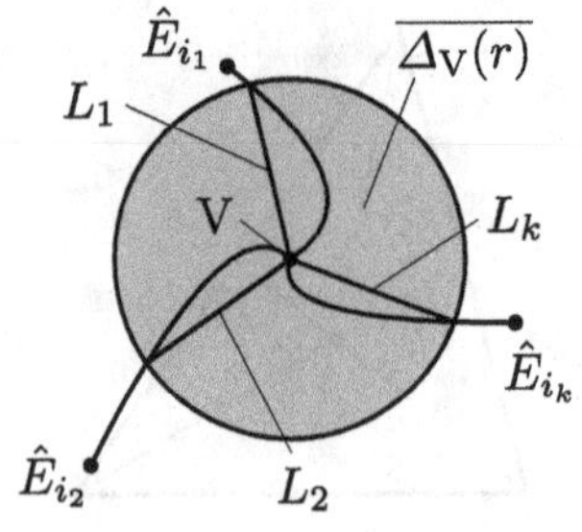

Fig. 19.2 Proof of Theorem 19.6.

we choose r sufficiently small, then no two of $L_1, \ldots, L_k$ ever meet each other (see Fig. 19.2).

We give the same operation for each vertex of the triangulation, and assume that the radius r is commonly chosen for all vertices. After this operation, both ends V_i, V'_i of the edge E_i $(i = 1, \ldots, m)$ are replaced by geodesic segments $\mathrm{V}_i\mathrm{A}_i$, $\mathrm{V}'_i\mathrm{B}_i$, where A_i (resp. B_i) are points on edges so that $d(\mathrm{A}_i, \mathrm{V}_i) = r$ (resp. $d(\mathrm{B}_i, \mathrm{V}'_i) = r$). (If $\mathrm{V} = \mathrm{V}_i$, then $\mathrm{V}_i\mathrm{A}_i$ coincides with one of the $L_1, \ldots, L_j$.) We denote by E'_i the subarc of E_i bounded by A_i and B_i. Since E'_i is compact and disjoint from the other edges, there exists a constant ρ_i such that $\Delta_{\mathrm{P}}(\rho_i)$ does not meet other edges for all $p \in E'_i$. We set

$$d := \min\left\{\frac{\rho_1}{2}, \ldots, \frac{\rho_m}{2}, d''_1, \ldots, d''_m\right\}.$$

We then divide E'_i into the union of subarcs all of length less than d. When replacing each of these subarcs by geodesic segments, we get a piecewise smooth curve as an approximation of E'_i. Moreover, adding $\mathrm{V}_i\mathrm{A}_i$, $\mathrm{V}'_i\mathrm{B}_i$, we get a piecewise smooth curve $\widetilde{E}_i$ bounded by V_i and V'_i as an approximation of E_i. By Lemma 19.2, each geodesic segment in $\widetilde{E}_i$ is expressed as a graph over the u-axis in the tubular neighborhood of E_i. So $\widetilde{E}_i$ has no self-intersections. Since $d < \delta'/2$, $\widetilde{E}_i$ never meets the other $\widetilde{E}_j$ $(i \neq j)$. In this way, we have constructed a geodesic polygonal division of the 2-manifold.

For one geodesic polygon Γ, we will write the vertices as $\mathrm{A}_1, \ldots, \mathrm{A}_m$ $(m \geq 4)$. Because Γ lies in one of the convex neighborhoods, it can be considered as a piecewise smooth simple closed curve in $\boldsymbol{R}^2$.

A vertex A_i of the geodesic polygon Γ is called *removable* if $\mathrm{A}_{i-1}, \mathrm{A}_i, \mathrm{A}_{i+1}$ lie on the same geodesic in S. A vertex which is not removable is called *essential.* We prove the following assertion.

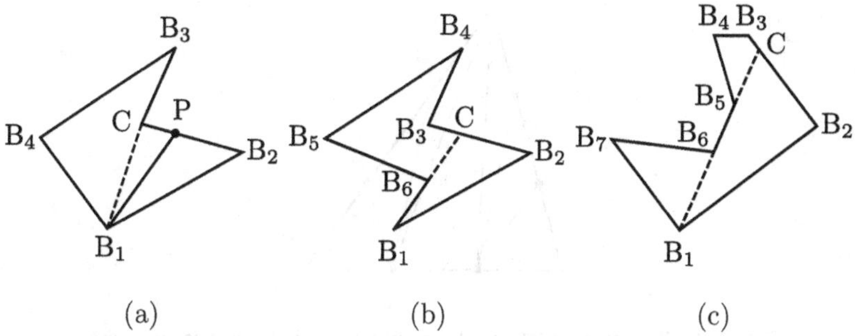

Fig. 19.3 Three cases for the position of C.

Lemma 19.7. *Suppose that the geodesic polygon Γ consisting of n essential vertices ($n \geq 4$). Then the interior region bounded by Γ can be divided into the union of bounded regions whose boundary are polygons having* 3 *essential vertices.*

Proof. We denote by $B_1, \ldots, B_n$ the essential vertices of Γ. Draw a geodesic from one vertex B_1 into the interior of Γ. At the first point P where this geodesic leaves the region inside Γ, if P is a vertex of Γ, this geodesic has separated Γ into two geodesic polygons with fewer edges.

If, on the other hand, P is not a vertex but lies in one of the edges of Γ, the geodesic segment B_1P lies in the interior region of Γ, and P does not lie in either of the edges B_1B_2, B_1B_n. If P lies neither in B_2B_3 nor $B_{n-1}B_n$, then Γ is divided into two polygons with fewer essential vertices. So we may assume that P lies in B_2B_3 or $B_{n-1}B_n$. Without loss of generality, we may assume that P lies in B_2B_3.

Moving the point P along the edge B_2B_3 toward B_3, at least for some time B_1P will remain a shortest geodesic in the region inside Γ. We consider the first point $P = C$ along the edge B_2B_3 where this property is lost. If C coincides with one of the vertices $B_3, \ldots, B_{n-1}$ and the geodesic segment B_1C does not meet Γ except at B_1, C, then Γ is divided into two polygons with fewer essential vertices (see Fig. 19.3(a)).

So we may assume the shortest geodesic B_1C contains the edge B_nB_1 or an edge B_iB_{i+1} for $i = 3, \ldots, n-1$. In the former case, Γ is divided into two polygons B_1B_2C and $B_nCB_3 \ldots B_{n-1}$ which both consist of fewer essential vertices (see Fig. 19.3(b)). In the latter case, Γ is divided into three polygons with fewer numbers of essential vertices, namely (see Fig. 19.3(c)) B_1B_2C, $CB_3 \ldots B_i$ and $B_{i+1} \ldots B_nB_1$. □

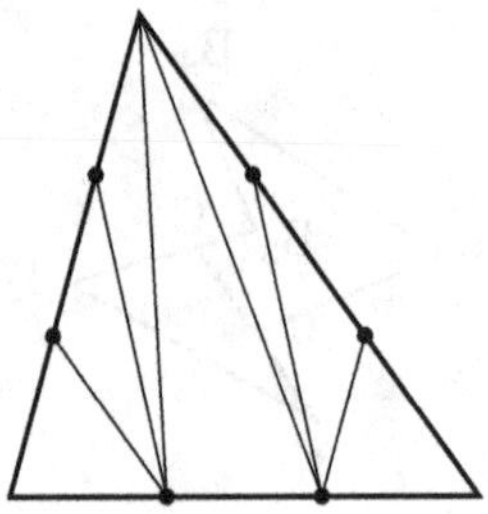

Fig. 19.4 The triangulation of a polygon having essential vertices.

Now, we return to the proof of Theorem 19.6. Applying Lemma 19.7 to each geodesic polygon, we get a geodesic polygonal division for the 2-manifold whose number of essential vertices is three on every polygon. Since each polygonal region with exactly three essential vertices can be divided into a union of geodesic triangular subregions as in Fig. 19.4, we finally get the desired geodesic triangulation.

Remark 19.8. We have obtained a geodesic triangulation consisting of regions bounded by three distinct geodesic segments. We fix such a polygonal region T arbitrarily. Let P be a vertex of T. Then the inverse of $\mathrm{Exp}_{\mathrm{P}}$ maps T to a triangular region in $T_{\mathrm{P}}S$ whose boundary consists of two line segments and one regular arc. So it is not difficult to show that the inverse image can be mapped into a linear triangle by a suitable diffeomorphsm. Hence, T is a triangular region in the sense of Definition 19.3.

Appendix A

Supplements

A.1. A review of calculus

In this section, we summarize some facts from calculus.

Taylor's formula. A function $f : I \to \boldsymbol{R}$ defined on an open interval I is said to be of *class* C^r if there is a continuous r'th derivative $f^{(r)}$. Moreover, the function f is said to be of class C^∞ if it is of class C^r for every positive integer r. Functions in this text are assumed to be of class C^∞ unless specified otherwise. The following Taylor's formula (cf. [37, Theorem 4 in Section 20]) is useful for investigating the local behavior of curves and surfaces. See Problem **3** in Section 1, equation (4.1) in Section 4, for example.

Theorem A.1.1 (Taylor's formula for functions of one variable). *Assume that a function $f(x)$ of one variable x is of class C^{r+1} on an interval containing a and $a+h$ $(h \neq 0)$. Then it holds that*

$$
\text{(A.1.1)} \quad f(a+h) = f(a) + \dot{f}(a)h + \frac{\ddot{f}(a)}{2}h^2 + \cdots + \frac{f^{(r)}(a)}{r!}h^r + \frac{f^{(r+1)}(a+\theta h)}{(r+1)!}h^{r+1}
$$

$$
\left(\dot{f} := \frac{df}{dx},\ \ddot{f} := \frac{d^2 f}{dx^2},\ f^{(k)} := \frac{d^k f}{dx^k}\right),
$$

where θ takes some value so that $0 < \theta < 1$.

The last term in equation (A.1.1) is called the *remainder*. For functions $\varphi(x)$ and $h(x)$, we write

$$
\varphi(x) = o(h(x)) \qquad (x \to a)
$$

if it holds that

$$\lim_{x \to a} \frac{\varphi(x)}{h(x)} = 0.$$

Namely, the function $\varphi(x)$ tends to 0 to an order higher than $h(x)$ does, as x approaches a. The symbol "o" is called *Landau's symbol.* Using this, the remainder of (A.1.1) is written as "$o(h^r)$ $(h \to 0)$". In this text, we write this simply as "$o(h^r)$", omitting "$(h \to 0)$".

A subset D of $\boldsymbol{R}^n$ (we mainly consider the case $n = 2$ or 3) is said to be (*pathwise*) *connected* if any two points in D can be connected by a continuous curve in D, and said to be an *open set* if a sufficiently small open disc centered at P is contained in D for each point P in D. A connected open set in $\boldsymbol{R}^n$ is called a *domain* or a *region.* A function $f(x_1, \ldots, x_n)$ of two variables defined on a domain D of $\boldsymbol{R}^n$ is said to be of *class* C^r if all partial derivatives up to order r exist and are continuous, and of class C^∞ if it is of class C^r for every positive integer r.

Taylor's formula for functions of two variables is expressed as follows (cf. [36, Theorem 2-11]). This formula is used on page 63 of Section 6, pages 68 and 70 of Section 7, the proof of Theorem 8.7 in Section 8, and the proof of Theorem 9.10 in Section 9, for example.

Theorem A.1.2 (Taylor's formula for functions of two variables). *Assume that a function $f(x, y)$ of two variables is of class C^∞ on a domain*

$$\{(x,y) \mid (x-a)^2 + (y-b)^2 < \varepsilon^2\} \subset \boldsymbol{R}^2 \qquad (\varepsilon > 0)$$

containing (a, b). Then for h, k satisfying $h^2 + k^2 < \varepsilon^2$, it holds that

$$\begin{aligned} f(a+h, b+k) = f(a,b) &+ \frac{\partial f}{\partial x}(a,b)h + \frac{\partial f}{\partial y}(a,b)k \\ &+ \frac{1}{2}\left(\frac{\partial^2 f}{\partial x^2}(a,b)h^2 + 2\frac{\partial^2 f}{\partial x \partial y}(a,b)hk + \frac{\partial^2 f}{\partial y^2}(a,b)k^2\right) \\ &+ R(h,k), \end{aligned}$$

where $\lim_{(h,k)\to(0,0)} \frac{R(h,k)}{h^2+k^2} = 0$, *that is,* $R(h,k) = o(h^2 + k^2)$.

L'Hôpital's rule. The following L'Hôpital's rule (cf. Theorem 9 and Problem 54 in [37, Section 11]) is used in Section 11 (pages 121 and 128) to investigate properties of the geodesic polar coordinate system.

Theorem A.1.3 (L'Hôpital's rule). *Assume two differentiable functions $f(x)$, $g(x)$ defined on an interval $(a, a+\varepsilon)$ $(\varepsilon > 0)$ satisfy*

$$\lim_{x \to a^+} f(x) = \lim_{x \to a^+} g(x) = 0,$$

where $\lim_{r\to a^+}$ *means the right-hand limit. Then*

$$\lim_{x\to a^+}\frac{f(x)}{g(x)}=\lim_{x\to a^+}\frac{\dot{f}(x)}{\dot{g}(x)}$$

holds, whenever the limit on the right-hand side exists. A similar formula holds for the left-hand limit $x\to a^-$ *when* $f(x)$ *and* $g(x)$ *are defined on* $(a-\varepsilon,a)$ $(\varepsilon>0)$.

The hyperbolic functions. The functions

$$\cosh t:=\frac{e^t+e^{-t}}{2},\qquad \sinh t:=\frac{e^t-e^{-t}}{2},\qquad \tanh t:=\frac{\sinh t}{\cosh t}=\frac{e^t-e^{-t}}{e^t+e^{-t}}$$

of one real variable t are called the *hyperbolic cosine*, *hyperbolic sine* and *hyperbolic tangent* of t, respectively. Furthermore, the identity

$$\cosh^2 t-\sinh^2 t=1$$

holds.

The triangle inequality for integrals. The inequality

$$\left|\int_a^b f(t)\,dt\right|\le\int_a^b |f(t)|\,dt \tag{A.1.2}$$

for a continuous function $f(t)$ on the interval $[a,b]$ follows immediately from the definition of the integral. Here, equality in (A.1.2) holds if and only if $f(t)$ does not change sign on the interval $[a,b]$. This inequality is called the *triangle inequality* (for integrals).

We shall extend this inequality to vector-valued functions. Consider a continuous function $\boldsymbol{f}=(f_1,f_2)\colon[a,b]\to\boldsymbol{R}^2$ defined on $[a,b]$ and taking values in $\boldsymbol{R}^2$, where $f_1(t)$, $f_2(t)$ are continuous functions on $[a,b]$. The integral of $\boldsymbol{f}$ is the vector defined by

$$\int_a^b \boldsymbol{f}(t)\,dt:=\left(\int_a^b f_1(t)\,dt,\int_a^b f_2(t)\,dt\right)\in\boldsymbol{R}^2.$$

Then the following triangle inequality, which is used in the proof of Theorem 3.3 in Section 3 and the proof of Theorem 4.5 in Section 4, holds for vector-valued functions.

Theorem A.1.4 (The triangle inequality for vector-valued functions). *It holds that*

$$\left|\int_a^b \boldsymbol{f}(t)\,dt\right|\le\int_a^b |\boldsymbol{f}(t)|\,dt \tag{A.1.3}$$

for a continuous vector-valued function $\boldsymbol{f}\colon[a,b]\to\boldsymbol{R}^2$ *defined on* $[a,b]$. *Equality holds in* (A.1.3) *if and only if there exists a constant vector* $\boldsymbol{a}$ *such that* $\boldsymbol{f}(t)=|\boldsymbol{f}(t)|\boldsymbol{a}$.

Proof. Identifying $\boldsymbol{R}^2$ with the complex plane $\boldsymbol{C}$, one can write $\int_a^b \boldsymbol{f}(t)\,dt = re^{i\theta}$ $(r \geq 0,\ \theta \in \boldsymbol{R})$. Using this, we define a new function $\tilde{\boldsymbol{f}}(t) := e^{-i\theta}\boldsymbol{f}(t)$. If one writes $\tilde{\boldsymbol{f}}(t) = \tilde{f}_1(t) + i\tilde{f}_2(t)$ in terms of real and imaginary parts, we have

$$\begin{aligned}\int_a^b \tilde{f}_1(t)\,dt + i\int_a^b \tilde{f}_2(t)\,dt &= \int_a^b \tilde{\boldsymbol{f}}(t)\,dt \\ &= e^{-i\theta}\int_a^b \boldsymbol{f}(t)\,dt = e^{-i\theta}(re^{i\theta}) = r,\end{aligned}$$

and hence $\int_a^b \tilde{f}_1(t)\,dt \geq 0$, $\int_a^b \tilde{f}_2(t)\,dt = 0$. Then we have the inequality

$$\begin{aligned}\left|\int_a^b \boldsymbol{f}(t)\,dt\right| &= \left|\int_a^b \tilde{\boldsymbol{f}}(t)\,dt\right| = \left|\int_a^b \tilde{f}_1(t)\,dt\right| \\ &\leq \int_a^b |\tilde{f}_1(t)|\,dt \leq \int_a^b |\tilde{\boldsymbol{f}}(t)|\,dt = \int_a^b |\boldsymbol{f}(t)|\,dt.\end{aligned}$$

Equality holds here if and only if $\tilde{f}_2(t)$ is identically zero and $\tilde{f}_1(t)$ does not change sign, which is equivalent to $\boldsymbol{f}(t) = |\boldsymbol{f}(t)|\boldsymbol{a}$ holding for some constant vector $\boldsymbol{a}$. □

The inverse and implicit function theorems. The implicit function theorem is necessary to show that an implicit function representation or a parametrization of a curve or surface actually represents a smooth figure (cf. page 6 in Section 1, pages 11 and 19 in Section 2, the proof of Theorem 8.7 in Section 8, and the proof of Proposition 11.7 in Section 11). Here, we introduce the inverse and implicit function theorems (cf. [36, Theorems 2-11 and 2-12]).

Theorem A.1.5 (The inverse function theorem).

(1) *Assume that a real-valued C^∞-function $f(x)$ defined on an interval including the value a satisfies $\dot{f}(a) \neq 0$, where $\dot{}$ means d/dx. Then there exists a unique C^∞ function $g(y)$ defined on an interval including $f(a)$ such that $g(f(x)) = x$ and $f(g(y)) = y$. Moreover, the derivative $\dot{g}$ of g satisfies*

$$\dot{g}(y) = \frac{1}{\dot{f}(g(y))}.$$

(2) *Assume that a C^∞-map*

$$\boldsymbol{f}\colon D \ni (u,v) \longmapsto \boldsymbol{f}(u,v) = (x(u,v), y(u,v)) \in \boldsymbol{R}^2$$

defined on a domain $D \subset \boldsymbol{R}^2$ containing the point $\mathrm{P} = (u_0, v_0) \in \boldsymbol{R}^2$ *satisfies*

$$\det \begin{pmatrix} \dfrac{\partial x}{\partial u}(u_0, v_0) & \dfrac{\partial x}{\partial v}(u_0, v_0) \\ \dfrac{\partial y}{\partial u}(u_0, v_0) & \dfrac{\partial y}{\partial v}(u_0, v_0) \end{pmatrix} \neq 0.$$

Then there exists a sufficiently small domain D' in $\boldsymbol{R}^2$ containing $\boldsymbol{f}(\mathrm{P})$ and a C^∞-map $\boldsymbol{g}\colon D' \to \boldsymbol{R}^2$ such that both $\boldsymbol{g} \circ \boldsymbol{f}$ and $\boldsymbol{f} \circ \boldsymbol{g}$ are the identity maps, that is, $\boldsymbol{g}$ is the inverse map of $\boldsymbol{f}$.

Theorem A.1.6 (The implicit function theorem). *Assume that a C^∞-function $F(x_1, \dots, x_n)$ defined on a domain D of $\boldsymbol{R}^n$ containing $\mathrm{P} = (p_1, \dots, p_n)$ satisfies*

$$F(\mathrm{P}) = 0, \quad \text{and} \quad \frac{\partial F}{\partial x_n}(\mathrm{P}) \neq 0.$$

Then there exists a neighborhood[1] *U of $(p_1, \dots, p_{n-1})$ in $\boldsymbol{R}^{n-1}$ and a C^∞-function f defined on U such that*

$$F(x_1, \dots, x_{n-1}, f(x_1, \dots, x_{n-1})) = 0, \qquad (x_1, \dots, x_{n-1}) \in U.$$

Moreover, the point $\mathrm{Q} := (x_1, \dots, x_n)$ satisfies $F(\mathrm{Q}) = 0$ on a sufficiently small neighborhood of P if and only if $x_n = f(x_1, \dots, x_{n-1})$. In addition,

$$\text{(A.1.4)} \qquad \frac{\partial f}{\partial x_k}(\hat{\mathrm{Q}}) = -\frac{\partial F}{\partial x_k}(\mathrm{Q}) \Big/ \frac{\partial F}{\partial x_n}(\mathrm{Q})$$

holds for each $k = 1, \dots, n-1$, where $\hat{\mathrm{Q}} := (x_1, \dots, x_{n-1})$.

This theorem is used on page 3 in Section 1 and page 60 in Section 6 to establish a criterion for the set of points satisfying $F(x, y) = 0$ or $F(x, y, z) = 0$ to form a smooth curve or smooth surface, respectively.

Change of variables in double integrals. As seen in Problem **5** in Section 6, the following formula for the change of variables in double integrals holds (cf. [36, Theorem 3-17]):

Theorem A.1.7 (Change of variables in double integrals.). *Let $\varphi\colon (\xi, \eta) \mapsto (u(\xi, \eta), v(\xi, \eta))$ be an injective C^1-map from a domain D' in the $\xi\eta$-plane into a domain D in the uv-plane. Let $\overline{\Omega'} \subset D'$ be a bounded closed domain*

[1] A domain of $\boldsymbol{R}^{n-1}$ containing a point is called a *neighborhood* of the point.

and set $\overline{\Omega} = \varphi(\overline{\Omega'})$. *Then, for an integrable function* $f(u,v)$ *on the domain* D *containing* $\overline{\Omega}$, *it holds that*

$$\iint_{\overline{\Omega}} f(u,v)\,du\,dv = \iint_{\overline{\Omega'}} f(u(\xi,\eta), v(\xi,\eta))|\det J|\,d\xi\,d\eta.$$

Here, $\det J$ *is the* Jacobian, *that is, the determinant of the* Jacobi matrix

$$J := \begin{pmatrix} u_\xi & u_\eta \\ v_\xi & v_\eta \end{pmatrix}$$

of the variable change φ, *and* $|\det J|$ *is the absolute value of the Jacobian.*

A.2. The fundamental theorems for ordinary differential equations

For a given n-tuple of functions of $(n+1)$-variables

$$f_i = f_i(t; x_1, \dots, x_n) \qquad (i = 1, \dots, n), \tag{A.2.1}$$

we consider the ordinary differential equation

$$\begin{cases} \dfrac{d}{dt} y_i(t) = f_i(t; y_1(t), \dots, y_n(t)), \\ y_i(t_0) = a_i \end{cases} \qquad (i = 1, \dots, n) \tag{A.2.2}$$

with respect to the unknown functions $y_1(t), \dots, y_n(t)$, where t_0, a_1, ..., a_n are given constants. Using the vector notation

$$\boldsymbol{y}(t) := (y_1(t), \dots, y_n(t)), \quad \boldsymbol{a} := (a_1, \dots, a_n), \quad \boldsymbol{f} := (f_1, \dots, f_n),$$

equation (A.2.2) is written in the form

$$\frac{d\boldsymbol{y}}{dt}(t) = \boldsymbol{f}(t, \boldsymbol{y}(t)), \qquad \boldsymbol{y}(t_0) = \boldsymbol{a}. \tag{A.2.3}$$

The following holds (cf. Theorem 1.1 in Chapter II and Theorem 4.1 in Chapter V of [9]):

Theorem A.2.1 (The fundamental theorem of ordinary differential equations). *Fix a constant* $t_0 \in \boldsymbol{R}$ *and a constant vector* $\boldsymbol{b} = (b_1, \dots, b_n) \in \boldsymbol{R}^n$, *and let* $f_i(t; x_1, \dots, x_n)$ $(i = 1, \dots, n)$ *be* C^∞*-functions defined on a domain*

$$\widetilde{D} := \{(t, \boldsymbol{x}) \in \boldsymbol{R}^{n+1} \mid |t - t_0| < \varepsilon, |x_i - b_i| < \rho \ (i = 1, \dots, n)\}$$

in $\boldsymbol{R}^{n+1}$ $(= \boldsymbol{R} \times \boldsymbol{R}^n)$, *where* $\rho > 0$, $\varepsilon > 0$, *and* $\boldsymbol{x} = (x_1, \dots, x_n)$. *Then there exist a neighborhood* W *of* $\boldsymbol{b}$ *in* $\boldsymbol{R}^n$ *and a positive constant* δ *satisfying the following: For any* $\boldsymbol{a} \in W$, *there exists a unique* $\boldsymbol{y}(t) = \boldsymbol{y}(t; \boldsymbol{a})$ *defined on an open interval* $(t_0 - \delta, t_0 + \delta)$ *satisfying* (A.2.3). *Moreover,* $\boldsymbol{y}(t; \boldsymbol{a})$ *is of class* C^∞ *on* $(t_0 - \delta, t_0 + \delta) \times W$.

This theorem is used to show the existence of geodesics in Section 10, and the properties of geodesics as in Lemmas 11.5 and 11.6 in Section 11.

The fundamental theorem for linear ordinary differential equations. Equation (A.2.2) is called a *linear ordinary differential equation* when the functions f_i have the special form

$$f_i(t; y_1, \ldots, y_n) = b_{i1}(t)\, y_1 + \cdots + b_{in}(t)\, y_n \qquad (i = 1, \ldots, n).$$

Here, $b_{ij}(t)$ $(i, j = 1, \ldots, n)$ are n^2 functions defined on an interval. In this case, (A.2.2) can be written as

$$\text{(A.2.4)} \qquad \dot{\boldsymbol{y}}(t) = \boldsymbol{y}(t)B(t), \qquad B(t) := \begin{pmatrix} b_{11}(t) & \ldots & b_{1n}(t) \\ \vdots & \ddots & \vdots \\ b_{n1}(t) & \ldots & b_{nn}(t) \end{pmatrix}.$$

Here we consider the vector $\boldsymbol{y}$ as a row vector.

The reason why equation (A.2.4) is said to be *linear* is that any linear combination $\alpha\boldsymbol{y}_1(t) + \beta\boldsymbol{y}_2(t)$ of solutions $\boldsymbol{y}_1(t)$, $\boldsymbol{y}_2(t)$ of (A.2.4) is also a solution. In addition to the general existence theorem (Theorem A.2.1), this guarantees that the domain of the solution (A.2.1), for a sufficiently small interval containing t, can be extended to any interval in which all of the b_{ij} are defined (cf. [9, Lemma 1.1 in Chapter IV]).

Theorem A.2.2 (The fundamental theorem of linear ordinary differential equations). *Let $b_{ij}(t)$ $(i, j = 1, \ldots, n)$ be n^2 C^∞-functions defined on an interval $[\alpha, \beta]$ containing $t_0 \in \boldsymbol{R}$, and $\boldsymbol{a} = (a_1, \ldots, a_n)$. Then there exists a unique n-tuple $\boldsymbol{y}(t) = (y_1(t), \ldots, y_n(t))$ of C^∞-functions defined on the interval $[\alpha, \beta]$ satisfying* (A.2.4) *and the initial condition $\boldsymbol{y}(t_0) = \boldsymbol{a}$.*

Moreover, suppose that each function b_{ij} is determined by r parameters $u_1, \ldots, u_r$ such that the vectors $\boldsymbol{u} = (u_1, \ldots, u_r)$ run over a domain in $\boldsymbol{R}^r$. More precisely, if

$$b_{ij} = b_{ij}(t; u_1, \ldots, u_r) \qquad (i, j = 1, \ldots, n)$$

are C^∞-functions on $[\alpha, \beta] \times U$, then the solution $\boldsymbol{y}(t) = \boldsymbol{y}(t; \boldsymbol{a}, \boldsymbol{u})$ of (A.2.4) *with the initial condition $\boldsymbol{y}(t_0) = \boldsymbol{a}$ is of class C^∞ on $[\alpha, \beta] \times \boldsymbol{R}^n \times U (\subset \boldsymbol{R}^{n+r+1})$.*

This theorem is used in the proof of the fundamental theorem for curves (Theorems 2.8 and 5.2 in Sections 2 and 5). Also, in the proof of the fundamental theorem of surface theory (Appendix B.10), the smoothness of the solutions with respect to the parameters will be applied.

A.3. Euclidean spaces

In this section, we review the fundamental properties of vectors in Euclidean spaces. For vectors $\boldsymbol{a} = (a_1, \ldots, a_n)$ and $\boldsymbol{b} = (b_1, \ldots, b_n)$ in $\boldsymbol{R}^n$ $(n \geq 2)$, the sum $\boldsymbol{a} + \boldsymbol{b}$ and the scalar multiplication $\lambda\boldsymbol{a}$ $(\lambda \in \boldsymbol{R})$ are defined, and by these operations, $\boldsymbol{R}^n$ is considered as a *vector space.*

The inner product. The *inner product* of vectors $\boldsymbol{a}$ and $\boldsymbol{b}$ is defined by

$$\boldsymbol{a} \cdot \boldsymbol{b} := a_1 b_1 + \cdots + a_n b_n.$$

The inner product is also represented by the matrix multiplication of a row vector and a column vector as follows:

$$\boldsymbol{a} \cdot \boldsymbol{b} = (a_1, \ldots, a_n) \begin{pmatrix} b_1 \\ \vdots \\ b_n \end{pmatrix}. \tag{A.3.1}$$

We then define the *norm* (i.e. the length) of $\boldsymbol{a}$ by

$$|\boldsymbol{a}| := \sqrt{\boldsymbol{a} \cdot \boldsymbol{a}}. \tag{A.3.2}$$

The following assertion holds.

Proposition A.3.1. *Let $\boldsymbol{a}$ and $\boldsymbol{b}$ be vectors in $\boldsymbol{R}^n$. If $\boldsymbol{b} \neq \boldsymbol{0}$, then*

$$\boldsymbol{a} \cdot \boldsymbol{b} \leq |\boldsymbol{a}||\boldsymbol{b}|. \tag{A.3.3}$$

Equality holds if and only if there exists a non-negative number c such that $\boldsymbol{a} = c\boldsymbol{b}$.

Proof. Consider a real quadratic polynomial

$$f(t) := |\boldsymbol{b}|^2 t^2 - 2(\boldsymbol{a} \cdot \boldsymbol{b})t + |\boldsymbol{a}|^2 = (\boldsymbol{a} - t\boldsymbol{b}) \cdot (\boldsymbol{a} - t\boldsymbol{b}) = |\boldsymbol{a} - t\boldsymbol{b}|^2$$

in t. Since $f(t) \geq 0$ for all t, the discriminant of f is non-positive, giving the inequality. The equality holds only when there exists a number $t = c$ such that $f(c) = 0$. Then $\boldsymbol{a} - c\boldsymbol{b} = \boldsymbol{0}$. Note that if $c < 0$, then $\boldsymbol{a} \cdot \boldsymbol{b}$ is negative and (A.3.3) obviously becomes a strict inequality, so we can conclude $c \geq 0$. □

Corollary A.3.2 (The Schwarz inequality). *Let $\boldsymbol{a}$ and $\boldsymbol{b}$ be vectors in $\boldsymbol{R}^n$. Then*

$$|\boldsymbol{a} \cdot \boldsymbol{b}| \leq |\boldsymbol{a}||\boldsymbol{b}|. \tag{A.3.4}$$

Proof. Replacing $\boldsymbol{b}$ in Proposition A.3.1 by $-\boldsymbol{b}$, we get $-\boldsymbol{a} \cdot \boldsymbol{b} \leq |\boldsymbol{a}||\boldsymbol{b}|$. □

As a consequence of the inequality, we can write

$$\boldsymbol{a} \cdot \boldsymbol{b} = |\boldsymbol{a}||\boldsymbol{b}| \cos\theta \qquad (0 \leq \theta \leq \pi), \tag{A.3.5}$$

where θ denotes the *angle* between these two vectors.

Positivity and negativity of bases. A family $\{\boldsymbol{a}_1, \dots, \boldsymbol{a}_n\}$ of linearly independent vectors in $\boldsymbol{R}^n$ forms a *basis* of the vector space $\boldsymbol{R}^n$. A basis of $\boldsymbol{R}^n$ is said to be *positive* or *positively oriented* if it can be deformed continuously to $\{\boldsymbol{e}_1, \dots, \boldsymbol{e}_n\}$ while keeping linear independency, where

$$\boldsymbol{e}_1 = \begin{pmatrix} 1 \\ 0 \\ \vdots \\ 0 \end{pmatrix}, \quad \boldsymbol{e}_2 = \begin{pmatrix} 0 \\ 1 \\ \vdots \\ 0 \end{pmatrix}, \quad \dots, \quad \boldsymbol{e}_n = \begin{pmatrix} 0 \\ 0 \\ \vdots \\ 1 \end{pmatrix}.$$

A non-positive basis is said to be *negative* or *negatively oriented*, see Fig. A.3.1, right.

When $n = 2$, we set $\boldsymbol{a} := \boldsymbol{a}_1$ and $\boldsymbol{b} := \boldsymbol{a}_2$. Then the sign of the basis $\{\boldsymbol{a}, \boldsymbol{b}\}$ is that of the determinant

$$\det(\boldsymbol{a}, \boldsymbol{b}) = \begin{vmatrix} a_1 & b_1 \\ a_2 & b_2 \end{vmatrix} = a_1 b_2 - b_1 a_2,$$

where $\boldsymbol{a} = (a_1, a_2)^T$, $\boldsymbol{b} = (b_1, b_2)^T$ are considered as column vectors.

Similarly, when $n = 3$, the sign of the basis $\{\boldsymbol{a}, \boldsymbol{b}, \boldsymbol{c}\}$ is that of the determinant

$$\det(\boldsymbol{a}, \boldsymbol{b}, \boldsymbol{c}) = \begin{vmatrix} a_1 & b_1 & c_1 \\ a_2 & b_2 & c_2 \\ a_3 & b_3 & c_3 \end{vmatrix}$$

$$= a_1 b_2 c_3 + b_1 c_2 a_3 + c_1 a_2 b_3 - a_1 c_2 b_3 - b_1 a_2 c_3 - c_1 b_2 a_3,$$

where $\boldsymbol{a} = (a_1, a_2, a_3)^T$, $\boldsymbol{b} = (b_1, b_2, b_3)^T$, $\boldsymbol{c} = (c_1, c_2, c_3)^T$ (see Fig. A.3.1). Here, A^T denotes the *transposition* of a matrix A.

A square matrix A whose determinant does not vanish is called a *regular matrix*. When $n = 2$ and $A = (\boldsymbol{a}, \boldsymbol{b})$ is regular, we set

$$\text{(A.3.6)} \qquad A^{-1} := \frac{1}{\det A} \begin{pmatrix} b_2 & -b_1 \\ -a_2 & a_1 \end{pmatrix}.$$

On the other hand, if $n = 3$ and $A = (\boldsymbol{a}, \boldsymbol{b}, \boldsymbol{c})$ is regular, we set

$$\text{(A.3.7)} \qquad A^{-1} := \frac{1}{\det A} \begin{pmatrix} b_2 c_3 - c_2 b_3 & c_2 a_3 - a_2 c_3 & a_2 b_3 - b_2 a_3 \\ b_3 c_1 - c_3 b_1 & c_3 a_1 - a_3 c_1 & a_3 b_1 - b_3 a_1 \\ b_1 c_2 - c_1 b_2 & c_1 a_2 - a_1 c_2 & a_1 b_2 - b_1 a_2 \end{pmatrix}^T.$$

The matrix A^{-1} is called the *inverse matrix* of A satisfying

$$\text{(A.3.8)} \qquad A^{-1} A = A A^{-1} = I \qquad (I \text{ is the identity matrix}).$$

If a matrix X satisfies $XA = I$ (resp. $AX = I$), then

$$X = X(AA^{-1}) = (XA)A^{-1} = A^{-1}$$
$$(\text{resp. } X = (A^{-1}A)X = A^{-1}(AX) = A^{-1}),$$

that is, X coincides with the matrix A^{-1}.

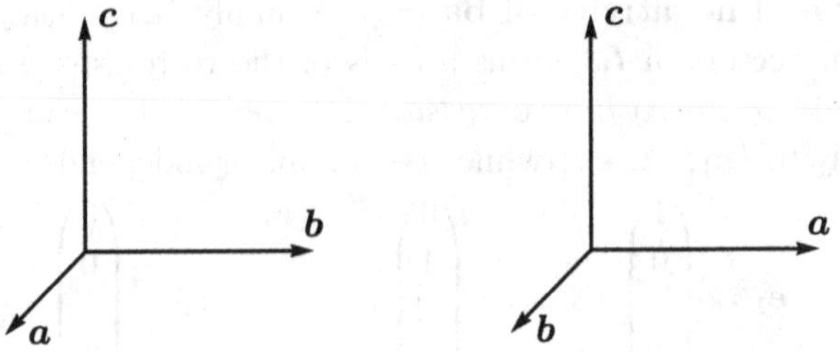

Positive basis. Negative basis.

Fig. A.3.1 The sign of a basis.

Congruent transformations. A transformation of $\boldsymbol{R}^n$ preserving the distance between any two points is called a *congruent transformation.* As seen below, a congruent transformation is represented using a parallel translation and an orthogonal matrix.

An $n \times n$-matrix with real components A is called an *orthogonal matrix* if the transposition A^T of A satisfies

$$A^T A = AA^T = I. \tag{A.3.9}$$

This implies that A^T is the inverse matrix of A. Decomposing A into n column vectors as $A = (\boldsymbol{a}_1, \dots, \boldsymbol{a}_n)$, A is an orthogonal matrix if and only if $\{\boldsymbol{a}_1, \dots, \boldsymbol{a}_n\}$ forms an orthonormal basis, that is,

$$\boldsymbol{a}_i \cdot \boldsymbol{a}_j = \delta_{ij} \qquad (i, j = 1, \dots, n),$$

where

$$\delta_{ij} := \begin{cases} 1 & \text{if } i = j, \\ 0 & \text{if } i \neq j \end{cases} \tag{A.3.10}$$

is the *Kronecker delta.* Taking the determinant of both sides of (A.3.9), one can see that (cf. Problem **1** in Section 2)

$$\det A = \pm 1 \tag{A.3.11}$$

holds for any orthogonal matrix. Here, we consider $\boldsymbol{a}$, $\boldsymbol{b} \in \boldsymbol{R}^n$ as column vectors. Since $\boldsymbol{a} \cdot \boldsymbol{b} = \boldsymbol{a}^T \boldsymbol{b}$ (cf. (A.3.1)), it holds that

$$A\boldsymbol{a} \cdot A\boldsymbol{b} = (A\boldsymbol{a})^T (A\boldsymbol{b}) = \left(\boldsymbol{a}^T A^T\right)(A\boldsymbol{b}) = \boldsymbol{a}^T (A^T A)\boldsymbol{b} = \boldsymbol{a} \cdot \boldsymbol{b} \tag{A.3.12}$$

for an arbitrary orthogonal matrix A and vectors $\boldsymbol{a}$, $\boldsymbol{b} \in \boldsymbol{R}^n$.

We define a map $\Phi \colon \boldsymbol{R}^n \to \boldsymbol{R}^n$ by

$$\Phi(\mathrm{P}) = A \begin{pmatrix} p_1 \\ \vdots \\ p_n \end{pmatrix} + \begin{pmatrix} a_1 \\ \vdots \\ a_n \end{pmatrix}, \qquad \mathrm{P} = \begin{pmatrix} p_1 \\ \vdots \\ p_n \end{pmatrix}, \tag{A.3.13}$$

where A is an orthogonal matrix. Since the distance $d(\mathrm{P},\mathrm{Q})$ between two points P, Q $\in \boldsymbol{R}^n$ is the norm of the vector $\overrightarrow{\mathrm{PQ}}$ with initial point P and terminal point Q, it holds that

$$d(\Phi(\mathrm{P}),\Phi(\mathrm{Q})) = d(\mathrm{P},\mathrm{Q}),$$

that is, the map Φ preserves the distance between the two points. In addition, by (A.3.12), the angle between two vectors is preserved by Φ. In other words, the map Φ in (A.3.13) is a congruent transformation. Conversely, an arbitrary congruent transformation of $\boldsymbol{R}^n$ can be represented as in the form in (A.3.13).

If a congruent transformation (A.3.13) satisfies (cf. (A.3.11)) $\det A = 1$, it is called an *orientation preserving congruent transformation.*

When $n = 2$, the following assertion holds (see Problem **1** in Section 2).

Proposition A.3.3. *If a 2×2-orthogonal matrix has positive determinant, then it gives a rotation about the origin. In particular, an orientation preserving congruent transformation consists of the composition of a rotation and a parallel translation.*

On the other hand, the following assertion holds for $n = 3$.

Proposition A.3.4. *If a 3×3-orthogonal matrix has positive determinant, then it gives a rotation about an axis that is a line passing through the origin. In particular, an orientation preserving congruent transformation consists of the composition of a rotation and a parallel translation.*

Proof. Since the characteristic equation of A has real coefficients and is of odd degree, A has at least one real eigenvalue, say α. Let $\boldsymbol{v}$ be an eigenvector of A with respect to the eigenvalue α. Then we have $A\boldsymbol{v} = \alpha\boldsymbol{v}$. Since A^T is the inverse matrix of A, we have

$$\alpha|\boldsymbol{v}|^2 = (\alpha\boldsymbol{v})\cdot\boldsymbol{v} = (A\boldsymbol{v})^T\boldsymbol{v} = \boldsymbol{v}^TA^T\boldsymbol{v} = \boldsymbol{v}^TA^{-1}\boldsymbol{v} = \boldsymbol{v}^T(\alpha^{-1}\boldsymbol{v}) = \alpha^{-1}|\boldsymbol{v}|^2,$$

that is, $\alpha^2 = 1$ holds. In particular, $\alpha = \pm 1$. Thus, if all of the three eigenvalues of A are real, then 1 is an eigenvalue of A because $\det A = 1$.

On the other hand, if A has a non-real eigenvalue $\beta \in \boldsymbol{C}$, the three eigenvalues of A can be written as

$$\lambda_1 = \alpha, \quad \lambda_2 = \beta, \quad \lambda_3 = \bar{\beta}, \qquad (\alpha \in \{1,-1\},\ \beta \in \boldsymbol{C})$$

because the coefficients of the characteristic polynomial are real. Since $1 = \det A = |\beta|^2\alpha$, we have $\alpha > 0$ and so $\alpha = 1$. So any 3×3-orthogonal matrix with positive determinant has an eigenvector $\boldsymbol{v}$ corresponding to the eigenvalue 1. By a suitable rotation of the axis of $\boldsymbol{R}^3$, we may assume that $\boldsymbol{v} = (0,0,1)^T$. Since $A\boldsymbol{v} = \boldsymbol{v}$ and $A^TA = I$, A can be written as

$$A = \left(\begin{array}{cc|c} \multicolumn{2}{c|}{\multirow{2}{*}{B}} & 0 \\ & & 0 \\ \hline 0 & 0 & 1 \end{array}\right),$$

where B is a 2×2 orthogonal matrix with positive determinant. By Proposition A.3.3 (cf. Problem **1** in Section 2), there exists a real number θ such that

$$B = \begin{pmatrix} \cos\theta & -\sin\theta \\ \sin\theta & \cos\theta \end{pmatrix},$$

which implies that A gives the rotation of angle θ with respect to the line through the origin with direction vector $\boldsymbol{v}$. □

Geometric meanings of determinants. We review here the geometric meanings of determinants of 2×2 and 3×3-matrices.

Let $\boldsymbol{a}, \boldsymbol{b} \in \boldsymbol{R}^2$ be planar vectors, and regard them as column vectors. Then the determinant $\det(\boldsymbol{a}, \boldsymbol{b})$ is equal to the area S of the parallelogram spanned by $\boldsymbol{a}$ and $\boldsymbol{b}$ up to sign (Fig. A.3.2). Here, $\det(\boldsymbol{a}, \boldsymbol{b}) > 0$ holds if and only if $\boldsymbol{b}$ points to the left side of $\boldsymbol{a}$, and $\det(\boldsymbol{a}, \boldsymbol{b}) < 0$ holds if $\boldsymbol{b}$ points to the right side of $\boldsymbol{a}$.

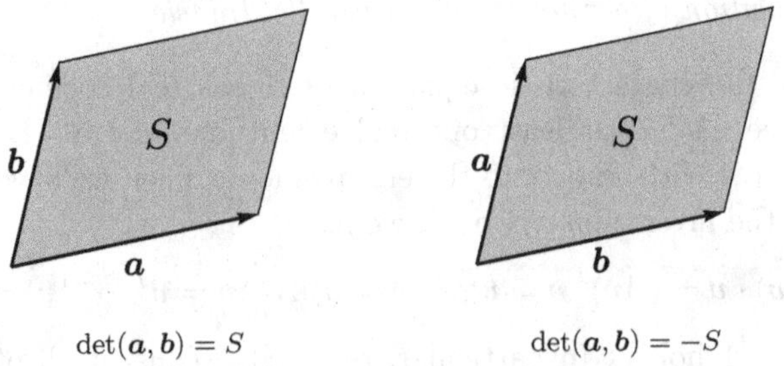

Fig. A.3.2 The area of a parallelogram.

Similarly, for spacial vectors $\boldsymbol{a}$, $\boldsymbol{b}$ and $\boldsymbol{c}$, the determinant $\det(\boldsymbol{a}, \boldsymbol{b}, \boldsymbol{c})$ is zero if and only if $\{\boldsymbol{a}, \boldsymbol{b}, \boldsymbol{c}\}$ are linearly dependent. Otherwise, the sign of this determinant coincides with that of the basis $\{\boldsymbol{a}, \boldsymbol{b}, \boldsymbol{c}\}$. Moreover, the absolute value of the determinant equals the volume V of the parallelepiped spanned by $\boldsymbol{a}$, $\boldsymbol{b}$ and $\boldsymbol{c}$ (Fig. A.3.3). This fact can be verified in the follow-

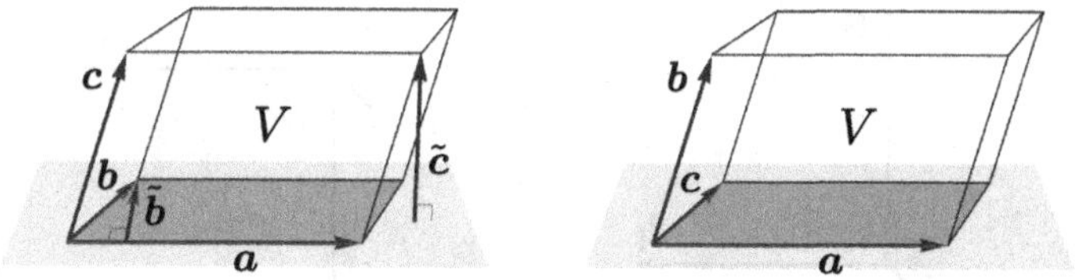

$\det(\boldsymbol{a}, \boldsymbol{b}, \boldsymbol{c}) = V$ $\qquad$ $\det(\boldsymbol{a}, \boldsymbol{b}, \boldsymbol{c}) = -V$

Fig. A.3.3 The volume of parallelepipeds and the determinant.

ing way: Take the vector $\tilde{\boldsymbol{b}}$ with initial point the foot of the perpendicular from the terminal point of $\boldsymbol{b}$ to $\boldsymbol{a}$, and with terminal point the same as that of $\boldsymbol{b}$. Then these vectors lie in the plane spanned by $\boldsymbol{a}$ and $\boldsymbol{b}$. Since $\boldsymbol{b} - \tilde{\boldsymbol{b}}$ is a scalar multiple of $\boldsymbol{a}$,

$$\det(\boldsymbol{a}, \boldsymbol{b}, \boldsymbol{c}) = \det(\boldsymbol{a}, \tilde{\boldsymbol{b}}, \boldsymbol{c})$$

holds. Next, let $\tilde{\boldsymbol{c}}$ be the vector whose initial point is the foot of the perpendicular from the terminal point of $\boldsymbol{c}$ to the plane spanned by $\boldsymbol{a}$ and $\boldsymbol{b}$, and whose terminal point is that of $\boldsymbol{c}$. Then the difference of $\boldsymbol{c}$ and $\tilde{\boldsymbol{c}}$ is a linear combination of $\boldsymbol{a}$ and $\boldsymbol{b}$. Hence we have

$$\det(\boldsymbol{a}, \tilde{\boldsymbol{b}}, \boldsymbol{c}) = \det(\boldsymbol{a}, \tilde{\boldsymbol{b}}, \tilde{\boldsymbol{c}}) = |\boldsymbol{a}|\,|\tilde{\boldsymbol{b}}|\,|\tilde{\boldsymbol{c}}| \det\left(\frac{\boldsymbol{a}}{|\boldsymbol{a}|}, \frac{\tilde{\boldsymbol{b}}}{|\tilde{\boldsymbol{b}}|}, \frac{\tilde{\boldsymbol{c}}}{|\tilde{\boldsymbol{c}}|}\right).$$

Here, $\{\boldsymbol{a}/|\boldsymbol{a}|, \tilde{\boldsymbol{b}}/|\tilde{\boldsymbol{b}}|, \tilde{\boldsymbol{c}}/|\tilde{\boldsymbol{c}}|\}$ is an orthonormal basis; the determinant of the orthogonal matrix corresponding to this basis is ± 1 (cf. (A.3.11)), and furthermore the sign of the determinant coincides with that of the basis $\{\boldsymbol{a}, \boldsymbol{b}, \boldsymbol{c}\}$. The product $|\boldsymbol{a}|\,|\tilde{\boldsymbol{b}}|$ is the area of the parallelogram spanned by $\boldsymbol{a}$ and $\boldsymbol{b}$; and $|\tilde{\boldsymbol{c}}|$ is the height of the parallelepiped with respect to the base plane spanned by $\boldsymbol{a}$ and $\boldsymbol{b}$. Thus, $|\boldsymbol{a}|\,|\tilde{\boldsymbol{b}}|\,|\tilde{\boldsymbol{c}}|$ is the volume of the parallelepiped.

The vector product. We define here the *vector product*, or the *cross product* $\boldsymbol{a} \times \boldsymbol{b}$, of two vectors $\boldsymbol{a}$, $\boldsymbol{b} \in \boldsymbol{R}^3$. This is used to define the binormal vectors for space curves in Section 5, and the unit normal vectors of surfaces in Section 6.

Firstly, the norm of $\boldsymbol{a} \times \boldsymbol{b}$ is defined to be the area of the parallelogram spanned by $\boldsymbol{a}$ and $\boldsymbol{b}$. In other words,

(A.3.14) $$|\boldsymbol{a} \times \boldsymbol{b}| = |\boldsymbol{a}||\boldsymbol{b}| \sin\theta \qquad (0 \le \theta \le \pi)$$

holds, where θ is the angle between the two vectors $\boldsymbol{a}$ and $\boldsymbol{b}$.

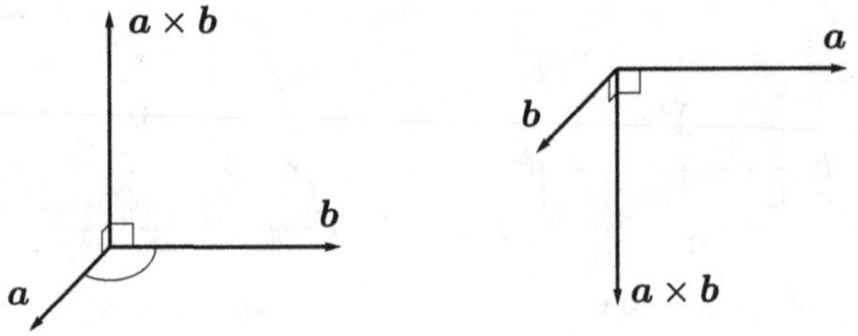

Fig. A.3.4 The vector product.

Next, the direction of $\boldsymbol{a} \times \boldsymbol{b}$ is defined to be perpendicular to the plane spanned by $\boldsymbol{a}$ and $\boldsymbol{b}$, so that $\{\boldsymbol{a}, \boldsymbol{b}, \boldsymbol{a} \times \boldsymbol{b}\}$ forms a positive basis (Fig. A.3.4).

When $\boldsymbol{a}$ and $\boldsymbol{b}$ are linearly dependent, we define $\boldsymbol{a} \times \boldsymbol{b} = \boldsymbol{0}$. Two non-zero vectors $\boldsymbol{a}$ and $\boldsymbol{b}$ are linearly independent if and only if $\boldsymbol{a} \times \boldsymbol{b} \neq \boldsymbol{0}$. By definition, it holds that

$$\boldsymbol{a} \times \boldsymbol{b} = -\boldsymbol{b} \times \boldsymbol{a}.$$

By (A.3.14), we have the following *Lagrange's identity*[1]:

$$|\boldsymbol{a} \times \boldsymbol{b}|^2 = |\boldsymbol{a}|^2|\boldsymbol{b}|^2 - (\boldsymbol{a} \cdot \boldsymbol{b})^2. \tag{A.3.15}$$

This identity is used on page 73 in Section 7 to express the area of surfaces in terms of the first fundamental forms.

As will be shown later, the vector product of vectors $\boldsymbol{a} = (a_1, a_2, a_3)$, $\boldsymbol{b} = (b_1, b_2, b_3)$ is calculated as

$$\boldsymbol{a} \times \boldsymbol{b} = \left(\begin{vmatrix} a_2 & b_2 \\ a_3 & b_3 \end{vmatrix}, - \begin{vmatrix} a_1 & b_1 \\ a_3 & b_3 \end{vmatrix}, \begin{vmatrix} a_1 & b_1 \\ a_2 & b_2 \end{vmatrix} \right). \tag{A.3.16}$$

The following identity for the *scalar triple product* is used to show the formula (5.7) for the torsion of space curves, and the formula (10.16) for the geodesic curvature.

Proposition A.3.5 (The scalar triple product). *It holds that*

$$(\boldsymbol{a} \times \boldsymbol{b}) \cdot \boldsymbol{c} = \det(\boldsymbol{a}, \boldsymbol{b}, \boldsymbol{c}), \tag{A.3.17}$$

for three vectors $\boldsymbol{a}$, $\boldsymbol{b}$, $\boldsymbol{c} \in \boldsymbol{R}^3$. *In particular,*

$$(\boldsymbol{a} \times \boldsymbol{b}) \cdot \boldsymbol{c} = (\boldsymbol{b} \times \boldsymbol{c}) \cdot \boldsymbol{a} = (\boldsymbol{c} \times \boldsymbol{a}) \cdot \boldsymbol{b} \tag{A.3.18}$$

holds.

[1] Lagrange, Joseph-Louis (1736–1813).

Proof. When $\boldsymbol{a}$, $\boldsymbol{b}$ are linearly dependent, both sides of (A.3.17) are $\boldsymbol{0}$, because $\boldsymbol{a} \times \boldsymbol{b} = \boldsymbol{0}$, and the conclusion follows.

Assume $\boldsymbol{a}$ and $\boldsymbol{b}$ are linearly independent. In this case, one can take the unique unit vector $\boldsymbol{n}$ perpendicular to the parallelogram spanned by $\boldsymbol{a}$, $\boldsymbol{b}$, so that $\{\boldsymbol{a}, \boldsymbol{b}, \boldsymbol{n}\}$ forms a positive basis. The absolute value $|\boldsymbol{n} \cdot \boldsymbol{c}|$ is the height of the parallelepiped spanned by $\boldsymbol{a}$, $\boldsymbol{b}$ and $\boldsymbol{c}$ with respect to the base plane spanned by $\boldsymbol{a}$ and $\boldsymbol{b}$. Since

$$(\boldsymbol{a} \times \boldsymbol{b}) \cdot \boldsymbol{c} = |\boldsymbol{a} \times \boldsymbol{b}|(\boldsymbol{c} \cdot \boldsymbol{n}),$$

$|(\boldsymbol{a} \times \boldsymbol{b}) \cdot \boldsymbol{c}|$ is the volume V of the parallelepiped spanned by $\boldsymbol{a}$, $\boldsymbol{b}$, $\boldsymbol{c}$, and we have

$$|(\boldsymbol{a} \times \boldsymbol{b}) \cdot \boldsymbol{c}| = V = |\det(\boldsymbol{a}, \boldsymbol{b}, \boldsymbol{c})|.$$

On the other hand, the sign of $\det(\boldsymbol{a}, \boldsymbol{b}, \boldsymbol{c})$ is that of the basis $\{\boldsymbol{a}, \boldsymbol{b}, \boldsymbol{c}\}$, which coincides with the sign of the inner product $\boldsymbol{c} \cdot \boldsymbol{n}$. The conclusion follows. □

The vector product $\boldsymbol{a} \times \boldsymbol{b}$ can be expressed as a linear combination of the orthonormal basis $\{\boldsymbol{e}_1, \boldsymbol{e}_2, \boldsymbol{e}_3\}$ by

$$\begin{aligned}\boldsymbol{a} \times \boldsymbol{b} &= (\boldsymbol{e}_1 \cdot (\boldsymbol{a} \times \boldsymbol{b}))\boldsymbol{e}_1 + (\boldsymbol{e}_2 \cdot (\boldsymbol{a} \times \boldsymbol{b}))\boldsymbol{e}_2 + (\boldsymbol{e}_3 \cdot (\boldsymbol{a} \times \boldsymbol{b}))\boldsymbol{e}_3 \\ &= \begin{vmatrix} 1 & a_1 & b_1 \\ 0 & a_2 & b_2 \\ 0 & a_3 & b_3 \end{vmatrix} \boldsymbol{e}_1 + \begin{vmatrix} 0 & a_1 & b_1 \\ 1 & a_2 & b_2 \\ 0 & a_3 & b_3 \end{vmatrix} \boldsymbol{e}_2 + \begin{vmatrix} 0 & a_1 & b_1 \\ 0 & a_2 & b_2 \\ 1 & a_3 & b_3 \end{vmatrix} \boldsymbol{e}_3.\end{aligned}$$

Hence we have the algebraic formula (A.3.16) for the vector product. Equation (A.3.16) can be expressed formally by

$$\begin{aligned}\boldsymbol{a} \times \boldsymbol{b} &= \begin{vmatrix} a_2 & b_2 \\ a_3 & b_3 \end{vmatrix} \boldsymbol{e}_1 - \begin{vmatrix} a_1 & b_1 \\ a_3 & b_3 \end{vmatrix} \boldsymbol{e}_2 + \begin{vmatrix} a_1 & b_1 \\ a_2 & b_2 \end{vmatrix} \boldsymbol{e}_3 \\ &= \begin{vmatrix} \boldsymbol{e}_1 & a_1 & b_1 \\ \boldsymbol{e}_2 & a_2 & b_2 \\ \boldsymbol{e}_3 & a_3 & b_3 \end{vmatrix} = \begin{vmatrix} \boldsymbol{e}_1 & \boldsymbol{e}_2 & \boldsymbol{e}_3 \\ a_1 & a_2 & a_3 \\ b_1 & b_2 & b_3 \end{vmatrix}.\end{aligned}$$

By (A.3.16), one can show the following distributive law for vector products:

$$(\alpha \boldsymbol{a} + \beta \boldsymbol{b}) \times \boldsymbol{c} = \alpha(\boldsymbol{a} \times \boldsymbol{c}) + \beta(\boldsymbol{b} \times \boldsymbol{c}), \tag{A.3.19}$$

where $\boldsymbol{a}$, $\boldsymbol{b}$, $\boldsymbol{c} \in \boldsymbol{R}^3$ and α, $\beta \in \boldsymbol{R}$. By Proposition A.3.5, we have:

Corollary A.3.6. *For arbitrary vectors $\boldsymbol{a}$, $\boldsymbol{b} \in \boldsymbol{R}^3$ and an orthogonal matrix A, it holds that*

$$A\boldsymbol{a} \times A\boldsymbol{b} = (\det A)A(\boldsymbol{a} \times \boldsymbol{b}).$$

In particular, $A\boldsymbol{a} \times A\boldsymbol{b} = A(\boldsymbol{a} \times \boldsymbol{b})$ holds if $\det A = 1$.

Proof. For an arbitrary vector $\boldsymbol{c} \in \boldsymbol{R}^3$, we set $\tilde{\boldsymbol{c}} = A^{-1}\boldsymbol{c}$. Then we have

$$\begin{aligned}(A\boldsymbol{a} \times A\boldsymbol{b}) \cdot \boldsymbol{c} &= (A\boldsymbol{a} \times A\boldsymbol{b}) \cdot A\tilde{\boldsymbol{c}} = \det(A\boldsymbol{a}, A\boldsymbol{b}, A\tilde{\boldsymbol{c}}) = \det(A(\boldsymbol{a}, \boldsymbol{b}, \tilde{\boldsymbol{c}})) \\ &= (\det A)\det(\boldsymbol{a}, \boldsymbol{b}, \tilde{\boldsymbol{c}}) = (\det A)(\boldsymbol{a} \times \boldsymbol{b} \cdot \tilde{\boldsymbol{c}}) \\ &= (\det A)(A(\boldsymbol{a} \times \boldsymbol{b}) \cdot A\tilde{\boldsymbol{c}}) = (\det A)(A(\boldsymbol{a} \times \boldsymbol{b})) \cdot \boldsymbol{c}.\end{aligned}$$

Since $\boldsymbol{c}$ is arbitrary, the conclusion follows. □

The following formula holds for the vector product of three vectors:

Proposition A.3.7 (The vector triple product)**.** *It holds that*

$$\boldsymbol{a} \times (\boldsymbol{b} \times \boldsymbol{c}) = (\boldsymbol{a} \cdot \boldsymbol{c})\boldsymbol{b} - (\boldsymbol{a} \cdot \boldsymbol{b})\boldsymbol{c}$$

for arbitrary vectors $\boldsymbol{a}$, $\boldsymbol{b}$, $\boldsymbol{c} \in \boldsymbol{R}^3$.

Proof. First, we shall prove

$$\boldsymbol{p} \times (\boldsymbol{p} \times \boldsymbol{q}) = (\boldsymbol{p} \cdot \boldsymbol{q})\boldsymbol{p} - |\boldsymbol{p}|^2\boldsymbol{q} \tag{A.3.20}$$

for arbitrary vectors $\boldsymbol{p}$, $\boldsymbol{q} \in \boldsymbol{R}^3$. In fact, if $\boldsymbol{p}$ and $\boldsymbol{q}$ are linearly dependent, both sides of (A.3.20) are $\boldsymbol{0}$. On the other hand, if $\boldsymbol{p}$ and $\boldsymbol{q}$ are linearly independent, $\boldsymbol{r} := \boldsymbol{p} \times (\boldsymbol{p} \times \boldsymbol{q})$ is perpendicular to $\boldsymbol{p} \times \boldsymbol{q}$. Here $\boldsymbol{p} \times \boldsymbol{q}$ is perpendicular to the plane spanned by $\boldsymbol{p}$ and $\boldsymbol{q}$. Thus, $\boldsymbol{r}$ can be expressed as a linear combination of $\boldsymbol{p}$ and $\boldsymbol{q}$. So we can write

$$\boldsymbol{p} \times (\boldsymbol{p} \times \boldsymbol{q}) = \alpha\boldsymbol{p} + \beta\boldsymbol{q} \qquad (\alpha, \beta \in \boldsymbol{R}). \tag{A.3.21}$$

Taking the inner product with $\boldsymbol{p}$ on both sides of (A.3.21), we have

$$(\boldsymbol{p} \times (\boldsymbol{p} \times \boldsymbol{q})) \cdot \boldsymbol{p} = \alpha|\boldsymbol{p}|^2 + \beta(\boldsymbol{p} \cdot \boldsymbol{q}).$$

The left-hand side of this is calculated by the formula for scalar triple products ((A.3.18) in Proposition A.3.5) as

$$\boldsymbol{p} \times (\boldsymbol{p} \times \boldsymbol{q}) \cdot \boldsymbol{p} = (\boldsymbol{p} \times \boldsymbol{p}) \cdot (\boldsymbol{p} \times \boldsymbol{q}) = 0.$$

Hence we have $\alpha|\boldsymbol{p}|^2 + \beta(\boldsymbol{p} \cdot \boldsymbol{q}) = 0$. Similarly, taking the inner product with $\boldsymbol{q}$ on both sides of (A.3.21),

$$\alpha(\boldsymbol{p} \cdot \boldsymbol{q}) + \beta|\boldsymbol{p}|^2 = |\boldsymbol{p} \times \boldsymbol{q}|^2 = |\boldsymbol{p}|^2|\boldsymbol{q}|^2 - (\boldsymbol{p} \cdot \boldsymbol{q})^2$$

by the scalar triple product and Lagrange's identity (A.3.15). So we obtain $\alpha = \boldsymbol{p} \cdot \boldsymbol{q}$ and $\beta = -|\boldsymbol{p}|^2$, proving (A.3.20).

We shall prove the conclusion using (A.3.20): The conclusion is obvious when $\boldsymbol{b}$ and $\boldsymbol{c}$ are linearly dependent. So we assume $\boldsymbol{b}$ and $\boldsymbol{c}$ are linearly independent. Then, in a similar way as above, one can see that $\boldsymbol{a} \times (\boldsymbol{b} \times \boldsymbol{c})$ is expressed as a linear combination of $\boldsymbol{b}$ and $\boldsymbol{c}$:

$$\boldsymbol{a} \times (\boldsymbol{b} \times \boldsymbol{c}) = \gamma \boldsymbol{b} + \delta \boldsymbol{c} \qquad (\gamma, \delta \in \boldsymbol{R}). \tag{A.3.22}$$

Here, by (A.3.20) and the scalar triple product, we have

$$\begin{aligned}(\boldsymbol{a} \times (\boldsymbol{b} \times \boldsymbol{c})) \cdot \boldsymbol{b} &= (\boldsymbol{b} \times \boldsymbol{a}) \cdot (\boldsymbol{b} \times \boldsymbol{c}) = -(\boldsymbol{b} \times (\boldsymbol{b} \times \boldsymbol{a})) \cdot \boldsymbol{c} \\ &= (-(\boldsymbol{b} \cdot \boldsymbol{a})\boldsymbol{b} + |\boldsymbol{b}|^2 \boldsymbol{a}) \cdot \boldsymbol{c} = -(\boldsymbol{b} \cdot \boldsymbol{a})(\boldsymbol{b} \cdot \boldsymbol{c}) + |\boldsymbol{b}|^2 (\boldsymbol{a} \cdot \boldsymbol{c}), \\ (\boldsymbol{a} \times (\boldsymbol{b} \times \boldsymbol{c})) \cdot \boldsymbol{c} &= (\boldsymbol{b} \times \boldsymbol{c}) \cdot (\boldsymbol{c} \times \boldsymbol{a}) = (\boldsymbol{c} \times (\boldsymbol{c} \times \boldsymbol{a})) \cdot \boldsymbol{b} \\ &= ((\boldsymbol{c} \cdot \boldsymbol{a})\boldsymbol{c} - |\boldsymbol{c}|^2 \boldsymbol{a}) \cdot \boldsymbol{b} = (\boldsymbol{c} \cdot \boldsymbol{a})(\boldsymbol{c} \cdot \boldsymbol{b}) - |\boldsymbol{c}|^2 (\boldsymbol{a} \cdot \boldsymbol{b}).\end{aligned}$$

So taking the inner product with $\boldsymbol{b}$ (resp. $\boldsymbol{c}$) on both sides of (A.3.22), we have

$$\begin{aligned}|\boldsymbol{b}|^2 \gamma + (\boldsymbol{b} \cdot \boldsymbol{c})\delta &= -(\boldsymbol{b} \cdot \boldsymbol{a})(\boldsymbol{b} \cdot \boldsymbol{c}) + |\boldsymbol{b}|^2 (\boldsymbol{a} \cdot \boldsymbol{c}), \\ (\boldsymbol{b} \cdot \boldsymbol{c})\gamma + |\boldsymbol{c}|^2 \delta &= (\boldsymbol{c} \cdot \boldsymbol{a})(\boldsymbol{c} \cdot \boldsymbol{b}) - |\boldsymbol{c}|^2 (\boldsymbol{a} \cdot \boldsymbol{b}).\end{aligned}$$

Hence we have $\gamma = \boldsymbol{a} \cdot \boldsymbol{c}$ and $\delta = -\boldsymbol{a} \cdot \boldsymbol{b}$, which is the conclusion. □

The following identity is easily obtained from the vector triple product:

Corollary A.3.8 (The Jacobi identity). *It holds that*

$$\boldsymbol{a} \times (\boldsymbol{b} \times \boldsymbol{c}) + \boldsymbol{b} \times (\boldsymbol{c} \times \boldsymbol{a}) + \boldsymbol{c} \times (\boldsymbol{a} \times \boldsymbol{b}) = \boldsymbol{0}$$

for arbitrary $\boldsymbol{a}$, $\boldsymbol{b}$, $\boldsymbol{c} \in \boldsymbol{R}^3$.

By Corollary A.3.8, the vector space $\boldsymbol{R}^3$ is considered as a *Lie algebra* (cf. [40]). In addition, we can see that *the associative law does not hold for the vector product.* In fact, since $\boldsymbol{a} \times (\boldsymbol{b} \times \boldsymbol{c}) = (\boldsymbol{a} \times \boldsymbol{b}) \times \boldsymbol{c} + \boldsymbol{b} \times (\boldsymbol{a} \times \boldsymbol{c})$, if $\boldsymbol{b}$ and $\boldsymbol{a} \times \boldsymbol{c}$ are linearly independent, then $\boldsymbol{a} \times (\boldsymbol{b} \times \boldsymbol{c}) \neq (\boldsymbol{a} \times \boldsymbol{b}) \times \boldsymbol{c}$.

Differentiation of matrix-valued functions. Let $A(t)$ be an $m \times n$-matrix

$$A(t) = \begin{pmatrix} a_{11}(t) & \dots & a_{1n}(t) \\ \vdots & \ddots & \vdots \\ a_{m1}(t) & \dots & a_{mn}(t) \end{pmatrix}$$

whose components are differentiable functions $a_{ij}(t)$ $(1 \leq i \leq m, 1 \leq j \leq n)$ of t. We define the *derivative* of $A(t)$ as

$$\dot{A}(t) = \frac{d}{dt} A(t) = \begin{pmatrix} \dot{a}_{11}(t) & \dots & \dot{a}_{1n}(t) \\ \vdots & \ddots & \vdots \\ \dot{a}_{m1}(t) & \dots & \dot{a}_{mn}(t) \end{pmatrix} \qquad \left(\dot{} = \frac{d}{dt} \right).$$

Proposition A.3.9 (The product rule). *Suppose* $A(t) = (a_{ij}(t))$ *and* $B(t) = (b_{ij}(t))$ *are an* $m \times k$*-matrix and a* $k \times n$*-matrix, respectively, whose components are all differentiable functions. Then*

$$\frac{d}{dt}(A(t)B(t)) = \dot{A}(t)B(t) + A(t)\dot{B}(t).$$

Proof. The product $C(t) := A(t)B(t)$ is an $m \times n$-matrix whose (i,j)-component is $c_{ij}(t) := \sum_{s=1}^{k} a_{is}(t)b_{sj}(t)$. Then we have

$$\dot{c}_{ij}(t) = \left(\sum_{s=1}^{k} \dot{a}_{is}(t)b_{sj}(t)\right) + \left(\sum_{s=1}^{k} a_{is}(t)\dot{b}_{sj}(t)\right).$$

Here, the first (resp. second) term of the right-hand side is the (i,j)-component of $\dot{A}(t)B(t)$ (resp. $A(t)\dot{B}(t)$). Thus we have the conclusion. □

Since the inner product of vectors can be expressed by matrix multiplication as in (A.3.1), Proposition A.3.9 implies the following corollary.

Corollary A.3.10 (The differentiation of inner products). *Suppose* $\boldsymbol{a}(t)$ *and* $\boldsymbol{b}(t)$ *are vector-valued functions whose components are all differentiable functions in* t*. Then*

$$\frac{d}{dt}(\boldsymbol{a}(t) \cdot \boldsymbol{b}(t)) = \dot{\boldsymbol{a}}(t) \cdot \boldsymbol{b}(t) + \boldsymbol{a}(t) \cdot \dot{\boldsymbol{b}}(t)$$

holds.

Moreover, the following derivation rule for the determinant holds:

Proposition A.3.11. *Let* $A(t)$ *be an* $n \times n$*-matrix-valued differentiable function, with column vectors* $\boldsymbol{a}_1(t), \ldots, \boldsymbol{a}_n(t)$*, that is,*

$$A(t) = (\boldsymbol{a}_1(t), \ldots, \boldsymbol{a}_n(t)).$$

Then

$$\begin{aligned}\frac{d}{dt}(\det A(t)) &= \det(\dot{\boldsymbol{a}}_1(t), \boldsymbol{a}_2(t), \ldots, \boldsymbol{a}_n(t)) \\ &\quad + \det(\boldsymbol{a}_1(t), \dot{\boldsymbol{a}}_2(t), \ldots, \boldsymbol{a}_n(t)) \\ &\quad + \cdots + \det(\boldsymbol{a}_1(t), \boldsymbol{a}_2(t), \ldots, \dot{\boldsymbol{a}}_n(t)).\end{aligned}$$

Proof. By multi-linearity of the determinant with respect to column vectors, it holds that

$$\begin{aligned}
&\frac{1}{h}(\det A(t+h) - \det A(t)) \\
&\quad = \frac{1}{h}\left\{\det(\boldsymbol{a}_1(t+h), \ldots, \boldsymbol{a}_n(t+h)) - \det(\boldsymbol{a}_1(t), \ldots, \boldsymbol{a}_n(t))\right\} \\
&\quad = \det\left(\frac{\boldsymbol{a}_1(t+h) - \boldsymbol{a}_1(t)}{h}, \boldsymbol{a}_2(t+h), \ldots, \boldsymbol{a}_n(t+h)\right) \\
&\qquad + \det\left(\boldsymbol{a}_1(t), \frac{\boldsymbol{a}_2(t+h) - \boldsymbol{a}_2(t)}{h}, \ldots, \boldsymbol{a}_n(t+h)\right) \\
&\qquad + \cdots + \det\left(\boldsymbol{a}_1(t), \boldsymbol{a}_2(t), \ldots, \frac{\boldsymbol{a}_n(t+h) - \boldsymbol{a}_n(t)}{h}\right)
\end{aligned}$$

for sufficiently small h. Letting $h \to 0$, we have the conclusion. □

Appendix B

Advanced Topics on Curves and Surfaces

B.1. Evolutes and the cycloid pendulum

Envelopes. Consider a family of curves $\{C_t\}_{t\in[a,b]}$, that is, for each value of the parameter t running over the interval $[a,b]$, there is a corresponding curve C_t. If there exists a curve σ tangent to C_t for all t, the curve σ is called an *envelope* of the family $\{C_t\}$. For example, let C_t be the line segment in the xy-plane joining $(t,0)$ and $(0,1-t)$ $(0 \le t \le 1)$. Drawing these segments, one can see the envelope σ, as in the left-side figure in Fig. B.1.1.

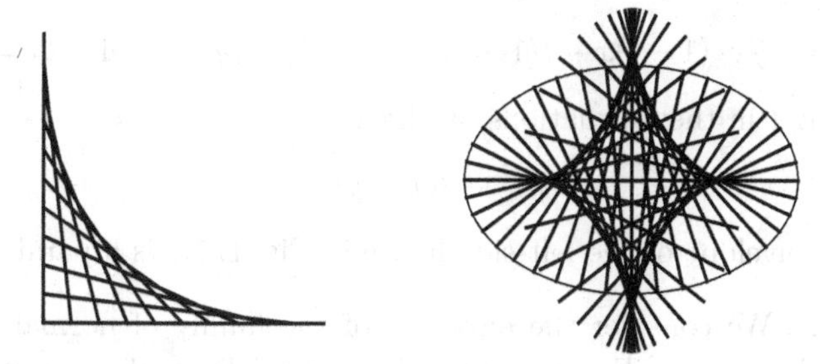

The envelope of a family of line segments. The envelope of the normal lines of an ellipse.

Fig. B.1.1 Envelopes.

If a family $\{C_t\}$ is given by an implicit function $F(x,y,t)=0$, then the system of equations

$$F(x,y,t)=0, \qquad F_t(x,y,t)=\frac{\partial F}{\partial t}(x,y,t)=0$$

gives an envelope. By eliminating the parameter t in these equations, or by solving these with respect to x, y, one can obtain the equation for the envelope.

In fact, assume that a curve $\sigma(t) = (x(t), y(t))$ is tangent to C_t for each t. Then $\sigma(t)$ lies on the curve C_t, that is, $F(x(t), y(t), t) = 0$ holds. Moreover, since σ is tangent to C_t at $\sigma(t)$, the formula for differentiation of implicit functions ((A.1.4) in Appendix A.1) yields

$$\frac{\dot{y}}{\dot{x}} = \frac{dy}{dx} = -\frac{F_x}{F_y}. \tag{B.1.1}$$

On the other hand, $F(x(t), y(t), t) = 0$ holds for all t. Differentiating this with respect to t, we have

$$F_t + F_x\dot{x} + F_y\dot{y} = 0.$$

Also $F_x\dot{x} + F_y\dot{y} = 0$ holds, because of (B.1.1). Hence we have $F_t = 0$.

Conversely, assume a curve $(x(t), y(t))$ satisfies

$$F(x(t), y(t), t) = F_t(x(t), y(t), t) = 0.$$

Then it holds that $F_x\dot{x} + F_y\dot{y} = 0$, so it is tangent to C_t at each t, that is, the curve is the envelope.

Recall the example in the left-side figure in Fig. B.1.1. Since the family of segments is given by $F(x, y, t) = ty + (1-t)x - t(1-t)$, the envelope is given by

$$(F =)ty + (1-t)x - t(1-t) = 0, \quad (F_t =)y - x - 1 + 2t = 0.$$

Eliminating t in these equations, we have

$$(x-y)^2 - 2(x+y) + 1 = 0.$$

Thus, the envelope of the left-side figure in Fig. B.1.1 is a parabola.

Evolutes. We consider the envelope of the family of normal lines of a given regular curve. For example, the normal lines of an ellipse induce the envelope as in the right-hand figure in Fig. B.1.1. The envelope is the locus of the centers of the osculating circles of the ellipse. In general, the following holds:

Theorem B.1.1. *The envelope $\sigma(s)$ of the normal lines of a given curve $\gamma(s)$ is the locus of the centers of the osculating circles:*

$$\sigma(s) := \gamma(s) + \frac{1}{\kappa(s)}\boldsymbol{n}(s), \tag{B.1.2}$$

where $\kappa(s)$ is the curvature and $\boldsymbol{n}(s)$ is the unit normal vector of $\gamma(s)$.

The curve $\sigma(s)$ is called the *evolute* (or *focal curve* or *caustic*) of $\gamma(s)$.

Proof. Assume the curve $\gamma(s) = (x(s), y(s))$ is parametrized by the arc-length s. Then the normal at s is given by the linear equation

$$F(X, Y, s) := x'(s)(X - x(s)) + y'(s)(Y - y(s)) = 0.$$

Since s is the arc-length of γ, we have

$$\frac{\partial F}{\partial s} = x''(s)(X - x(s)) + y''(s)(Y - y(s)) - 1.$$

Solving $F = \partial F/\partial s = 0$ with respect to X and Y, we have

$$X = x - \frac{y'}{x'y'' - x''y'} = x - \frac{y'}{\kappa}, \qquad Y = y + \frac{x'}{x'y'' - x''y'} = y + \frac{x'}{\kappa}.$$

Thus, the envelope $\sigma(s) = (X(s), Y(s))$ is expressed as in (B.1.2). □

Differentiating (B.1.2), we have $\sigma' = (1/\kappa)'\boldsymbol{n}$ by the Frenet formula (2.15) in Section 2. This means that the velocity vector $\sigma'(s)$ of the evolute vanishes where the corresponding point on the original curve $\gamma(s)$ is a vertex. Generically, such a point on the evolute is a cusp point (see page 261). In particular, for a non-circular ellipse, four cusp points appear on the evolute, as shown in the right-side figure of Fig. B.1.1.

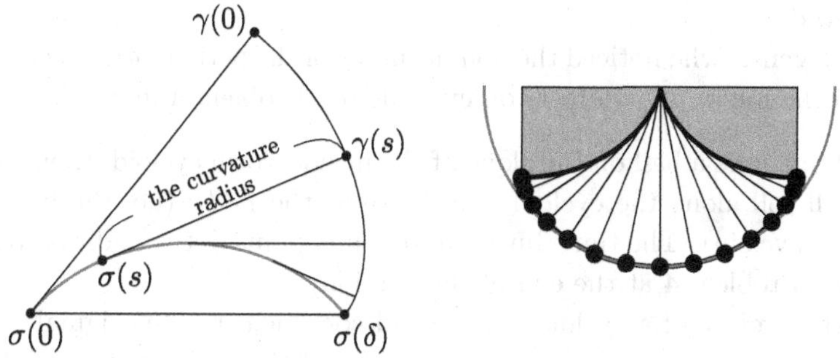

The involute. The cycloid pendulum.

Fig. B.1.2 The involute and the cycloid pendulum.

As an inverse problem, one can find the curve whose evolute is the given curve, using the fact that the evolute is the envelope of the normal lines: Take a board whose border is the curve $\sigma(s)$, and lay out a string along the border between $\sigma(0)$ and $\sigma(\delta)$. Then start to release the string, fixing one end $\sigma(0)$ as shown in the left-side figure in Fig. B.1.2. Let $\gamma(s)$ be the end of the string when the string is straightened out from $\sigma(s)$ onward

$(0 \leq s \leq \delta)$. If s is the arc-length of the curve σ, we have (cf. Fig. B.1.2, left)

$$\gamma(s) = \sigma(s) + (\delta - s)\sigma'(s) \qquad (0 \leq s \leq \delta), \tag{B.1.3}$$

and then $\gamma'(s) = (\delta - s)\sigma''(s)$. Since s is the arc-length of σ, we infer that $\sigma'(s)$ is perpendicular to $\gamma'(s)$. That is, the straight part of the string is the normal direction of γ. Therefore, by Theorem B.1.1, $\sigma(s)$ is the involute of the newly obtained curve $\gamma(s)$ (cf. Problem **2** at the end of this section).

The curve $\gamma(s)$ obtained as above is called the *involute* of $\sigma(s)$. The radius of the osculating circle of the involute $\gamma(s)$ is the length of the straight part of the string $\delta - s$. Note that δ (> 0) can be chosen arbitrarily. That is, the involute of a given curve is not uniquely determined.

The cycloid pendulum. Galilei[1] found that the period of the swing of a pendulum of small amplitude depends only the length of the pendulum, and does not depend on the weight and the amplitude. This property is called *isochronism*. In fact, the period T is expressed by the length l of the pendulum and the acceleration of gravity g as $T \approx 2\pi\sqrt{l/g}$, where "$\approx$" means "approximately equal". Though this formula is approximately correct for small amplitude, it is known that the period is longer for a larger amplitude. So how can one make a pendulum with period independent of amplitude?

Huygens,[2] who noticed the non-accuracy of the period formula of Galilei, found the following two facts on cycloids (cf. Problem **4** in Section 1):

(1) If one puts a ball on the slope of the upside-down cycloid, then the ball will roll along the cycloid and arrive at the *nadir* (i.e. the bottom of the cycloid). The time duration from any point to the nadir is constant (cf. Problem **4** at the end of this section).
(2) The evolute of a cycloid is a cycloid congruent to the original cycloid. In particular, for an appropriate choice of the position, the involute of the cycloid is congruent to the original cycloid (cf. Problem **3** of this section).

Based on these facts, consider the machinery as shown in the right-side figure in Fig. B.1.2. The borders of the two shaded boards are both half of a cycloid. By the property (2), a weight rolling along the left side of the board draws the left half of a cycloid as the string releases until it turns

[1] Galilei, Galileo (1564–1642).
[2] Huygens, Christiaan (1629–1695).

into a vertical segment. It then moves to the right side of the board as the weight travels along the right half of the cycloid. In this way, the weight draws a period of the cycloid. Hence by the property (1), the period T of such a cycloid pendulum does not depend on the amplitude; in fact, the equality $T = 2\pi\sqrt{l/g}$ holds precisely (cf. Problem **4** of this section). Here, l is indeed a radius of the gray circle in the right-side figure in Fig. B.1.2.

The usual (circular) pendulum moves along a circle, and the locus of the usual pendulum and that of the cycloid pendulum almost coincide for small amplitude. In fact, at the nadir, the cycloid and the circle are tangent to the third order (cf. page 16). This shows that Galilei's isochronism of the circular pendulum is a good approximation.

Cycloids also have the property of *brachistochrone* curves. That is, consider a time duration of a mass starting at a point A and traveling along a curve, pulled only by gravity, to another point B lower than A (Fig. B.1.3). It is known that the path minimizing the travel time is the cycloid passing through A with horizontal base line. A proof of this fact is given in Section 18.

Fig. B.1.3 The brachistochrone curve.

Exercises B.1

1 Regard the x-axis and the y-axis as a floor and a wall, respectively, and consider a bar leaning against the wall as a line segment joining $(0, 0)$ and $(0, 1)$. When the bar slips down from the wall to the floor, show that the envelope of the bar is the astroid (cf. Example 1.3 in Section 1).

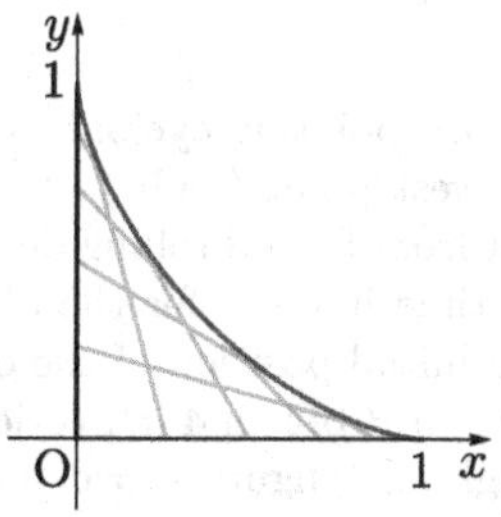

2 Let $\sigma(s)$ $(0 \le s \le l)$ be a curve parametrized by arc-length s and fix δ $(0 < \delta \le l)$. Verify by a direct computation that $\sigma(s)$ is the evolute of

$$\gamma(s) := \sigma(s) + (\delta - s)\sigma'(s) \qquad (0 \le s \le \delta).$$

3 Show that the evolute of a cycloid is a cycloid congruent to the original one. Using this fact, show that the involute of a cycloid, setting δ equal to the half-length of one period of the cycloid as in the figure below, is a cycloid congruent to the original one.

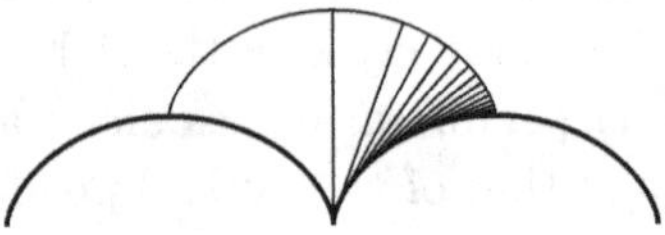

4 (1) Let $\gamma(w) := (u(w), v(w))$ $(b \leq w \leq c)$ be a curve on the uv-plane such that the initial point $\gamma(b)$ lies on the u-axis, and the other points lie in the lower half-plane $\{(u,v) \in \boldsymbol{R}^2 \mid v < 0\}$. Consider an object (a mass point) of mass m starting at $\gamma(b)$ at time $t = 0$ slipping down the slope $\gamma(w)$, pulled only by the gravity $(0, -mg)$ (g is the acceleration of gravity). Show that the time T for the object to arrive at the point $\gamma(c)$ is given by (see the left-side figure below)

$$T = \frac{1}{\sqrt{2g}} \int_b^c \frac{1}{\sqrt{|v|}} \sqrt{\left(\frac{du}{dw}\right)^2 + \left(\frac{dv}{dw}\right)^2}\, dw$$
$$= \frac{1}{\sqrt{2g}} \int_b^c \sqrt{\frac{\dot{u}^2 + \dot{v}^2}{|v|}}\, dw.$$

(Hint: Apply the principle of conservation of mechanical energy.)

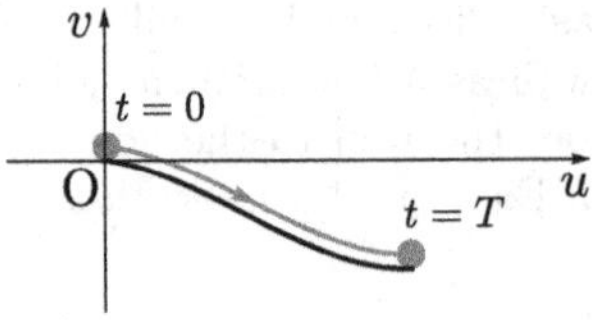

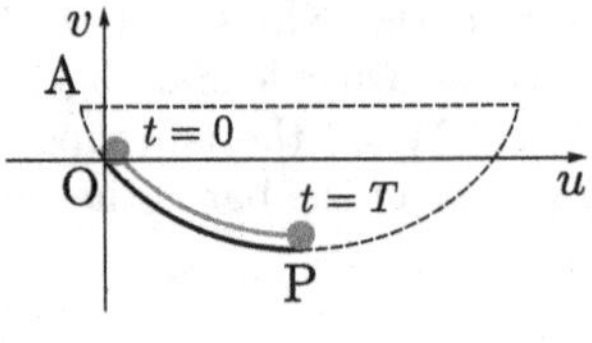

(2) Let C be a downward-pointing cycloid as in the right-side figure above, and let P be its lowest point (nadir). Consider an object starting at a point of C different from P, and rolling down along C with initial velocity $\mathbf{0}$. Show that the time it takes for the object to arrive at P is $\pi\sqrt{a/g}$, independent of the initial position of the object, where a is the radius of the turning circle as in Problem **4** in Section 1. (Hint: Set the coordinate system as in the right-side figure above, and let b be the length of the arc AO of the cycloid, and then parametrize the cycloid by the arc-length with initial point A.)

(3) The distance between Tokyo and Osaka is about 403 km. Assume the two cities are joined by an underground cycloid tunnel, and start with a ball in Tokyo. Compute the time duration for the ball to arrive in Osaka.

B.2. Convex curves and curves of constant width

Characterizations of convex curves. In this section, we prove the characterization of convex curves as in Theorem B.2.1. This theorem is used in the proof of the four-vertex theorem for a general simple closed regular curve (not necessarily a convex curve, Theorem 4.4 in Section 4).

Theorem B.2.1. *Let $\gamma(s)$ $(0 \leq s \leq l)$ be a simple closed regular curve, and D the open region (not including boundary points) bounded by the curve. Then the following five assertions are mutually equivalent:*

(1) *The curvature function of γ does not change sign.*
(2) *For each oriented line d, there exists a unique tangent line $\tilde{d}$ to the curve γ which is parallel to d in the same direction. Moreover, the set of the tangent points of the curve γ and the line $\tilde{d}$ consists of either a single point or a line segment.*
(3) *There is no intersection of D with any tangent line of γ. In particular, D lies on one side of any tangent line.*
(4) *Each line segment joining two points in D lies in D.*
(5) *Each line segment joining two points of $\overline{D}$ lies in $\overline{D}$, where $\overline{D}$ is the closure of D, that is, the union of D and its boundary points. In particular, $\gamma(s)$ is a convex curve in the sense of Section 2.*

Proof. Without loss of generality we may assume that D lies on the left-side of the curve.

(1)⇒(2): By Theorems 3.2 and 3.3 in Section 3, a simple closed regular curve can be deformed to the circle continuously. As we assumed that D lies on the left-side of the curve, γ turns in the counterclockwise direction. Since the rotation index does not change during the deformation, we deduce that the rotation index of γ is 1. Then $\kappa \geq 0$ because the curvature function $\kappa(s)$ does not change sign by the assumption (1). Here, $\gamma(s)$ can be expressed as (2.19) in Section 2 using the curvature function $\kappa(s)$, where s is the arc-length parameter of γ. Then the function

$$\theta(s) = \int_0^s \kappa(u)\, du$$

gives the angle between $\gamma'(s)$ and $\gamma'(0)$, because of (3.1) in page 29. Since $\kappa \geq 0$, the angle $\theta(s)$ simply increases from 0 to 2π as s runs over $0 \leq s \leq l$. Hence there exists the desired line $\tilde{d}$ in (2), by the intermediate value theorem.

On the other hand, assume that the tangents of the curve at $\gamma(s_1)$ and $\gamma(s_2)$ are parallel to the line d. Since $\theta(s)$ does not increase on the interval $[s_1, s_2]$, the two tangents are identical, that is, we have proved the uniqueness of $\tilde{d}$. Hence $\gamma([s_1, s_2])$ is a line segment on the tangent line $\tilde{d}$, which proves (2).

(2)⇒(3): Take a point P on the curve γ, and let $\tilde{d}$ be the tangent line of γ at P. Translating a line d from far away toward the image of γ, let Q be the first tangent point of the line and γ. Then d is the tangent of the curve at Q, and there are no intersections of D and d. Here, the tangent $\tilde{d}$ of γ parallel to d must coincide with d. Hence $\tilde{d}$ does not intersect D.

(3)⇒(4): For two points P, Q in D, assume the line segment $\overline{\mathrm{PQ}}$ intersects the complement of D. Then there exists a point R on the segment which is the intersection point of the segment and the curve γ. Since $\mathrm{P}, \mathrm{Q} \in D$, $\mathrm{R} \neq \mathrm{P}, \mathrm{Q}$. Let d_{R} be the tangent of the curve at R (cf. Fig. B.2.1).

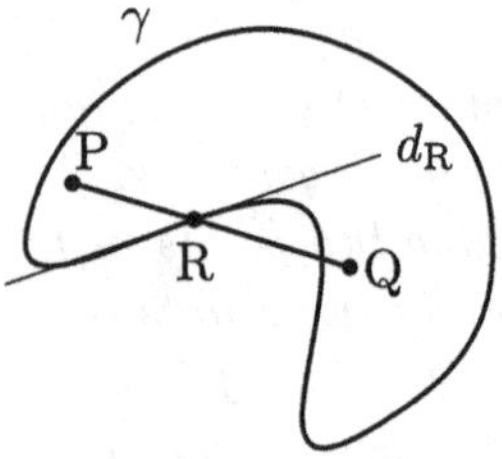

Fig. B.2.1 Proof of Theorem B.2.1.

If $\overline{\mathrm{PQ}}$ lies on d_{R}, then P, Q are the common points of D and d_{R}, contradicting the assumption (3). So $\overline{\mathrm{PQ}}$ intersects d_{R} at R, that is, P and Q lie on different sides of d_{R}. This contradicts that D lies to one side of the tangent line.

(4)⇒(5): Let P, Q be two distinct points in $\overline{D}$. Then one can take sequences $\{\mathrm{P}_n\}$, $\{\mathrm{Q}_n\}$ in D such that

$$\lim_{n\to\infty} \mathrm{P}_n = \mathrm{P}, \qquad \lim_{n\to\infty} \mathrm{Q}_n = \mathrm{Q}.$$

Since the line segments $\overline{\mathrm{P}_n\mathrm{Q}_n}$ $(n = 1, 2, 3, \dots)$ lie in D, the limit $\overline{\mathrm{PQ}}$ lies in $\overline{D}$.

(5)⇒(1): If $\kappa(s) < 0$, the osculating circle, which is a clockwise turning circle, is an approximation of the curve to second order (cf. page 16). Then the line segment joining two points on the curve near $\gamma(s)$ lies in the right-hand side region of the curve, as does the clockwise turning circle. This contradicts that D is on the left side of the curve. Hence $\kappa(s) \geq 0$ holds. □

We defined strict convexity of simple closed curves on page 24 in Section 2.

Corollary B.2.2. *Let $\gamma(s)$ $(0 \leq s \leq l)$ be a closed convex curve whose curvature function does not have any zeros. Then γ is strictly convex.*

Proof. Let P and Q be two distinct points on the curve γ, and L the line segment bounded by P and Q. We denote by D the interior region of γ. Suppose that there exists a point X $(\neq \mathrm{P}, \mathrm{Q})$ on L such that X $\notin D$. By (5), X lies in $\overline{D}$, and L must be the tangent line of γ at X. Then by (3) and (5), there is no intersection of D with L, and L is contained in the image of the curve γ. In particular the curvature of γ at X is equal to zero, a contradiction. □

Moreover, the following theorem holds.

Theorem B.2.3. *A closed regular curve with rotation index ± 1 is a convex curve (resp. a strictly convex curve) if its curvature function does not change sign (has no zeros).*

Proof. Changing the direction of the curve if necessary, we may assume that the curvature is non-negative (resp. positive) and the rotation index is 1. By Theorem B.2.1 and Corollary B.2.2, it is sufficient to show that such a curve has no self-intersections. So, we assume a closed curve $\gamma(s)$ $(0 \leq s \leq l)$ parametrized by the arc-length has a self-intersection. We may assume $\gamma(0)$ is the self-intersection without loss of generality.

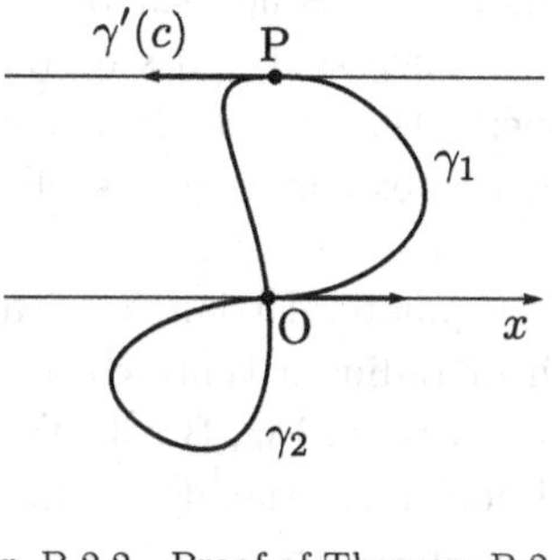

Fig. B.2.2 Proof of Theorem B.2.3.

Since the curvature is non-negative, the Gauss map $\gamma'(s)$ as in (3.2) in Section 3 rotates once on the unit circle in the counterclockwise direction.

By a rotation and a translation if necessary, we may assume

(B.2.1) $\gamma(0) = \mathrm{O} = (0,0), \quad \gamma'(0) = (\pm 1, 0)$ (the direction of the x-axis).

Moreover, we may assume that the curve starting at $\gamma(0)$ travels on the upper half-plane $y > 0$ until it returns to the self-intersection point $\gamma(0)$. (In fact, if not, we replace γ by $-\gamma$.) We denote this part of the curve by γ_1.

Take a line parallel to the x-axis that is far above the x-axis, and translate it downwards. Let $\mathrm{P} = \gamma(c)$ $(0 < c < l)$ be the first contact of such a line with γ_1 (Fig. B.2.2). The tangent vector of the curve at P points in the direction of the positive or negative x-axis. If it coincides with $\gamma'(0)$, then γ' does not change when traveling from O to P, that is, $\gamma(s)$ is a line segment on this interval. In particular, by (B.2.1), this line segment lies on the x-axis, which contradicts that P lies on the upper-half plane. Hence the tangent vector at P points in the opposite direction of $\gamma'(0)$, and the angle of γ' increases by π between O and P. The angle also increases when traveling back from P to O, and the rotation angle of the Gauss map of γ_1 is greater than π. By the same argument with respect to another part of the curve starting at the self-intersection O (i.e., γ_2 in Fig. B.2.2), the rotation angle of such a part is also greater than π. Thus, the rotation angle of the Gauss map of γ is greater than 2π, contradicting the fact that the rotation index is 1. □

One important class of convex curves is the curves of constant width.

Curves of constant width. The *width* of a convex curve γ with respect to the direction d is the height of the curve assuming the tangent line parallel to d is the ground. (If γ is non-smooth, a tangent line of γ at P is a line passing through P such that γ lies on one side of it.) The circle is a curve with the property that *the width is constant for all directions.* There are infinitely many convex curves with such a property, called *curves of constant width.*

For example, take an equilateral triangle with edges of length a, and draw circular arcs, each of radius a centered at a vertex, which form a closed curve with three corners (cf. Fig. B.2.3(a)). This curve has constant width a and known as Reuleaux's triangle. Next, draw circular arcs with radius $a + \varepsilon$, each centered at a vertex of the equilateral triangle, and join them by circular arcs of radius ε centered at the vertices. Then one obtains a curve of constant width without corners as shown in Fig. B.2.3(b). This curve is C^1-regular but not C^2. By using similar methods, one can get

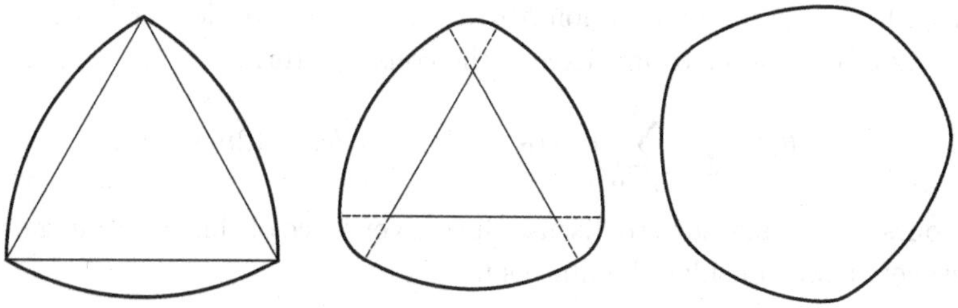

(a) Reuleaux's triangle. (b) (c)

Fig. B.2.3 Curves of constant width.

regular C^1 curves of constant width with n symmetric parts for each odd number n, by joining the circular arcs centered at vertices of a regular n-gon.

Construction of curves of constant width. We show that there are infinitely many curves of constant width that are of class C^∞, not merely class C^1. Let

$$\boldsymbol{e}(\theta) := (\cos\theta, \sin\theta), \qquad \boldsymbol{n}(\theta) := (-\sin\theta, \cos\theta) \qquad (\theta \in \boldsymbol{R}).$$

Then $\boldsymbol{e}(\theta)$ is a unit vector whose angle against the positive direction of the x-axis is θ, and $\boldsymbol{n}(\theta)$ is the unit vector perpendicular to $\boldsymbol{e}(\theta)$ pointing to the left-side of it. For a smooth function $h(\theta)$ of period 2π with $\ddot{h} + h > 0$ ($\dot{} = d/d\theta$), we set

$$\gamma(\theta) := \dot{h}(\theta)\boldsymbol{e}(\theta) - h(\theta)\boldsymbol{n}(\theta) \qquad (0 \le \theta \le 2\pi).$$

Then the velocity vector $\dot{\gamma} = (\ddot{h} + h)\boldsymbol{e}$ of the curve γ does not vanish and is parallel to $\boldsymbol{e}$. Moreover, the arc-length parameter s of the curve γ satisfies $ds/d\theta = \ddot{h} + h > 0$, and the curvature is computed as $\kappa = 1/(\ddot{h} + h)$. Hence the total curvature is

$$\int_\gamma \kappa(\theta)\, ds = \int_0^{2\pi} d\theta = 2\pi,$$

that is, the rotation index of γ is 1. Since the curvature function is positive, the closed curve γ is a convex curve, because of Theorem B.2.3. Here, the tangents of γ at θ and at $\theta + \pi$ are parallel, because $\boldsymbol{e}(\theta + \pi) = -\boldsymbol{e}(\theta)$. In addition, if the function is taken to be $h(\theta) + h(\theta + \pi) = a$ for a positive constant a, the line segment joining $\gamma(\theta)$ and $\gamma(\theta + \pi)$ is perpendicular to the tangent lines of γ at θ and $\theta + \pi$ whose length is a. Summing up, the curve γ is a curve of constant width a. To create a curve of constant width, it is sufficient to take a function $h(\theta)$ such that its Fourier series expansion consists of odd-order terms, except the constant term:

$$h(\theta) = \frac{a}{2} + \sum_{n=0}^{\infty} (a_n \cos(2n+1)\theta + b_n \sin(2n+1)\theta).$$

Of course, it is necessary to discuss the convergence of this series, in general. However, if we consider the finite sum

$$h(\theta) = \frac{a}{2} + \sum_{n=0}^{N} (a_n \cos(2n+1)\theta + b_n \sin(2n+1)\theta)$$

for a positive integer N, and adjusting the coefficients (mainly a) so that $\ddot{h} + h > 0$, one can obtain infinitely many curves of constant width. The curve in (c) of Fig. B.2.3 is the curve of constant width corresponding to the function

$$h(\theta) = 35 + \cos 3\theta + \sin 5\theta.$$

B.3. Line integrals and the isoperimetric inequality

In this section, we will introduce Green's formula on line integrals, and, as an application, we prove the isoperimetric inequality (B.3.3) of Section 1 (page 9).

Green's formula. Let $f(x, y)$ and $g(x, y)$ be two continuous functions defined on a domain D of the xy-plane $\boldsymbol{R}^2$. Then the formal sum

$$\alpha := f(x, y)\, dx + g(x, y)\, dy \tag{B.3.1}$$

is called a *differential form*, or a *one form*. Let $\gamma(t) = (x(t), y(t))$ be a smooth curve (not necessarily a regular curve) on D defined on the interval $[a, b]$. Then we define the *line integral* of the differential form (B.3.1) along γ as follows:

$$\int_\gamma \alpha := \int_a^b \left(f(x(t), y(t)) \frac{dx}{dt}(t) + g(x(t), y(t)) \frac{dy}{dt}(t) \right) dt. \tag{B.3.2}$$

It can be easily checked that this integral does not depend on the choice of parametrization of the curve (see Problem **1** at the end of this section).

So we consider $C := \gamma([a, b])$ as the image of the curve with the orientation given by γ (see the left side of Fig. B.3.1), and denote the line integral by

$$\int_C \alpha \qquad (\alpha := f(x, y)\, dx + g(x, y)\, dy).$$

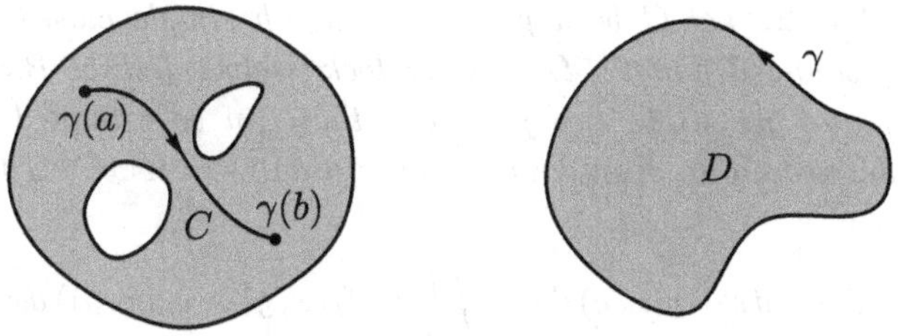

Fig. B.3.1

Example B.3.1. Let C be a counterclockwise oriented circle of radius r centered at the origin of the plane, and consider the line integral

$$I := \int_C (-y\,dx + x\,dy)\,.$$

We parametrize C by $\gamma(t) := r(\cos t, \sin t)$ $(0 \le t \le 2\pi)$. Since $\dot{x} = -r\sin t$ and $\dot{y} = r\cos t$, we have

$$\begin{aligned} I &= \int_0^{2\pi} (-y(t)\dot{x}(t) + x(t)\dot{y}(t))\,dt = \int_0^{2\pi} r^2(\sin^2 t + \cos^2 t)\,dt \\ &= r^2 \int_0^{2\pi} dt = 2\pi r^2, \end{aligned}$$

which is a twice of the area of the disc bounded by C, as we sill see (cf. (B.3.4) and (B.3.5)).

A continuous curve $\gamma : [a,b] \to D$ is said to be *piecewise smooth* if there is a division

$$a = t_0 < t_1 < \cdots < t_{n-1} < t_n = b$$

of the interval $[a,b]$ such that, for each $i = 1, \ldots, n$, γ is C^∞ on (t_{i-1}, t_i) and the velocity vector $\dot{\gamma}(t)$ $(t_{i-1} < t < t_i)$ can be extended as a continuous vector-valued function on the closed interval $[t_{i-1}, t_i]$. We can generalize the definition of line integral for piecewise smooth curves as follows: Let γ be a piecewise smooth curve as above, then the image $C := \gamma([a,b])$ is obtained by the union of subarcs

$$C = C_1 \cup \cdots \cup C_n, \qquad (C_i := \gamma([t_{i-1}, t_i]),\ i = 1, \ldots, n),$$

which is frequently denoted by $(C =)C_1 + \cdots + C_n$. The line integral is defined by

$$\int_C \alpha \left(= \int_{C_1 + \cdots + C_n} \alpha \right) := \sum_{i=1}^n \int_{C_i} \alpha.$$

Now, the well-known *Green's formula* is stated as follows (cf. [24]).

Theorem B.3.2. *Let C be a piecewise smooth simple closed curve surrounding a bounded domain Ω counterclockwisely (cf. Fig. B.3.1, right). Let $f(x, y)$ and $g(x, y)$ be two functions that can be extended as smooth functions defined on an open domain containing $\overline{\Omega} := \Omega \cup C$. Then it holds that*

$$\int_C f(x,y)\,dx + g(x,y)\,dy = \iint_{\overline{\Omega}} (-f_y(x,y) + g_x(x,y))\,dx\,dy.$$

The isoperimetric inequality. As mentioned in Section 1 (page 9), the *isoperimetric inequality*

$$4\pi A \leq l^2 \qquad \text{(B.3.3)}$$

holds for any smooth simple closed regular curve γ, where l is the length of the curve, and A is the area of the domain surrounded by γ. Taking the *convex hull* (i.e., the intersection of all convex domains containing γ) of the curve (cf. Fig. B.3.2 and Problem **3** at the end of this section), one can restrict to the case of convex curves (see Appendix B.2). In this section, we prove (B.3.3) for smooth convex curves. (However, in the following discussions, C^2-regularity is sufficient to prove the assertion. A proof for general convex curves having low regularity is given in [14].)

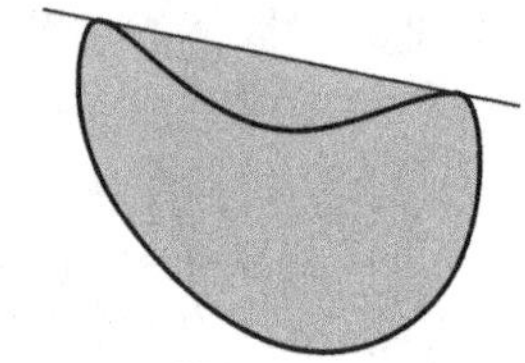

Fig. B.3.2

The explicit statement that we would like to prove is as follows.

Theorem B.3.3. *Let C be a smooth convex regular curve in $\boldsymbol{R}^2$. Then the inequality* (B.3.3) *holds, and equality holds if and only if C is a circle.*

Our proof is a modification of the proof originally given by E. Schmidt [34]. We prepare the following proposition.

Proposition B.3.4. *Under the assumption of Theorem B.3.3, there is a pair of parallel lines such that*

- *the curve C lies between the two lines,*
- *and each of the two lines meet the curve at a point where the curvature function of the curve does not vanish.*

Proof. We fix a point P on C where the curvature function of C does not vanish, and denote by L_1 the tangent line of the curve at P. Take a line L_2 which is parallel to L_1 such that C lies between L_1 and L_2. By translating

L_2 toward the curve, we may assume that L_2 first meets C at a point Q. Let $\gamma(t)$ $(t \in \boldsymbol{R})$ be a smooth curve such that

$$\gamma(t+l) = \gamma(t) \qquad (t \in \boldsymbol{R}),$$

whose restriction to the closed interval $[0, l]$ gives a parametrization of the curve C such that the interior domain lies on the left-hand side of the curve. Then the curvature function $\kappa(t)$ of γ takes a non-negative value for every t, because of (1) in Theorem B.2.1. Without loss of generality we may set

$$\gamma(0) = \mathrm{P}, \quad \gamma(b) = \mathrm{Q} \qquad (0 < b < l).$$

If $\kappa(b) > 0$, then the pair of lines (L_1, L_2) satisfies the desired properties (see Fig. B.3.3, left).

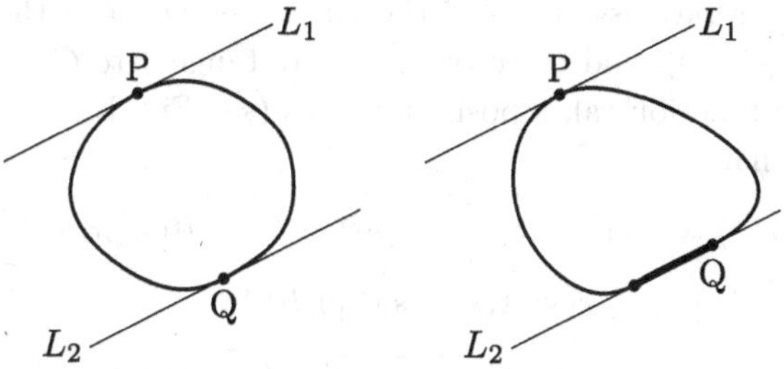

Fig. B.3.3

So we assume $\kappa(b) = 0$. Then the set (see Fig. B.3.3, right)

$$Z := \{t \in [b, l] \mid \kappa(s) = 0 \text{ for each } s \text{ satisfying } b \le s \le t\}$$

is equal to $\{b\}$, or Z is a closed subinterval of $[b, l]$, and we can replace b by

$$b := \max(Z).$$

The idea of the remainder of this proof is to show that we can shift $\gamma(0)$ and $\gamma(b)$ slightly to $\gamma(\varepsilon_1)$ and $\gamma(b + \varepsilon_2)$ for real numbers $\varepsilon_1, \varepsilon_2$ close to 0, and show both $\kappa(\varepsilon_1)$ and $\kappa(b + \varepsilon_2)$ do not vanish. We set $t_n := b + (1/n)$ for $n = 1, 2, 3, \ldots$, then $t_n \notin Z$ whenever $b + (1/n) < l$, and so there exists a point $s_n \in [b, t_n]$ such that $\kappa(s_n) > 0$ for each positive integer n. Let $L_2(n)$ be the tangent line of C at $\gamma(s_n)$. Then there exists a point $s'_n \in (s_n < s'_n < s_n + l)$ such that the tangent line $L_1(n)$ at $\gamma(s'_n)$ is parallel to $L_2(n)$ but does not coincide with $L_2(n)$. Since the sequence $\{s'_n\}_{n=1,2,3,\ldots}$ is bounded in $\boldsymbol{R}$, taking a subsequence, we may assume that s'_n converges to a point s'_∞. Since $L_2(n)$ converges to L_2, $L_1(n)$ must converge to L_1, and so we can conclude that $s'_\infty = 0$. Since $\kappa(0) > 0$, we may assume that $\kappa(s'_n) > 0$ for sufficiently large n. Then for such n, the pair $(L_1(n), L_2(n))$ of parallel lines attains the desired properties. $\square$

Proof of Theorem B.3.3. We may assume that t is the arc-length parameter of the curve $\gamma(s) := (x(s), y(s))$ $(0 \leq s \leq l)$ such that the interior domain D is on the left side of the curve. Then the curvature function $\kappa(s)$ is non-negative. By Green's formula (Theorem B.3.2), we have

$$\text{(B.3.4)} \qquad \int_0^l x(s)y'(s)\,ds = \int_C x\,dy = \iint_{\overline{D}} dx\,dy = A,$$

and

$$\text{(B.3.5)} \qquad -\int_0^l y(s)x'(s)\,ds = -\int_C y\,dx = \iint_{\overline{D}} dx\,dy = A.$$

We now apply Proposition B.3.4: Rotating and translating the xy-coordinates, we may assume that the curve lies between the two vertical lines $x = \pm\delta$ $(\delta > 0)$, and these two lines are tangent to C at points where the curvature function takes positive values (see Fig. B.3.4). We may further assume that

$$x(0) = x(l) = -\delta, \qquad x(m) = \delta \qquad (0 < m < l).$$

Define a curve $\mathit{\Gamma}(s) := (x(s), z(s))$ $(s \in [0, l])$ by

$$z(s) := \begin{cases} -\sqrt{\delta^2 - x(s)^2} & (s \in [0, m]), \\ \sqrt{\delta^2 - x(s)^2} & (s \in [m, l]), \end{cases}$$

see Fig. B.3.4. Then the curve $\mathit{\Gamma}$ parametrizes the circle of radius δ centered at the origin.

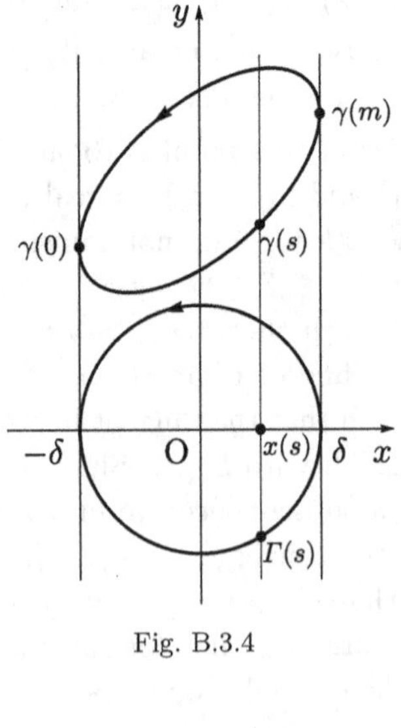

Fig. B.3.4

Moreover, we can prove the following lemma.

Lemma B.3.5. *$\Gamma(s)$ $(0 \le s \le l)$ gives a piecewise smooth parametrization of the circle.*

Proof. Since $\gamma(s)$ is positively curved at $s = 0$ and $s = m$, $0 < x(s) < \delta$ holds whenever $0 < s < m$ or $m < s < l$. Thus, $\Gamma(s)$ gives a smooth parametrization on $(0, m)$ and (m, l). So it is sufficient to show that $\Gamma(s)$ is piecewise smooth at $x = \pm\delta$, i.e., the points where $s = 0$, m and l. In fact, we can show furthermore that the $\Gamma'(s)$ is continuous at $s = 0$, m and l. We consider the case $x = \delta$ (the case $x = -\delta$ is similar). The curvature function $\kappa(s)$ of the curve γ is positively curved at $s = 0, m$. Since $x'(m) = 0$ and $y'(m) = 1$ (see Fig. B.3.4), we have

$$0 < \kappa(m) = \det \begin{pmatrix} x'(m) & x''(m) \\ y'(m) & y''(m) \end{pmatrix} = \det \begin{pmatrix} 0 & x''(m) \\ 1 & y''(m) \end{pmatrix} = -x''(m).$$

Using L'Hôpital's rule (cf. Theorem A.1.3 in Appendix A.1), we have

$$\lim_{s \to m} \frac{(x')^2}{\delta^2 - x^2} = \left(\lim_{s \to m} \frac{1}{\delta + x} \right) \left(\lim_{s \to m} \frac{(x')^2}{\delta - x} \right) = \frac{1}{2\delta} \lim_{s \to m} \frac{2x'x''}{-x'} = \frac{-x''(m)}{\delta}.$$

Since $x'(m) = 0$ and $x''(m) < 0$, we have $x'(s) > 0$ (resp. $x'(s) < 0$) if s $(< m)$ (resp. s $(> m)$) is sufficiently close to m. So it holds that

$$\lim_{s \to m^{\pm}} \frac{x'(s)}{\sqrt{\delta^2 - x(s)^2}} = \mp \sqrt{\frac{\kappa(m)}{\delta}},$$

where $\lim_{s \to m^{\pm}}$ denotes the rightward and leftward limits, respectively. In particular,

$$\lim_{s \to m^{\pm}} z'(s) = \frac{\mp x(s) x'(s)}{\sqrt{\delta^2 - x(s)^2}} = \sqrt{\kappa(m)\delta}$$

holds, and we have $\lim_{s \to m} \Gamma'(s) = (0, \sqrt{\kappa(m)\delta})$. □

Proof of Theorem B.3.3, continued. By Lemma B.3.5, we can apply the formula (B.3.5) to $\Gamma(s)$, and get

$$\pi\delta^2 = -\int_0^l z(s)\, x'(s)\, ds.$$

This equality with (B.3.4) yields that

$$\begin{aligned} A + \pi\delta^2 &= \int_0^l x(s) y'(s)\, ds - \int_0^l z(s) x'(s)\, ds \\ &= \int_0^l (x(s) y'(s) - z(s) x'(s))\, ds. \end{aligned}$$

Applying the Schwartz inequality (cf. Proposition A.3.1)

$$xy' - zx' = (x, z)\begin{pmatrix} y' \\ -x' \end{pmatrix} \leq \sqrt{(x^2 + z^2)(x'^2 + y'^2)} = \delta,$$

we have

$$A + \pi\delta^2 \leq l\delta. \tag{B.3.6}$$

By applying the relationship between the arithmetic mean and geometric mean, we have

$$2\sqrt{A\pi}\,\delta \leq A + \pi\delta^2,$$

which proves (B.3.3). We next examine the equality condition: The equality of (B.3.6) holds if and only if $(x(s), z(s))$ is proportional to $(-y'(s), x'(s))$ (cf. Proposition A.3.1), that is, there exists a non-negative valued function $\mu(s)$ such that

$$x(s) = \mu(s)y'(s), \quad z(s) = -\mu(s)x'(s) \qquad (0 \leq s \leq l). \tag{B.3.7}$$

Since

$$\delta^2 = x(s)^2 + z(s)^2 = \mu^2(y'(s)^2 + x'(s)^2) = \mu^2,$$

we have $\mu(s) = \delta$. Differentiating (B.3.7), we have $x''(s) = \delta\, y''(s)$. On the other hand, $y''(s) = \kappa(s)x'(s)$ holds, by the Frenet formula (cf. (2.15) in Section 2). So we have $\kappa(s) = 1/\delta$, that is, γ is a circle of radius δ. In particular $A = \pi\delta^2$ holds. □

Exercises B.3

1 Show that the definition of line integral given in (B.3.2) does not depend on the choice of parametrization of the curve.

2 In general, the values of line integrals are not determined by the starting point and the ending point: for example, compute the line integrals

$$\int_{C_1} x\,dy, \qquad \int_{C_2+C_3} x\,dy,$$

where C_1, C_2, C_3 are line segments starting from $(0,0)$, $(0,0)$, $(0,1)$ and terminating at $(1,1)$, $(0,1)$, $(1,1)$, respectively.

3 Given (B.3.3) holds for convex curves, explain why a simple closed regular curve (not necessarily convex) satisfies the isoperimetric inequality (B.3.3), using Fig. B.3.2.

4 Let $\mathcal{T}$ be the set of triangles whose total perimeter length is 1. Prove that the equilateral triangle in $\mathcal{T}$ attains the maximum area. (Hint: The area of the triangle can be computed as $\sqrt{s(s-a)(s-b)(s-c)}$ (Heron's formula), where a, b and c are the lengths of the sides and $s = (a+b+c)/2$.)

B.4. First fundamental forms and maps

As we have already seen in Section 7, a parametrization of a surface by the parameter (u, v) can be regarded as drawing a *map* of the surface on the uv-plane. In this section, we consider the world maps of our earth.

It requires rather high technology to draw a world map, because the shape of the earth is nearly a sphere. Although circular cylinders and circular cones can be easily developed to the plane, a sphere cannot be developed to the plane without expansion and contraction. That is, it is impossible to draw the distance-angle-preserving map in the sense of Proposition 7.4 in Section 7. So we use various types of world maps, which necessarily lose some properties of the sphere, depending on our specified purposes. In this section, we introduce several concrete examples of such world maps. For simplicity, we assume that the earth is a sphere of radius one.[1]

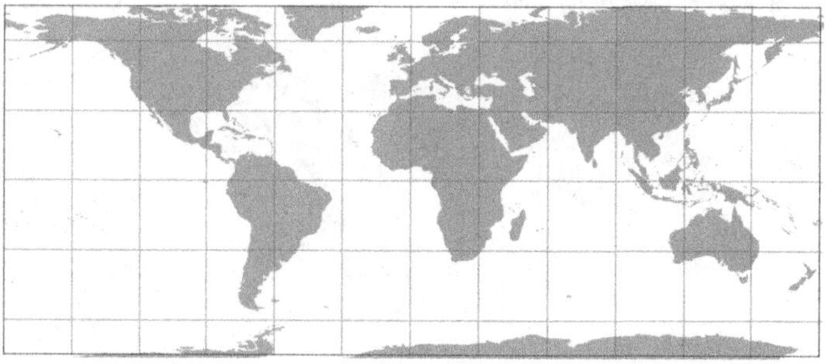

Fig. B.4.1 Development to the cylinder.

[1]The earth is a sphere of radius approximately 6,400 km (cf. Problem **3**-(2) in Section 1). This value comes from the original definition of the unit "meter", which is one ten millionth of the length of a meridian from the north pole to the equator. The modern definition of the "meter", however, is based on the speed of the light.

Development to the cylinder. One natural method to draw the world map, such that the latitudes and the meridians remain perpendicular, is the following: Consider the circular cylinder circumscribing the earth at the equator, and map each meridian to the generating line of the cylinder tangent to the meridian at the equator.

We describe such a map analytically: Letting u and v be the longitude and latitude of a point (x, y, z) on the sphere, respectively, we have

$$x = \cos u \cos v, \quad y = \sin u \cos v, \quad z = \sin v \qquad \left(-\pi < u < \pi,\ -\frac{\pi}{2} < v < \frac{\pi}{2}\right). \tag{B.4.1}$$

Then the map to the uv-plane obtained by this is the above mentioned map, which preserves the length of the meridians (Fig. B.4.1). The higher the latitude, the more the length expands for that latitude's image in the uv-plane. The first fundamental form of this coordinate system is expressed as

$$ds^2 = (\cos^2 v)\, du^2 + dv^2. \tag{B.4.2}$$

The coefficient of dv^2 (i.e. the coefficient G of the first fundamental form) is 1 because the length of meridians is preserved by this map.

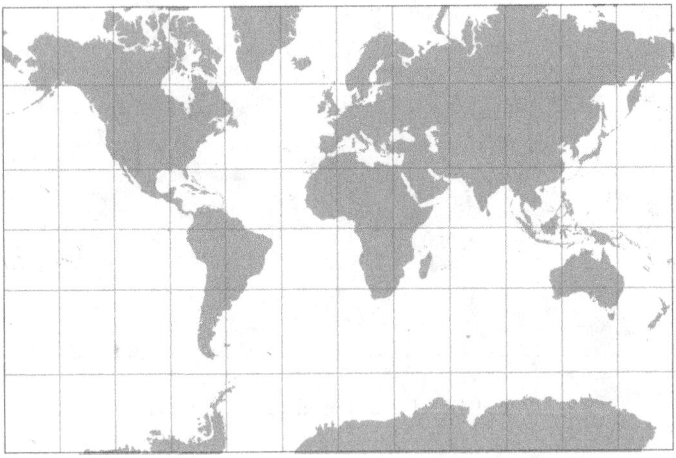

Fig. B.4.2 Mercator's world map.

Mercator's world map. A disadvantage of the previous example is that the ratio of vertical and horizontal lengths is far from 1 at the further points from the equator. If one takes the function $\eta = \eta(v)$ satisfying

$$dv = \cos v \, d\eta, \tag{B.4.3}$$

and takes a new coordinate system (ξ, η) as

$$\xi = u, \qquad \eta = \eta(v),$$

the first fundamental form is written as

$$ds^2 = (\cos^2 v)\,(d\xi^2 + d\eta^2),$$

that is, the ratio of the vertical and the horizontal lengths turns out to be 1. In fact, the function η satisfying this can be computed by (B.4.3), resulting in

$$\eta = \int_0^v \frac{dt}{\cos t} = \log\left(\frac{\cos v}{1 - \sin v}\right).$$

Then

$$ds^2 = \frac{1}{\cosh^2 \eta}(d\xi^2 + d\eta^2)$$

holds, see Problem **1** in this section. The map corresponding to the $\xi\eta$-plane is called *Mercator's*[2] *world map* (Fig. B.4.2). In this map, a point (ξ, η) on the map corresponds to the point

$$\left(\frac{\cos\xi}{\cosh\eta}, \frac{\sin\xi}{\cosh\eta}, \tanh\eta\right) \qquad (-\pi < \xi < \pi, \quad -\infty < \eta < \infty)$$

on the sphere. The coefficients of the first fundamental form with respect to (ξ, η) satisfy $E = G$ and $F = 0$. That is, it is an isothermal coordinate system, as in page 74 and Section 16. Thus, by (7.19) in Section 7, angles on the map coincide with actual angles on the sphere. In other words, the angle between the north direction on the compass and the direction on board during a voyage coincide with the angle between the route and the meridians on Mercator's map. The north and south poles in Mercator's world map diverge to infinity, because $\lim_{u\to\pm\pi/2} \eta = \pm\infty$. Ever since the era of Grand Voyages, Mercator's map has been widely used as a chart.

[2]Mercator, Gerardus (1512–1594).

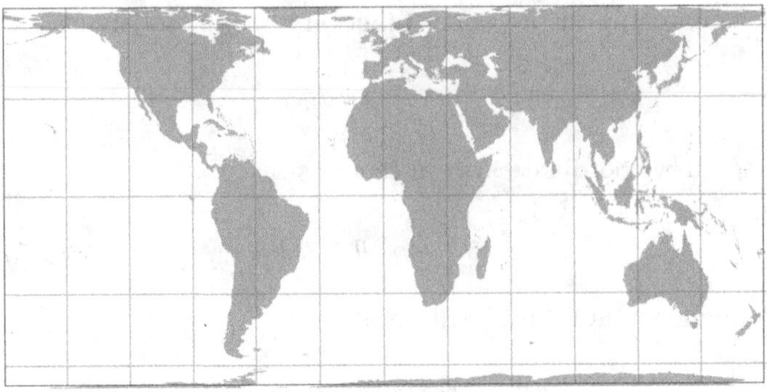

Fig. B.4.3 An equi-area map.

Equi-area maps. In the correspondence between the sphere and the cylinder as in (B.4.1), we set $\xi = u$ (as in Mercator's map) and set ζ to be the height from the plane containing the equator, that is,

$$\xi = u, \qquad \zeta = \sin v.$$

By this coordinate change, we have

$$\begin{aligned} ds^2 &= (\cos^2 v)du^2 + dv^2 = (\cos^2 v)\, d\xi^2 + \frac{1}{\cos^2 v}\, d\zeta^2 \\ &= (1-\zeta^2)\, d\xi^2 + \frac{1}{1-\zeta^2}\, d\zeta^2, \end{aligned}$$

since $d\zeta = (\cos v)\, dv$. The point on the sphere corresponding to (ξ, ζ) is given by

$$(\sqrt{1-\zeta^2}\cos\xi,\ \sqrt{1-\zeta^2}\sin\xi,\ \zeta) \qquad (-\pi < \xi < \pi,\ -1 < \zeta < 1).$$

The coefficients of the first fundamental form satisfy $EG - F^2 = 1$. Then by (7.14) in Section 7, the area of the domain on the sphere coincides with the area of the corresponding domain on the $\xi\zeta$-plane (Fig. B.4.3). As a disadvantage, the ratio between the lengths of the horizontal and vertical directions grows larger near the poles on this map. On the other hand, the ξ-curves and η-curves are orthogonal on the sphere, since $F = 0$.

A map which represents the area correctly, as in this example, is called an *equi-area projection*, or an *equi-area map*. Since the map is equi-area if and only if $EG - F^2 = 1$, there are many ways to draw equi-area maps. The projection introduced here is called *Lambert's cylindrical projection.*[3]

[3]Lambert, Johan Heinrich (1728–1777).

The stereographic projection. Let P be a point different from the north pole on the sphere of radius 1 centered at the origin, and let Q be the point on the xy-plane which is the intersection point of the half line starting at the north pole and passing through P. Then the map $\pi\colon \mathrm{P} \mapsto \mathrm{Q}$ is called *stereographic projection* (Fig. B.4.4, left). Let $(X, Y, 0)$ be the

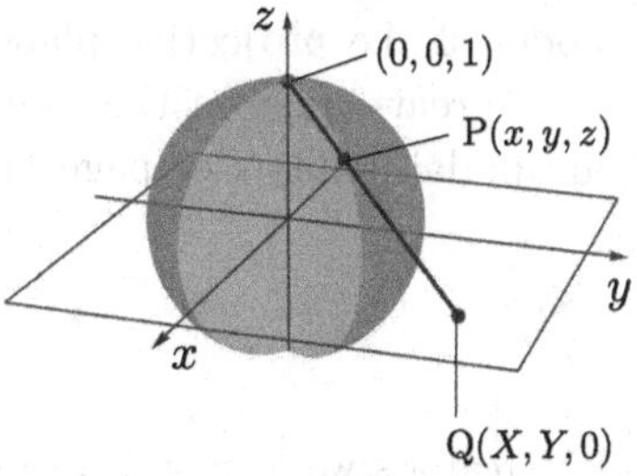

The stereographic projection.

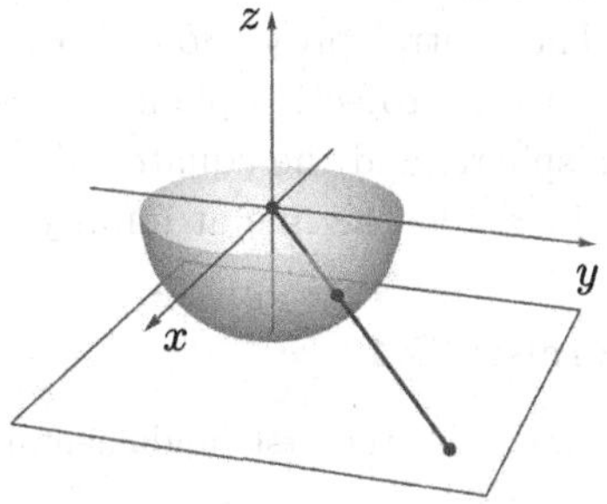

The central projection.

Fig. B.4.4 The stereographic projection and the central projection.

point corresponding to the point (x, y, z) on the sphere by stereographic projection. By definition, we have

$$X = \frac{x}{1-z}, \qquad Y = \frac{y}{1-z}.$$

Noticing that $x^2 + y^2 + z^2 = 1$, these equations can be solved with respect to x, y, z as

$$x = \frac{2X}{1+X^2+Y^2}, \quad y = \frac{2Y}{1+X^2+Y^2}, \quad z = \frac{X^2+Y^2-1}{X^2+Y^2+1}.$$

Then the first fundamental form of $\boldsymbol{p}(X, Y) = (x, y, z)$ can be computed as

$$ds^2 = \frac{4}{(1+X^2+Y^2)^2}(dX^2 + dY^2).$$

Since the coefficients of the first fundamental form satisfy $E = G$ and $F = 0$, this map represents angles correctly, like Mercator's map. By this projection, a circle on the sphere is mapped to a circle or a line in the plane. Stereographic projection is used to identify the complex plane union the point at infinity with the sphere.

The central projection. The *central projection* is a correspondence which maps a point of the southern hemisphere to the projection from the origin of the point onto the tangent plane of the sphere at the south pole

(Fig. B.4.4, right). Though such a map preserve neither the angle nor the area, great circles (curves on the sphere obtained as intersections of planes through the origin and the sphere) are mapped to lines (see Problem **4** of this section). In particular, a line segment joining two points on the map corresponds to the shortest path on the sphere. This property is convenient for finding shortest routes.

The central projection also gives a model of the projective plane. A line in the projective plane corresponds to a great circle on the southern hemisphere, and the equator identifying antipodal points (see page 115) is considered as the line at infinity.

Exercises B.4

1 Show that the first fundamental form of Mercator's world map is expressed as

$$ds^2 = \frac{1}{\cosh^2 \eta}(d\xi^2 + d\eta^2)$$

in terms of ξ and η.

2 Find an equi-area map of the torus in Problem **1** in Section 6.

3 Find a parametrization of the cone $z = \sqrt{x^2 + y^2}$ which preserves the length and the angle.

4 Show that a great circle is mapped to a line by the central projection.

B.5. Curvature line coordinates and asymptotic line coordinates

In this section, we fix a point P on a surface S in $\boldsymbol{R}^3$, and let

$$\text{(B.5.1)} \qquad \boldsymbol{p}\colon D \ni (s,t) \longmapsto \boldsymbol{p}(s,t) \in \boldsymbol{R}^3$$

be a parametrization of S on a neighborhood of the point P, where D is a domain of $\boldsymbol{R}^2$. We set

$$\text{(B.5.2)} \qquad \mathrm{P} = \boldsymbol{p}(s_0, t_0) \qquad ((s_0, t_0) \in D).$$

Although parameters for surfaces were usually written as (u, v) in Chapter II, we use (s, t) for the initial parametrization in this section, because we will then consider a change of parameters $(s, t) \mapsto (u, v)$.

Curvature lines. A regular curve on the surface is said to be a *curvature line* if its velocity vector gives a principal direction (cf. page 93). A curvature line can be regarded as the curve which travels along the surface in "the most curved" direction. So drawing a net consisting of curvature lines on the surface, one may have a parametrization of the surface which best represents how the surface bends.

Theorem B.5.1. *Let* P *be a non-umbilic point of the surface* S *with parametrization* $\mathbf{p}(s,t)$ *as in* (B.5.1) *and* (B.5.2). *Then there exists a parameter change*

$$D' \ni (s,t) \longmapsto (u,v) \in \widetilde{D}$$

on a neighborhood D' ($\subset D$) *of* (s_0, t_0) *such that, under the new parametrization* $\mathbf{p}(u,v)$, *the* u*-curves and* v*-curves of* $\mathbf{p}(u,v)$ *are all curvature lines. Such a parameterization is called a* curvature line coordinate system. *Under the parametrization* $\mathbf{p}(u,v)$, *the first and second fundamental forms are expressed as*

$$ds^2 = E\,du^2 + G\,dv^2, \qquad II = L\,du^2 + N\,dv^2. \tag{B.5.3}$$

Conversely, if a parametrization $\mathbf{p}(u,v)$ *of the surface satisfies* (B.5.3), *then* (u,v) *is a curvature line coordinate system.*

Proof. Existence of the curvature line coordinate system is shown on page 239.

We show that (B.5.3) holds for a parametrization $\mathbf{p}(u,v)$ of the surface by the curvature line coordinate system (u,v). We set

$$ds^2 = E\,du^2 + 2F\,du\,dv + G\,dv^2, \quad II = L\,du^2 + 2M\,du\,dv + N\,dv^2. \tag{B.5.4}$$

The vectors $\boldsymbol{p}_u$ and $\boldsymbol{p}_v$ are perpendicular, because of Lemma 9.5 in Section 9. Then $F = \boldsymbol{p}_u \cdot \boldsymbol{p}_v = 0$. Moreover, since the u-directions and v-directions correspond to the principal directions, letting $\theta = 0,\ \pi/2$ in Proposition 9.2 in Section 9, we have that the principal curvatures are

$$\lambda_1 = \frac{L}{E}, \qquad \lambda_2 = \frac{N}{G}.$$

By the definition of Gaussian curvature (8.7) and its relationship (9.10) with the principal curvatures,

$$\frac{LN}{EG} = \lambda_1\lambda_2 = K = \frac{LN - M^2}{EG - F^2} = \frac{LN - M^2}{EG}$$

holds, and hence $M = 0$.

Conversely, assume $F = M = 0$. Then by (8.7) in Section 8, we have

$$\lambda_1 = \frac{L}{E}, \qquad \lambda_2 = \frac{N}{G}.$$

On the other hand, by Proposition 9.2 in Section 9, the normal curvature κ_n of the direction of angle θ from the u-axis on the uv-plane is

$$\kappa_n = \frac{L\cos^2\theta + N\sin^2\theta}{E\cos^2\theta + G\sin^2\theta}.$$

Hence the normal curvatures with respect to the directions of the u and v axes (the directions for $\theta = 0,\ \pi/2$) coincide with λ_1 and λ_2, respectively, that is, (u, v) is a curvature line coordinate system. □

Asymptotic lines. A curve on the surface is said to be an *asymptotic line* if the velocity vector gives the asymptotic direction (cf. page 96) at each point of the curve. If the surface contains a straight line, it is an asymptotic line. In fact, both the normal curvature and the geodesic curvature of the line are 0. We consider a parametrization of the surface by the net generated by asymptotic curves.

Theorem B.5.2. *Let* P *be a hyperbolic point (that is, a point where the Gaussian curvature K is negative) of the surface S with parametrization $\mathbf{p}(s,t)$ as in* (B.5.1) *and* (B.5.2). *Then there exists a parameter change $D'(\subset D) \ni (s,t) \mapsto (u,v) \in \widetilde{D}$ on a neighborhood of* P *such that, under the new parametrization $\mathbf{p}(u,v)$, the u-curves and v-curves are asymptotic lines. Such a parameter (u, v) is called an* asymptotic line coordinate system. *Under this parametrization, the second fundamental form satisfies*

$$II = 2M\,du\,dv. \tag{B.5.5}$$

Conversely, a parametrization $\mathbf{p}(u,v)$ of the surface satisfying (B.5.5) *is an asymptotic line coordinate system.*

Proof. The existence of the asymptotic line coordinate system will be shown later, together with the existence of the curvature line coordinate system.

For a given parametrization $\boldsymbol{p}(u,v)$ of the surface, we write the fundamental forms as in (B.5.4). By letting $\theta = 0,\ \pi/2$ in Proposition 9.2 in Section 9, we have

$$\kappa_1 = \frac{L}{E}, \qquad \kappa_2 = \frac{N}{G},$$

where κ_1 and κ_2 are the normal curvatures with respect to the u-direction and v-direction, respectively. These are asymptotic directions if and only if $\kappa_1 = \kappa_2 = 0$, by definition. Here, since $EG \neq 0$, this is equivalent to $L = N = 0$. □

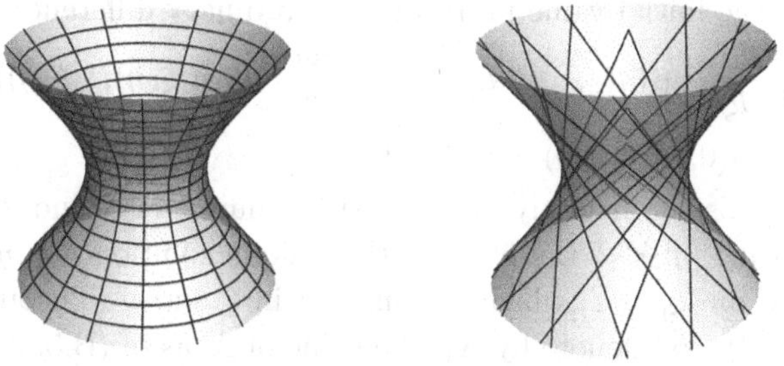

Curvature line coordinates. Asymptotic line coordinates.

Fig. B.5.1 Curvature line and asymptotic line coordinate systems.

Figure B.5.1 shows the curvature line net (left) and the asymptotic line net (right) on the same hyperboloid of one sheet. Notice the different visual impressions given by the two coordinate systems.

Existence of curvature and asymptotic line coordinate systems. We give the proof for existence of the coordinate systems in Theorems B.5.1 and B.5.2. When a tangent vector is assigned to each point on the surface, we call this correspondence a *vector field* of the surface. For a given parametrized surface $\boldsymbol{p}(s,t)$, each tangent vector of the surface at $\boldsymbol{p}(s,t)$ can be expressed as a linear combination of $\boldsymbol{p}_s(s,t)$ and $\boldsymbol{p}_t(s,t)$. So a vector field X of the surface can be expressed in the form[1].

(B.5.6) $$X = X(s,t) = \alpha(s,t)\boldsymbol{p}_s(s,t) + \beta(s,t)\boldsymbol{p}_t(s,t),$$

where α and β are functions in (s,t). The vector field X is said to be *smooth* if α and β are of class C^∞.

Lemma B.5.3. *Let S be a surface parametrized as in* (B.5.1) *and* (B.5.2) *on a neighborhood of an arbitrarily given point* P *of the surface. Let X be a smooth vector field on the surface without zeros, and $\gamma(\tau) = p(s(\tau), t(\tau))$ a regular curve on the surface passing through* P *with* $\mathrm{P} = \gamma(0)$, *that is,* $(s(0), t(0)) = (s_0, t_0)$ (*cf.* (B.5.2)). *If X at* P *and $\dot\gamma(0)$ are linearly independent, there exists a coordinate system (ξ, η) such that* P *corresponds to the origin, and at each point, the ξ-curve is tangent to X, and the image of the η-axis is γ.*

Proof. We express the vector field X as in (B.5.6). Then, for each fixed η,

[1] From the viewpoint of manifold theory, this vector field on the surface can be identified with the vector field $\alpha(s,t)\partial/\partial s + \beta(s,t)\partial/\partial t$ on the manifold that is the source space of the map $\boldsymbol{p}$ (cf. (13.20) in Section 13).

we consider the initial value problem of an ordinary differential equation

$$\begin{cases} \dfrac{da}{d\xi} = \alpha(a(\xi,\eta), b(\xi,\eta)), & \dfrac{db}{d\xi} = \beta(a(\xi,\eta), b(\xi,\eta)), \\ (a(0,\eta), b(0,\eta)) = (s(\eta), t(\eta)). \end{cases}$$

Then there exist sufficiently small positive numbers δ and ε such that: for each η with $|\eta| < \delta$ there exists the unique solution $a(\xi,\eta)$, $b(\xi,\eta)$ of the equation on $|\xi| < \varepsilon$, which are smooth in ξ and η (Theorem A.2.1 in Appendix A.2). We denote by X_{P} the value of X as in (B.5.6) at P. Since X_{P} and

$$\dot\gamma(0) = \dot s(0)\boldsymbol{p}_s(s_0,t_0) + \dot t(0)\boldsymbol{p}_t(s_0,t_0)$$

are linearly independent at P,

$$\left(\frac{\partial a}{\partial \xi}(0,0), \frac{\partial b}{\partial \xi}(0,0)\right) = (\alpha(s_0,t_0), \beta(s_0,t_0)),$$

$$\left(\frac{\partial a}{\partial \eta}(0,0), \frac{\partial b}{\partial \eta}(0,0)\right) = (\dot s(0), \dot t(0))$$

are linearly independent as vectors in $\boldsymbol{R}^2$. Then the inverse function theorem (Theorem A.1.5 in Appendix A.1) yields that the map $(\xi,\eta) \mapsto (a(\xi,\eta), b(\xi,\eta))$ is a diffeomorphism on a neighborhood of $(a,b) = (u_0,v_0)$. Hence we have a new coordinate system (ξ,η) of a neighborhood P on the surface given by

$$s = a(\xi,\eta), \qquad t = b(\xi,\eta).$$

Since $a(0,0) = s_0$ and $b(0,0) = t_0$, the origin corresponds to the point P.

Moreover, since

$$\text{(B.5.7)} \qquad \begin{cases} \dfrac{\partial \boldsymbol{p}}{\partial \xi} = \dfrac{\partial s}{\partial \xi}\dfrac{\partial \boldsymbol{p}}{\partial s} + \dfrac{\partial t}{\partial \xi}\dfrac{\partial \boldsymbol{p}}{\partial t} = \alpha \boldsymbol{p}_s + \beta \boldsymbol{p}_t = X, \\ \dfrac{\partial \boldsymbol{p}}{\partial \eta}(0,\eta) = \dfrac{\partial s}{\partial \eta}\dfrac{\partial \boldsymbol{p}}{\partial s} + \dfrac{\partial t}{\partial \eta}\dfrac{\partial \boldsymbol{p}}{\partial t} = \dot s(\eta)\boldsymbol{p}_s + \dot t(\eta)\boldsymbol{p}_t = \dot\gamma(\eta), \end{cases}$$

each ξ-curve is tangent to X. On the other hand,

$$\boldsymbol{p}(\xi,\eta)|_{(\xi,\eta)=(0,0)} = \gamma(0) = \mathrm{P}$$

and the second equation in (B.5.7) imply that $\boldsymbol{p}(0,\eta) = \gamma(\eta)$. □

Lemma B.5.3 yields the following lemma.

Lemma B.5.4. *Let X and Y be two smooth vector fields on the surface which are linearly independent at each point. Then for an arbitrary point* P *on the surface, there exists a new coordinate system (u,v) on a neighborhood of* P *such that* P *corresponds to the origin, and each u-curve (resp. v-curve) is tangent to X (resp. Y).*

Proof. Let $\gamma_1(\tau)$ and $\gamma_2(\tau)$ be curves on the surface satisfying $\gamma_1(0) = \gamma_2(0) = \mathrm{P}$, $\dot\gamma_1(0) = Y_\mathrm{P}$ and $\dot\gamma_2(0) = X_\mathrm{P}$, where X_P and Y_P denote the values of vector fields X and Y, respectively, at the point P. Applying Lemma B.5.3 to the vector field X and the curve γ_1, we get a coordinate system (ξ_1, η_1) on a neighborhood of P such that each ξ_1-curve is tangent to X, and the η_1-curve passing through the origin coincides with γ_1. Similarly, applying Lemma B.5.3 to the vector field Y and the curve γ_2, we get a coordinate system (ξ_2, η_2) on a neighborhood of P such that each ξ_2-curve is tangent to Y, and the η_2-curve passing through the origin coincides with γ_2. Changing the coordinate system (ξ_1, η_1) to (ξ_2, η_2), ξ_1 and η_1 can be expressed as functions in ξ_2 and η_2. We now prove that (η_1, η_2) is a new coordinate system of the surface on a neighborhood of P. To do this, it is sufficient to show that the Jacobian (the determinant of the Jacobi matrix) of $(\xi_2, \eta_2) \mapsto (\eta_1, \eta_2)$ does not vanish at the origin, which is equivalent to having $\partial\eta_1/\partial\xi_2 \neq 0$ at the origin. Replacing ξ and η by ξ_1 and η_1, respectively, in the equation (B.5.7), it holds that

$$\begin{aligned} Y_\mathrm{P} = \boldsymbol{p}_{\xi_2}(0,0) &= \frac{\partial\xi_1}{\partial\xi_2}(0,0)\,\boldsymbol{p}_{\xi_1}(0,0) + \frac{\partial\eta_1}{\partial\xi_2}(0,0)\,\boldsymbol{p}_{\eta_1}(0,0) \\ &= \frac{\partial\xi_1}{\partial\xi_2}(0,0)X_\mathrm{P} + \frac{\partial\eta_1}{\partial\xi_2}(0,0)Y_\mathrm{P}. \end{aligned}$$

Here we used the relation $\dot\gamma_1(0) = Y_\mathrm{P}$. Thus, we have

$$\frac{\partial\xi_1}{\partial\xi_2}(0,0) = 0, \qquad \frac{\partial\eta_1}{\partial\xi_2}(0,0) = 1,$$

in particular, by the inverse function theorem (cf. Theorem A.1.5 in Appendix A.1), (η_1, η_2) gives a coordinate system on a neighborhood of P. An η_2-curve of this coordinate system is given as $\eta_1 =$ constant, which is a ξ_1-curve in the coordinate system (ξ_1, η_1). Then such a curve is tangent to X. Similarly, each η_1-curve is tangent to Y. Thus, we have the desired coordinate system $(u, v) = (\eta_2, \eta_1)$. □

We apply Lemma B.5.4 to prove the existence of the coordinate systems in Theorems B.5.1 and B.5.2. We fix a point P on the surface S and take a parametrization $\boldsymbol{p}(s,t)$ of the surface as in (B.5.1) and (B.5.2), and denote the coefficients of the first and second fundamental forms by E, F, G and L, M, N, respectively, with respect to the parameters (s,t).

Proof of the existence of curvature line coordinate systems. The two principal curvatures λ_1, λ_2 are mutually distinct except at umbilic points.

Hence the matrix

$$\begin{pmatrix} L-\lambda_j E & M-\lambda_j F \\ M-\lambda_j F & N-\lambda_j G \end{pmatrix} \qquad (j=1,2)$$

as in page 92 is of rank 1. Then, for each $j=1,2$, either $(L-\lambda_j E, M-\lambda_j F)$ or $(M-\lambda_j F, N-\lambda_j G)$ is not a zero vector, and at least one of two vector fields

$$X_1 = (-M+\lambda_1 F)\boldsymbol{p}_s + (L-\lambda_1 E)\boldsymbol{p}_t,$$
$$X_2 = (-N+\lambda_1 G)\boldsymbol{p}_s + (M-\lambda_1 F)\boldsymbol{p}_t$$

on the surface does not vanish at P. We denote by X one of the X_1 or X_2 such that $X_{\mathrm{P}} \neq \boldsymbol{0}$. Then by (9.8) in Section 9, X gives the principal direction with respect to λ_1 at each point of a neighborhood of P. Similarly, we take Y as one of

$$Y_1 = (-M+\lambda_2 F)\boldsymbol{p}_s + (L-\lambda_2 E)\boldsymbol{p}_t,$$
$$Y_2 = (-N+\lambda_2 G)\boldsymbol{p}_s + (M-\lambda_2 F)\boldsymbol{p}_t$$

such that $Y_{\mathrm{P}} \neq \boldsymbol{0}$, which gives the principal direction with respect to λ_2. Applying Lemma B.5.4, we get a new coordinate system, which is a curvature line coordinate system. □

Proof of the existence of asymptotic line coordinate systems. For a point P where the Gaussian curvature is negative, by changing coordinate if necessary, we may assume that $L \neq 0$ at P. In fact, if $L=0$ and $N \neq 0$ (resp. $L=N=0$), the replacement (s,t) by $(t,-s)$ (resp. (s,t) by $(s-t, s+t)$) gives the desired property. Then the asymptotic direction $\alpha\boldsymbol{p}_s + \beta\boldsymbol{p}_t$ satisfies

$$L\alpha^2 + 2M\alpha\beta + N\beta^2 = \frac{1}{L}\left((L\alpha+M\beta)^2 + (LN-M^2)\beta^2\right) = 0.$$

Hence, if we set $LN-M^2 = -k^2$, $(-M\pm k, L)$ corresponds to the asymptotic direction. A similar procedure as in the case of curvature line coordinate systems for vector fields

$$X = -(M+k)\boldsymbol{p}_s + L\boldsymbol{p}_t, \quad Y = -(M-k)\boldsymbol{p}_s + L\boldsymbol{p}_t$$

gives us an asymptotic line coordinate system. □

As another application of Lemma B.5.4, we show the fact that a curve having a corner can be rounded off to be a smooth curve, which is used in Theorem 3.5 in Section 3.

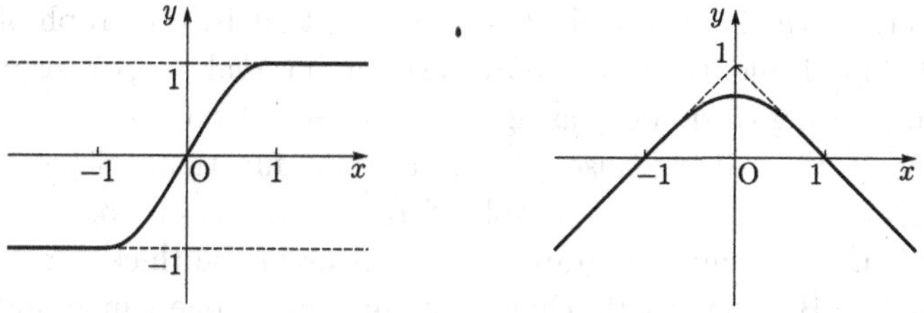

The graph of φ. The graph of f.

Fig. B.5.2

Proposition B.5.5. *Suppose that two plane curves $\gamma_1(t)$, $\gamma_2(t)$ intersect at $\mathrm{P} = \gamma_1(0) = \gamma_2(0)$ transversally. Then the resulting corner* P *can be rounded off smoothly (as a C^∞-regular curve).*

Proof. Consider a family of curves obtained by translating $\gamma_1(t)$ in the direction $\dot\gamma_2(0)$, and let X be a vector field consisting of the unit tangent vectors for the family of curves. Similarly, consider the family of curves obtained by translating $\gamma_2(t)$ in the direction of $\dot\gamma_1(0)$, and define a vector field Y as the unit tangent vectors of such a family of curves. Applying Lemma B.5.4 for these X and Y, we have a new coordinate system (ξ, η), such that P corresponds to the origin, and the ξ-axis and η-axis correspond to the curves γ_1 and γ_2, respectively. Thus, if one can round off the corner of two coordinate axes at the origin, we have the desired curve, by representing it in the original coordinate system.

We show how to round off the corner of two orthogonal half-lines (Fig. B.5.2, right). Let

$$\varphi(x) := \begin{cases} \left(\displaystyle\int_0^x e^{\frac{1}{u^2-1}}\,du\right) \Big/ \left(\displaystyle\int_0^1 e^{\frac{1}{u^2-1}}\,du\right) & (|x| < 1), \\ x/|x| & (|x| \geq 1), \end{cases}$$

see Fig. B.5.2, left. The integrand for $|x| < 1$ is nothing but the function ρ as in (10.5) in the proof of Theorem 10.5 in Section 10 for $c = 0$ and $\delta = 1$. Moreover, if we set

$$f(x) = -\int_{-1}^{x} \varphi(u)\,du,$$

its graph gives the desired curve, as shown in the thick curve in Fig. B.5.2, right.

In fact, $-\varphi(u)$ is identically 1 on $x \leq -1$, that is, the graph of $f(x)$ is a line of slope 1 and $f(-1) = 0$. On the open interval $(-1, 0)$ (resp. $(0, 1)$), $f(x)$ is increasing (resp. decreasing) because $-\varphi(u) > 0$ (resp. $-\varphi(u) < 0$). Moreover, $f(1) = 0$, because $\varphi(x)$ is an odd function. Since $-\varphi(u)$ is identically -1 on $x \geq 1$, the graph of $f(x)$ is a line of slope -1 on this interval. Summing up, the graph of $f(x)$ is as in the thick curves in Fig. B.5.2, right. By homothetic change of this curve, one can round off the curve on an arbitrarily small neighborhood of the corner. □

Exercise B.5

1 Consider a map $T\colon \boldsymbol{R}^3 \setminus \{\mathrm{O}\} \to \boldsymbol{R}^3$ which maps each point P in $\boldsymbol{R}^3$ different from the origin O to the point Q in $\boldsymbol{R}^3$ satisfying $\overrightarrow{\mathrm{OQ}} = \overrightarrow{\mathrm{OP}}/|\overrightarrow{\mathrm{OP}}|^2$. Such a map is called an *inversion* of $\boldsymbol{R}^3$ (the inversion with respect to the sphere centered at the origin with radius one). Show that the curvature lines of a surface $\boldsymbol{p} = \boldsymbol{p}(u, v)$ which do not pass through the origin correspond to the curvature lines of the inverted surface $T \circ \boldsymbol{p}$. (Hint: Show that the curvature line coordinate system is invariant under the inversion.)

B.6. Surfaces with $K = 0$

A surface whose Gaussian curvature K vanishes identically is called a *flat surface*. It is natural to ask if there are flat surfaces other than planes, circular cones, and circular cylinders.

Ruled surfaces. By bending a thin plastic board, one can get a surface which is isometric to the plane because the bending creates no expansion/contraction. This is one of the simplest examples of a flat surface. Typical examples of such surfaces are obtained by motions of lines, and called *ruled surfaces*, and the family of lines are said to be *generating lines*. As will be shown below, almost all flat surfaces are ruled surfaces.

Ruled surfaces are parametrized as follows: Let $\gamma(u)$ be a regular space curve, and let $\xi(u)$ a vector-valued smooth function which does not become zero. For each u, we consider $\xi(u)$ as a non-zero vector with initial point $\gamma(u)$. Introducing a new parameter v, the surface

$$\boldsymbol{p}(u, v) := \gamma(u) + v\xi(u) \qquad \text{(B.6.1)}$$

is a ruled surface. For example, a hyperbolic paraboloid $z = \frac{x^2}{a^2} - \frac{y^2}{b^2}$

$(a, b > 0)$ is expressed as a ruled surface in the following two ways:

$$p(u,v) = (au, 0, u^2) + v(a, \pm b, 2u).$$

Also, a hyperboloid of one sheet $\frac{x^2}{a^2} + \frac{y^2}{b^2} - \frac{z^2}{c^2} = 1$ $(a, b, c > 0)$ can be expressed as a ruled surface in the following two ways (Fig. B.6.1):

$$p(u,v) = (a\cos u, b\sin u, 0) + v(-a\sin u, b\cos u, \pm c).$$

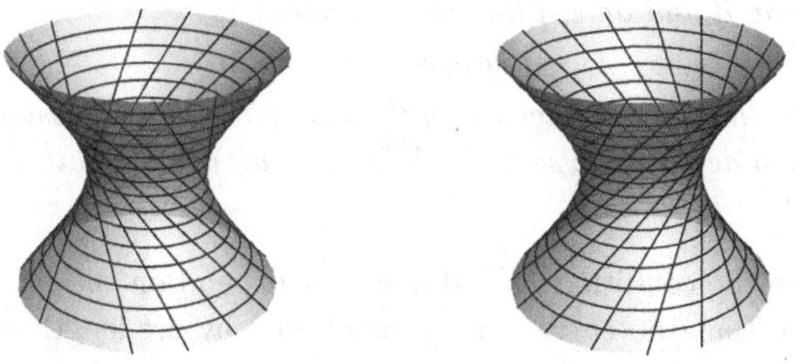

Fig. B.6.1

Developable surfaces. A *developable surface* is a regular ruled surface whose Gaussian curvature K is identically zero. We give a necessary and sufficient condition for a ruled surface to be developable. Consider a ruled surface as in (B.6.1). Then

$$\text{(B.6.2)} \qquad p_u = \dot{\gamma} + v\dot{\xi}, \qquad p_v = \xi \qquad \left(\dot{} = \frac{d}{du}\right),$$

and we have

$$p_{uu} = \ddot{\gamma} + v\ddot{\xi}, \qquad p_{uv} = \dot{\xi}, \qquad p_{vv} = \mathbf{0}.$$

We choose the unit normal vector $\nu(u,v)$ as in (6.9). Then the coefficients of the second fundamental forms are

$$L = p_{uu} \cdot \nu, \quad M = p_{uv} \cdot \nu = \dot{\xi} \cdot \nu, \quad N = p_{vv} \cdot \nu = 0,$$

and thus by (8.7), we have

$$(EG - F^2)K = LN - M^2 = -M^2 = -(\dot{\xi} \cdot \nu)^2.$$

This means that the Gaussian curvature of regular ruled surfaces is non-positive.

On the other hand, since (cf. Proposition A.3.5 in Appendix A.3)

$$\dot{\xi}\cdot\nu = \frac{\dot{\xi}\cdot(\boldsymbol{p}_u\times\boldsymbol{p}_v)}{|\boldsymbol{p}_u\times\boldsymbol{p}_v|} = \frac{\det(\dot{\xi},\boldsymbol{p}_u,\boldsymbol{p}_v)}{|\boldsymbol{p}_u\times\boldsymbol{p}_v|} = \frac{\det(\dot{\gamma},\xi,\dot{\xi})}{|\boldsymbol{p}_u\times\boldsymbol{p}_v|},$$

a necessary and sufficient condition for the surface to be developable is $\dot{\xi}\cdot\nu = 0$, that is,

$$\det(\dot{\gamma},\xi,\dot{\xi}) = 0. \qquad \text{(B.6.3)}$$

We give the following application of this fact for curvature lines.

Theorem B.6.1. *A regular space curve $\gamma(t)$ on the surface $\mathbf{p}(u,v)$ is a curvature line if and only if the ruled surface*

$$\boldsymbol{q}(t,w) := \gamma(t) + w\nu(t)$$

generated by the normal lines along the curve has zero Gaussian curvature, that is, $\boldsymbol{q}$ is a developable surface. Here, $\nu = \nu(t)$ is the unit normal vector of $\mathbf{p}$ at $\gamma(t)$.

Proof. Since the condition $K = 0$ does not depend on the parameters, we may assume that the curve γ is parametrized by arc-length s: $\boldsymbol{q}(s,w) = \gamma(s)+w\nu(s)$. By the condition (B.6.3) the Gaussian curvature of $\boldsymbol{q}$ vanishes identically if and only if

$$\det(\gamma',\nu,\nu') = 0 \qquad \left({}' = \frac{d}{ds}\right),$$

which is equivalent to the existence of a triple $(a(s), b(s), c(s))$ of functions satisfying

$$a\gamma' + b\nu + c\nu' = \mathbf{0}, \qquad (a,b,c) \neq \mathbf{0}. \qquad \text{(B.6.4)}$$

Taking the inner product of (B.6.4) with ν, one deduces that $b = 0$. Then $c \neq 0$ holds. In fact, $c = 0$ implies $\gamma' = \mathbf{0}$, contradicting that γ is parametrized by the arc-length. On the other hand, taking the inner product of (B.6.4) and $\boldsymbol{p}_u$ and noticing that $b = 0$, we have

$$\begin{aligned} 0 &= \boldsymbol{p}_u\cdot(a\gamma' + c\nu') \\ &= a\boldsymbol{p}_u\cdot(\boldsymbol{p}_u u' + \boldsymbol{p}_v v') + c\boldsymbol{p}_u\cdot(\nu_u u' + \nu_v v') \\ &= (aE - cL)u' + (aF - cM)v', \end{aligned}$$

where we set $\gamma(s) = \mathbf{p}(u(s), v(s))$. Taking the inner product with $\boldsymbol{p}_v$ similarly, we have $(aF - cM)u' + (aG - cN)v' = 0$. Then, noticing that $c \neq 0$, we obtain that the Gaussian curvature of $\boldsymbol{q}(s,w)$ vanishes identically if and only if

$$\left(L - \frac{a}{c}E\right)u' + \left(M - \frac{a}{c}F\right)v' = 0, \quad \left(M - \frac{a}{c}F\right)u' + \left(N - \frac{a}{c}G\right)v' = 0.$$

By (9.8) in Section 9, this is equivalent to a/c being a principal curvature with corresponding principal curvature direction (u', v'). □

Summing up this fact and the condition (B.6.3), we have the following corollary.

Corollary B.6.2. *A regular space curve $\gamma(t)$ on the surface is a curvature line if and only if*

$$\det(\dot{\gamma}, \nu, \dot{\nu}) = 0 \tag{B.6.5}$$

at each point of the curve, where $\nu = \nu(t)$ is the unit normal vector of the surface at $\gamma(t)$, and the overhead dot denotes d/dt.

The equation (B.6.5) is called the *differential equation for curvature lines.*

We now return to the discussion of developable surface, and prove the following assertion.

Theorem B.6.3. *The following three classes of regular ruled surfaces are developable.*

Cones: Surfaces generated by lines passing through one fixed point. A representative example is the circular cone.

Cylinders: Surfaces generated by parallel lines. A representative example is the circular cylinder.

Tangent developables: Surfaces given by $\boldsymbol{p}(u, v) = \gamma(u) + v\dot{\gamma}(u)$ for a space curve $\gamma(u)$.[1]

Conversely, a developable surface is obtained by joining these classes of surfaces smoothly, that is, there exists an open dense subset of the surface each of whose connected components is in one of these three classes.

Proof. When the surface is a cylinder or a cone, one can easily find a parametrization so that the equality (B.6.3) holds. Also, when the surface is a tangent developable, it is easy to see that (B.6.3) holds.

So it is sufficient to show the second assertion. Consider a ruled surface as in (B.6.1). Since ξ is a non-vanishing vector field along γ, we can write

$$\boldsymbol{p} = \gamma(u) + r\xi_1(u), \qquad \xi_1(u) := \frac{\xi(u)}{|\xi(u)|},$$

where $r = v|\xi(u)|$. We fix a point $(u, r) = (u_0, r_0)$ at which $\boldsymbol{p}_u$ and $\boldsymbol{p}_r$ are linearly independent. Then there exist open intervals I, J such that $(u_0, r_0) \in I \times J$ and $\boldsymbol{p}(u, r)$ gives a regular surface on $I \times J$. We set

$$U := \{u \in I \,|\, \dot{\xi}_1(u) \neq \boldsymbol{0}\}, \qquad A_1 := I \setminus U.$$

[1]A surface in this class does not satisfy the condition for a regular surface ((6.7) in Section 6) at $v = 0$. In fact, the image of the curve $\gamma(u)$ is the singular set.

We denote by W_1 the set of interior points of A_1 in I. W_1 may be empty. But if it is not an empty set, take a point $u_1 \in W_1$ arbitrarily. Let $I_1(\subset W_1)$ be the open interval defined as the connected component of W_1 containing u_1. Then $\dot{\xi}_1 = \mathbf{0}$ on I_1, and then $\xi_1(u)$ is a constant vector on I_1. Thus

$$\boldsymbol{p} = \gamma(u) + r\xi_1(u_1)$$

holds on $I_1 \times J$. Thus $\boldsymbol{p} = \boldsymbol{p}(u,r)$ is a cylinder on $I_1 \times J$. Since u_1 is chosen arbitrary, we can conclude that $\boldsymbol{p}$ consists of cylinders on $W_1 \times J$.

Differentiating $\xi_1(u) \cdot \xi_1(u) = 1$, we have $\xi_1(u) \cdot \dot{\xi}_1(u) = 0$. For $u \in U$, $\dot{\xi}_1(u) \neq \mathbf{0}$ holds, and so ξ_1 and $\dot{\xi}_1$ are linearly independent on U. So there are unique smooth functions $a(u)$ and $b(u)$ on U such that

$$\dot{\gamma}(u) = a(u)\xi_1(u) + b(u)\dot{\xi}_1(u) \qquad (u \in U).$$

We now set

$$\sigma(u) := \gamma(u) - b(u)\xi_1(u).$$

Then $\boldsymbol{p}$ can be rewritten as

$$\text{(B.6.6)} \qquad \boldsymbol{p} = \sigma(u) + s\xi_1(u) \qquad (s := r + b(u))\,.$$

By the definition of the curve σ,

$$\dot{\sigma}(u) = (a(u) - \dot{b}(u))\xi_1(u).$$

We then set

$$W_2 := \{u \in U \,|\, a(u) \neq \dot{b}(u)\}, \qquad A_2 := U \setminus W_2.$$

We denote by W_3 the set of interior points of A_2 in I. W_3 may be empty. But if it is not an empty set, take $u = u_0 \in W_3$ arbitrarily. Let $I_3(\subset W_3)$ be the open interval containing u_0 defined as the connected component of W_3. Then $\dot{\sigma} = 0$ on I_3, and $\sigma(u)$ is a constant vector on I_2. Thus $\boldsymbol{p}(u,s)$ is a cone on $I_3 \times J$. Since u_0 is chosen arbitrary, we can conclude that $\boldsymbol{p}$ consists of cones on $W_3 \times J$.

On the other hand, for $u \in W_2$, we can write

$$\text{(B.6.7)} \qquad \boldsymbol{p} = \sigma(u) + t\dot{\sigma}(u) \qquad \left(t := \frac{r}{a(u) - \dot{b}(u)}\right).$$

Thus $\boldsymbol{p}$ can be considered as a tangential developable on $W_2 \times J$. Since

$$(W_1 \times J) \cup (W_2 \times J) \cup (W_3 \times J)$$

is an open dense subset of $I \times J$, we get the assertion.

We remark that the sets W_1 and W_2 are usually empty, that is, the surface is a tangent developable in most cases. □

The geometric meaning of developability of surfaces. In general, a surface with $K = 0$ is locally (in the sense of Section 7) isometric to the plane, that is, there exists a parametrization with $E = G = 1$ and $F = 0$ (see Lemma 16.2 in Chapter III). Further, we are able to construct a local deformation of a given developable surface into a plane that preserves length and angles, as in the following Theorem B.6.4.

Theorem B.6.4. *Let* P *be a point on a developable surface* S *in* $\boldsymbol{R}^3$. *Then there exists a neighborhood* U *of* P *such that the restriction of the surface to* U *can be continuously deformed, preserving the first fundamental form, into a domain in a plane.*

Proof. A sufficiently small neighborhood of the point P can be parametrized as

$$\boldsymbol{q}(u,v) = \gamma(u) + v\xi(u) \qquad (|u|, |v| < \varepsilon) \tag{B.6.8}$$

without loss of generality, where ε is a positive number. Since $\boldsymbol{q}$ is a regular surface, $\dot{\gamma}$ and ξ are linearly independent. Since $\boldsymbol{q}$ is developable, the unit normal vector ν at $\boldsymbol{q}(u,v)$ depends only on u (cf. Problem **1** at the end of this section). So we denote it $\nu = \nu(u)$.

Moreover, we may assume $|\dot{\gamma}(u)| = 1$ and $|\xi(u)| = 1$. We denote by $\kappa_g(u)$ the geodesic curvature of the curve $\gamma(u)$ within the surface $\boldsymbol{q}(u,v)$. Then

$$\boldsymbol{n}_g := \nu(u) \times \dot{\gamma}(u)$$

gives the conormal vector, that is, a unit tangent vector of the surface to the left side of $\dot{\gamma}$ and perpendicular to $\dot{\gamma}$. The unit vector $\xi(u)$ can then be written as

$$\xi(u) = \cos\theta(u)\,\dot{\gamma}(u) + \sin\theta(u)\,\boldsymbol{n}_g(u). \tag{B.6.9}$$

By the definitions of the geodesic curvature κ_g and normal curvature κ_n (cf. Section 10), we have

$$\ddot{\gamma}(u) = \kappa_g(u)\boldsymbol{n}_g(u) + \kappa_n(u)\nu(u).$$

Also, $(\dot{\gamma}(u), \boldsymbol{n}_g(u), \nu(u))$ forms an orthonormal basis of $\boldsymbol{R}^3$, and thus $\dot{\boldsymbol{n}}_g(u)$ can be written as a linear combination of those basis vectors. Because $|\boldsymbol{n}_g(u)| = 1$ and $\boldsymbol{n}_g(u) \cdot \dot{\gamma}(u) = 0$,

$$\dot{\boldsymbol{n}}_g = -\kappa_g\dot{\gamma} + (\dot{\boldsymbol{n}}_g \cdot \nu)\nu$$

holds at u. Using this and differentiating (B.6.9), we have

$$\dot{\xi} = -(\dot{\theta} + \kappa_g)\sin\theta\,\dot{\gamma} + (\dot{\theta} + \kappa_g)\cos\theta\,\boldsymbol{n}_g + (\kappa_n\cos\theta + (\dot{\boldsymbol{n}}_g \cdot \nu)\sin\theta)\nu.$$

By the developability condition (B.6.3), we have

$$0 = \det(\dot\gamma, \xi, \dot\xi) = (\kappa_n \cos\theta + (\dot{\boldsymbol{n}}_g \cdot \nu)\sin\theta)\sin\theta.$$

Since $\dot\gamma$ and ξ are linearly independent, $\sin\theta \neq 0$ holds. Then we have

$$\text{(B.6.10)} \qquad \dot{\boldsymbol{n}}_g \cdot \nu = -\frac{\kappa_n \cos\theta}{\sin\theta}.$$

In particular, the coefficient of ν in the above representation of $\dot\xi$ must be $\mathbf{0}$, and so

$$\dot\xi(u) = (\dot\theta(u) + \kappa_g(u))(-\sin\theta(u)\,\dot\gamma(u) + \cos\theta(u)\,\boldsymbol{n}_g(u)).$$

Using this and noting that

$$\boldsymbol{q}_u(u,v) = \dot\gamma(u) + v(\dot\theta(u) + \kappa_g(u))(-\sin\theta(u)\,\dot\gamma(u) + \cos\theta(u)\,\boldsymbol{n}_g(u)),$$
$$\boldsymbol{q}_v(u,v) = \xi(u) = \cos\theta(u)\,\dot\gamma(u) + \sin\theta(u)\,\boldsymbol{n}_g(u),$$

the first fundamental form for q is

$$\text{(B.6.11)} \qquad E = 1 - 2v\sin\theta(\dot\theta + \kappa_g) + v^2, \qquad F = \cos\theta, \qquad G = 1.$$

For each $t \in [0,1]$, we consider the following ordinary differential equation (cf. Appendix A.2)

$$\text{(B.6.12)} \qquad \frac{d}{du}(\boldsymbol{e}_t(u), \boldsymbol{n}_t(u), \nu_t(u)) = (\boldsymbol{e}_t(u), \boldsymbol{n}_t(u), \nu_t(u)) \begin{pmatrix} 0 & -\kappa_g(u) & -t\kappa_n(u) \\ \kappa_g(u) & 0 & -tB(u) \\ t\kappa_n(u) & tB(u) & 0 \end{pmatrix},$$

where (see (B.6.10))

$$B(u) := -\frac{\kappa_n \cos\theta}{\sin\theta}.$$

By setting $(\boldsymbol{e}_t(0), \boldsymbol{n}_t(0), \nu_t(0))$ equal to the identity matrix as an initial condition, the vector-valued functions $\boldsymbol{e}_t(u)$, $\boldsymbol{n}_t(u)$, $\nu_t(u)$ satisfying (B.6.12) are uniquely determined. Moreover, using the same argument as in the proof of Theorem 5.2 in Section 5, $\boldsymbol{e}_t(u), \boldsymbol{n}_t(u), \nu_t(u)$ gives an orthonormal frame for each u. We set

$$\gamma_t(u) := \int_0^u \boldsymbol{e}_t(s)\,ds$$

and

$$\text{(B.6.13)} \qquad \boldsymbol{q}_t(u,v) := \gamma_t(u) + v\xi_t(u),$$
$$\xi_t(u) := (\cos\theta(u))\,\dot\gamma_t(u) + (\sin\theta(u))\,\boldsymbol{n}_t(u).$$

Then by definition, it holds that

$$\dot\xi_t = -\sin\theta(\dot\theta + \kappa_g)\dot\gamma_t + \cos\theta(\dot\theta + \kappa_g)\boldsymbol{n}_t + t(\kappa_n \cos\theta + B(u)\sin\theta)\nu_t,$$

and it can be easily checked that $\det(\dot\gamma_t, \xi_t, \dot\xi_t)$ vanishes, that is, the surface q_t $(0 \le t \le 1)$ is developable. By definition, $\boldsymbol{q}_1$ is congruent to $\boldsymbol{q}$. Since the unit normal vector $\nu_1(u)$ is a constant vector, the image of $\boldsymbol{q}_0$ lies in a plane in $\boldsymbol{R}^3$ perpendicular to ν_1. One can also easily check that the first fundamental form of $\boldsymbol{q}_t(u,v)$ is the same as that of $\boldsymbol{q}$. □

General flat surfaces. There exist a flat surfaces which are not ruled. See [17, page 68], for example.

Exercise B.6

1 Let $\boldsymbol{q}(u,v)=\gamma(u)+v\xi(u)$ be a developable surface, where $\gamma(u)$ is a regular curve in $\boldsymbol{R}^3$ and $\xi(u)$ is a vector-valued function in u such that $\dot\gamma(u)$ and $\xi(u)$ are linearly independent for each u. Prove that

$$\nu=\frac{\dot\gamma\times\xi}{|\dot\gamma\times\xi|}$$

is a unit normal vector field of $\boldsymbol{q}$.

B.7. A relationship between surfaces with constant Gaussian curvature and with constant mean curvature

We show in this section that surfaces of constant mean curvature can be obtained from surfaces of constant positive Gaussian curvature.

First, we consider a property of plane curves: Take a regular plane curve $\gamma(s)$ parametrized by the arc-length s, and let $\boldsymbol{n}(s)$ and $\kappa(s)$ be the leftward unit normal and the curvature, respectively. For a real constant t (which is not considered as a parameter of the curve), we define the new curve

$$\tilde\gamma(s):=\gamma(s)+t\boldsymbol{n}(s), \tag{B.7.1}$$

which is called the *parallel curve* of signed distance t of γ.

Lemma B.7.1. *The curvature $\tilde\kappa$ of the parallel curve $\tilde\gamma$ in* (B.7.1) *is related to the curvature κ of γ by*

$$\frac{1}{\tilde\kappa}=\frac{1}{\kappa}-t. \tag{B.7.2}$$

If $\tilde\gamma$ has a singular point, the leftward unit normal vector $\boldsymbol{n}$ of γ may change from the leftward normal to the rightward normal of $\tilde\gamma$ at the singular point. In this case, $\tilde\kappa$ as in (B.7.2) may differ by its sign from the usual curvature of $\tilde\gamma$.

Proof. Since the unit normal vector of $\tilde\gamma(s)$ is $\boldsymbol{n}(s)$, two curves γ and $\tilde\gamma$ have the same envelope of the normal lines. This envelope is the evolute of both γ and $\tilde\gamma$, that is, the locus of the centers of the osculating circles (see

Appendix B.1). Since the difference of the radii of the osculating circles of γ and $\tilde{\gamma}$ is t, we have (B.7.2). □

Similar to the case of plane curves, a surface obtained by translating the surface $\boldsymbol{p}(u,v)$ in the normal direction out to distance t is

$$\tilde{\boldsymbol{p}}(u,v) := \boldsymbol{p}(u,v) + t\,\nu(u,v), \tag{B.7.3}$$

called a *parallel surface* of $\boldsymbol{p}(u,v)$, where ν is the unit normal vector field of $\boldsymbol{p}$. A parallel surface may contain singular points, that is, points which do not satisfy (6.7) in Section 6. The following corresponds to (B.7.2) for the case of surfaces:

Theorem B.7.2. *The principal curvatures $\tilde{\lambda}_1$ and $\tilde{\lambda}_2$ of the surface $\tilde{\boldsymbol{p}}$ as in* (B.7.3) *are given in terms of the principal curvatures λ_1 and λ_2 of the surface $\boldsymbol{p}$ as*

$$\frac{1}{\tilde{\lambda}_j} = \frac{1}{\lambda_j} - t \qquad (j = 1, 2). \tag{B.7.4}$$

In particular, the Gaussian curvature $\widetilde{K}$ and mean curvature $\widetilde{H}$ of $\tilde{\boldsymbol{p}}$ can be expressed as

$$\widetilde{K} = \frac{K}{1 - 2tH + t^2K}, \qquad \widetilde{H} = \frac{H - tK}{1 - 2tH + t^2K}$$

using the Gaussian curvature K and mean curvature H of $\boldsymbol{p}$.

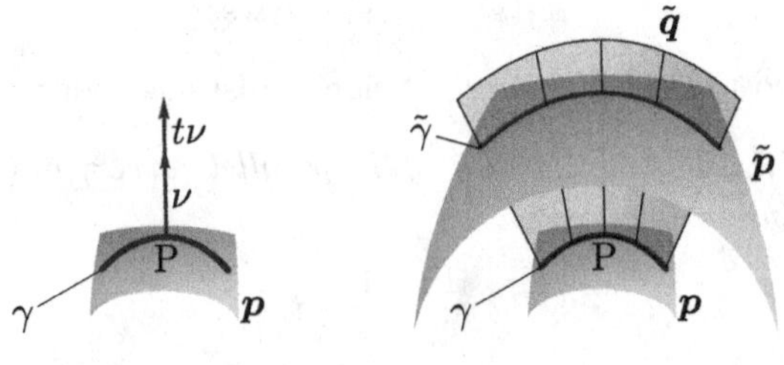

Fig. B.7.1 The proof of Theorem B.7.2.

Proof. It can be easily checked that ν_u and ν_v are perpendicular to ν, by the Weingarten formula (Proposition 8.5 in Section 8). Then the unit normal vector of $\tilde{\boldsymbol{p}}$ coincides with (the parallel translation of) ν. Consider a curvature line $\gamma(s) = \boldsymbol{p}(u(s), v(s))$ on $\boldsymbol{p}(u,v)$ passing through P, where s

is the arc-length parameter with $\gamma(0) = \mathrm{P}$ (Fig. B.7.1). By Theorem B.6.1 in Appendix B.6, the ruled surface

$$\boldsymbol{q}(s, w) := \gamma(s) + w\nu(u(s), v(s))$$

along γ is developable. Let t be a constant in (B.7.3), and set $w = t$. Then

$$\tilde{\gamma}(s) := \boldsymbol{q}(s, t)$$

is a curve on the surface $\tilde{\boldsymbol{p}}$, and the ruled surface generated by the normal lines (with the direction ν) coincides with $\boldsymbol{q}$, which is developable. Then by Theorem B.6.1 again, $\tilde{\gamma}(s)$ is a curvature line of $\tilde{\boldsymbol{p}}$.

Since the ruled surface $\boldsymbol{q}(s, w)$ is developable, and the Gaussian curvature is identically zero, there exists a correspondence from the surface $\boldsymbol{q}(s, w)$ to the plane preserving length (cf. such a surface can be realized as $\boldsymbol{q}_1(u, v)$ in (B.6.13)). Let σ and $\tilde{\sigma}$ be the corresponding plane curves of γ and $\tilde{\gamma}$. Then the curve $w \mapsto \boldsymbol{q}(s, w)$ corresponds to the line perpendicular to $\sigma(s)$, and then $\tilde{\sigma}$ is the parallel curve of σ at distance t. The normal curvature κ_n of $\gamma(s)$ as a curve on the surface $\boldsymbol{p}$ given by

$$\kappa_n = \gamma'' \cdot \nu = -\gamma' \cdot \nu'$$

is the geodesic curvature of γ as a curve on the surface $\boldsymbol{q}$. Moreover, the geodesic curvature depends only on the first fundamental form (cf. Section 14), and κ_n is the curvature of σ as a plane curve. Similarly, the normal curvature of $\tilde{\gamma}$ coincides with the curvature of $\tilde{\sigma}$. Hence by the relation (B.7.2) of the curvatures of parallel curves, the principal curvature $\tilde{\lambda}_j$ of $\tilde{\boldsymbol{p}}$ and λ_j satisfy (B.7.4). □

The proof of Theorem B.7.2 is incomplete in the strict sense. In fact, since the curvature line does not exist at an umbilic point of $\boldsymbol{p}$, the formula (B.7.4) has not been proved for such a point. However, even if P is an umbilic point, the formula holds if P is a limit of a sequence of non-umbilic points because of continuity of H and K. On the other hand, if all points on the surface inside the ball of ε centered at P are umbilic, then the surface is a part of a plane or a sphere (Proposition 9.7 in Section 9). If this is the case, the formula can be proved easily. So, we arrive at a complete proof.

Theorem B.7.2 can be proved by direct calculations as follows.

An alternative proof of Theorem B.7.2. On a neighborhood of a non-umbilic point, we assume that the surface is parametrized by a curvature line coordinate system as $\boldsymbol{p} = \boldsymbol{p}(u, v)$. Denote the coefficients of the first and second fundamental forms by E, F, G and L, M, N. Then by Theorem B.5.1 in Appendix B.5, $F = M = 0$ holds. Then the differentials of the unit normal vector ν are written as

$$\nu_u = -(L/E)\boldsymbol{p}_u, \qquad \nu_v = -(N/G)\boldsymbol{p}_v, \tag{B.7.5}$$

as a special case of the Weingarten formula (Proposition 8.5 in Section 8). Using these, the first and second fundamental forms of $\boldsymbol{p} + t\nu$ can be computed directly. □

In the proof above, the first and second fundamental matrices of $\boldsymbol{p} + t\nu$ are both diagonal, that is, (u, v) is a curvature line coordinate system of $\boldsymbol{p} + t\nu$ as well as of $\boldsymbol{p}$. So we have the following corollary.

Corollary B.7.3. *Let $\mathbf{p}(u, v)$ be a regular surface parametrized by a curvature line coordinate system, and let $\nu(u, v)$ the unit normal vector field of $\mathbf{p}$. Then $\nu(u, v)$ is also the unit normal vector field of the parallel surface $\tilde{\mathbf{p}}(u, v) := \mathbf{p}(u, v) + t\nu(u, v)$, and (u, v) is also a curvature line coordinate system of $\tilde{\mathbf{p}}(u, v)$. In particular, the curvature lines (on the uv-plane) are preserved by the operation of taking parallel surfaces.*

We give here a proof of the following fact, which was mentioned in page 81 in Section 8.

Proposition B.7.4. *A regular surface whose Gaussian and mean curvatures are both constant is a part of a plane, a circular cylinder or a sphere.*

Proof. If the Gaussian and mean curvatures are both constant, the principal curvatures λ_1 and λ_2 are also constant.

If $\lambda_1 = \lambda_2$, then all points of the surface are umbilic points, and the surface is a part of a plane or a sphere (cf. Exercise **3** in Section 9). So it is sufficient to consider the case that $\lambda_1 \neq \lambda_2$. Then we may assume $\lambda_1 \neq 0$. By replacing the unit normal vector field ν by $-\nu$ if necessary, we may assume $\lambda_1 > 0$ without loss of generality. We fix a point on a surface. Since $\lambda_1 \neq \lambda_2$, we can take a curvature line coordinate system as in Theorem B.5.1. Then $\lambda_1 = L/E$ and $\lambda_2 = N/G$ hold. Differentiating (B.7.5), we have

$$-\lambda_2 \boldsymbol{p}_{vu} = (\nu_v)_u = \nu_{uv} = (\nu_u)_v = -\lambda_1 \boldsymbol{p}_{uv}.$$

Since $\lambda_1 \neq \lambda_2$, we have $\boldsymbol{p}_{uv} = \mathbf{0}$. Then there exist two smooth vector-valued functions $\boldsymbol{a}(u)$, $\boldsymbol{b}(v)$ of one variable such that

$$\boldsymbol{p}(u, v) = \boldsymbol{a}(u) + \boldsymbol{b}(v). \tag{B.7.6}$$

Since

(B.7.7) $\boldsymbol{a}'(u)\cdot\boldsymbol{a}'(u) = E\,(>0), \quad \boldsymbol{a}'(u)\cdot\boldsymbol{b}'(v) = 0, \quad \boldsymbol{b}'(v)\cdot\boldsymbol{b}'(v) = G\,(>0),$

the two space curves $\boldsymbol{a}(u)$, $\boldsymbol{b}(v)$ are regular. So we can take u, v to be the arc-length parameters of $\boldsymbol{a}(u)$, $\boldsymbol{b}(v)$, respectively. After this change of parameters, we have

(B.7.8) $\boldsymbol{a}'(u)\cdot\boldsymbol{a}'(u) = 1, \quad \boldsymbol{a}'(u)\cdot\boldsymbol{b}'(v) = 0, \quad \boldsymbol{b}'(v)\cdot\boldsymbol{b}'(v) = 1.$

Then $\nu(0,0) = \boldsymbol{a}'(0)\times\boldsymbol{b}'(0)$ gives a unit normal vector of the surface $\boldsymbol{p}$ at $(0,0)$. By differentiating (B.7.8), $\boldsymbol{a}''(0)$ is perpendicular to $\boldsymbol{a}'(0)$ and $\boldsymbol{b}'(0)$, and so it is proportional to $\nu(0,0)$. Thus $|\boldsymbol{a}''(0)| = \lambda_1$ because $\lambda_1 > 0$. Since $\boldsymbol{a}''(0) \neq \boldsymbol{0}$, $\boldsymbol{a}'(0)$ and $\boldsymbol{a}'(u)$ are linearly independent whenever u $(\neq 0)$ is sufficiently small. Since

$$\boldsymbol{b}'(v)\cdot\boldsymbol{a}'(0) = \boldsymbol{b}'(v)\cdot\boldsymbol{a}'(u) = 0,$$

$\boldsymbol{b}'(v)$ is perpendicular to $\boldsymbol{a}'(0)$ and $\boldsymbol{a}'(u)$. Since $\boldsymbol{b}'(v)$ is a unit vector, we can conclude that $\boldsymbol{b}'(v)$ is a constant unit vector $\boldsymbol{c}$. So we have the expression

$$\boldsymbol{p}(u,v) = \boldsymbol{a}(u) + v\boldsymbol{c},$$

where $\boldsymbol{a}'(u)$ is perpendicular to $\boldsymbol{c}$ for all u. This surface p is a cylinder over the plane curve $\boldsymbol{a}(u)$, and so $\lambda_2 = 0$. Moreover the constancy of λ_1 implies that $\boldsymbol{a}(u)$ consists of a circle of radius $1/\lambda_1$. □

Theorem B.7.2 implies that, if the Gaussian curvature K of the surface $\boldsymbol{p}$ is a positive constant, then setting $t = 1/\sqrt{K}$, the parallel surface $\tilde{\boldsymbol{p}} = \boldsymbol{p} + (1/\sqrt{K})\nu$ is of constant mean curvature. On the other hand, if the mean curvature H is a non-zero constant, then letting $t = 1/(2H)$, a surface of positive constant Gaussian curvature is obtained.

In Section 8, we construct surfaces of revolution of constant Gaussian curvature (cf. page 84). Using these, one can construct surfaces of revolution of constant mean curvature (Fig. B.7.2). The left-side figure of Fig. B.7.2 is called an *unduloid*, and the right-side figure is called a *nodoid*. These surfaces of revolution are named *Delaunay*[1] *surfaces.*

[1]Delaunay, Charles Eugène (1816–1872).

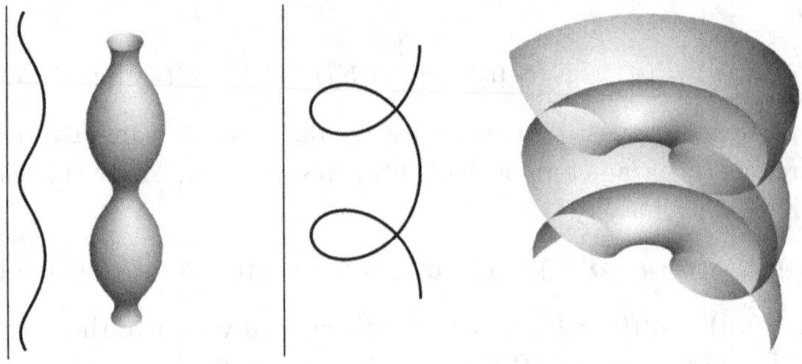

Fig. B.7.2 Surfaces of revolution of constant mean curvature.

Plotting Delaunay surfaces. Delaunay discovered that the generating curves of surfaces of revolution of constant mean curvature are obtained by rotating conics as follows:

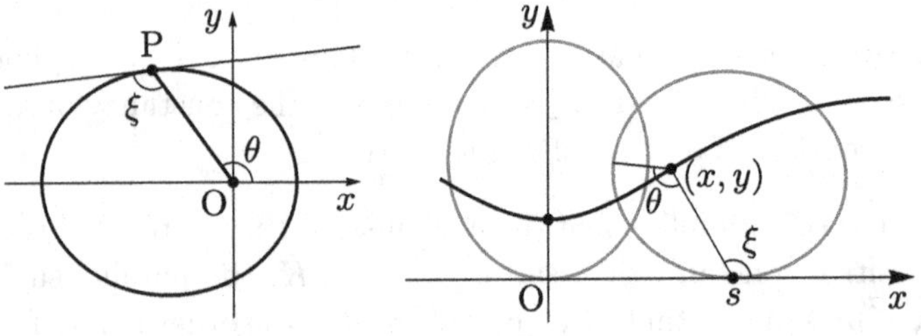

Fig. B.7.3 Plotting the generating curve of the unduloid.

The generating curve of an unduloid is obtained as the locus of a focal point of an ellipse while rolling it without slippage along a given line. In fact, the surface obtained by rotating this curve about the given line has constant mean curvature. We show this fact: As a preliminary, we notice that

$$r = r(\theta) = \frac{a}{1 + \varepsilon \cos\theta} \qquad (a > 0,\ 0 < \varepsilon < 1) \tag{B.7.9}$$

on the plane with respect to the polar coordinate system (r, θ) represents an ellipse such that O is one of the focal points, and ε is its eccentricity (cf. Problem **3** in Section 1). Since this ellipse can be parametrized as

$\gamma(\theta) := r(\theta)(\cos\theta, \sin\theta)$, the tangent vector at $\mathrm{P} = \gamma(\theta)$ is computed as

$$\dot{\gamma}(\theta)\left(=\frac{d\gamma}{d\theta}\right) = \left(\frac{-a\sin\theta}{(1+\varepsilon\cos\theta)^2}, \frac{a(\varepsilon+\cos\theta)}{(1+\varepsilon\cos\theta)^2}\right).$$

Let ξ be the angle between the vector $\overrightarrow{\mathrm{PO}}$ and the tangent of the ellipse at P (Fig. B.7.3, left). Then we have

$$\cos\xi = \frac{-\varepsilon\sin\theta}{\sqrt{1+2\varepsilon\cos\theta+\varepsilon^2}}, \qquad \sin\xi = \frac{1+\varepsilon\cos\theta}{\sqrt{1+2\varepsilon\cos\theta+\varepsilon^2}}.$$

We rotate the ellipse along the x-axis as in Fig. B.7.3, right. When the ellipse has rotated angle θ about the focal point, then the point tangent to the x-axis has traveled the distance $s(\theta)$, which is the arc-length of $\gamma(\theta)$, that is,

$$s(\theta) = \int_0^\theta \frac{a\sqrt{1+2\varepsilon\cos t+\varepsilon^2}}{(1+\varepsilon\cos t)^2}\,dt.$$

Then the focal point of the ellipse is represented as

$$(x, y) = (x(\theta), y(\theta)) = (s(\theta) + r(\theta)\cos\xi(\theta), r(\theta)\sin\xi(\theta)).$$

Here, the mean curvature of the surface obtained by rotating the curve $(x(\theta), y(\theta))$ around the x-axis is computed as

$$H = \frac{\dot{y}\ddot{x} - \ddot{y}\dot{x}}{2(\sqrt{\dot{x}^2+\dot{y}^2})^3} + \frac{\dot{x}}{2y\sqrt{\dot{x}^2+\dot{y}^2}},$$

by Meusnier's theorem (Problem **1** in Section 9). Here,

$$\dot{x} = \frac{a(1+\varepsilon\cos\theta)}{\Delta^3}, \quad \dot{y} = \frac{a\varepsilon\sin\theta}{\Delta^3}, \quad \sqrt{\dot{x}^2+\dot{y}^2} = \frac{a}{\Delta^2},$$

where $\Delta := \sqrt{1+2\varepsilon\cos\theta+\varepsilon^2}$. Hence we have $H = (1-\varepsilon^2)/(2a)$, which is a constant.

On the other hand, consider rotating a hyperbola along a line as in Fig. B.7.4.

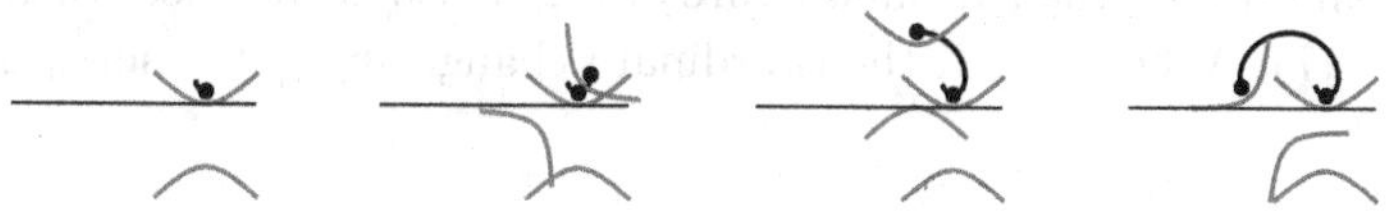

Fig. B.7.4 Plotting the generating curve of the nodoid.

Continuing the rotation, the tangent intersection of the hyperbola and the line moves out to infinity, and the line tends to the asymptotic line of the hyperbola. From this limit state, we continue rotating the other component of the hyperbola along the given line. Repeating these over and over again, the locus of the focal point of the hyperbola is the generating curve of a nodoid. In fact, the polar equation of the ellipse (B.7.9), used in the case of an unduloid, also represents a hyperbola when $\varepsilon > 1$. Then the mean curvature of the rotated surface can be computed similarly, which is constant.

Letting $\varepsilon = 1$, (B.7.9) represents a parabola whose focal point is the origin. The locus of the focal point by rotating the parabola along a line is a catenary (cf. Example 1.4 in Section 1) and the corresponding surface of revolution is a catenoid (Problem **7** in Section 8), which is a minimal surface.

By Theorem B.7.2, appropriate parallel surfaces of surfaces of revolution of constant mean curvature are surfaces of revolution of positive constant Gaussian curvature. They are smooth extensions of the surfaces of revolution of constant Gaussian curvature classified in Section 8 (see Fig. 8.2).

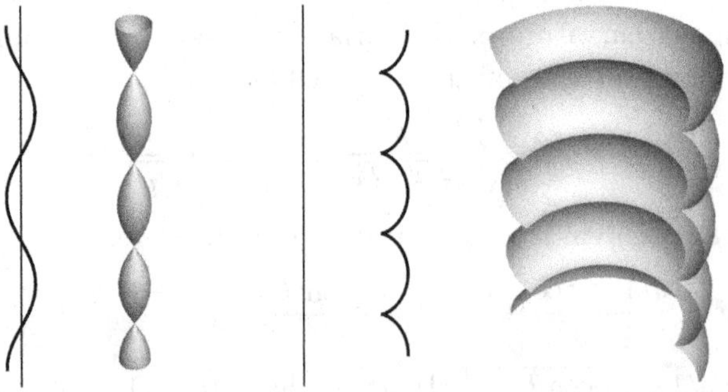

Fig. B.7.5 Extensions of surfaces of revolution of positive constant Gaussian curvature.

In fact, for the surface of revolution obtained in Section 8 by $x(u) = a\cos u$ and (8.17), the left-hand figure of Fig. B.7.5 is obtained when $a < 1$, from (8.17). When $a > 1$, the coordinate change $\sin w/2 = \sin\theta_0/2 \cdot \sin u$ yields

$$x(w) = a\left(1 - 2\sin^2\frac{\theta_0}{2}\cdot\sin^2 w\right),$$

$$z(w) = 2\sin\frac{\theta_0}{2}\int_0^w \cos^2 t\sqrt{\frac{1-\tan^2(\theta_0/2)\cdot\sin^2 t}{2-\sin^2(\theta_0/2)\cdot\sin^2 t}}\,dt,$$

where $a = 1/\sin\theta_0$ $(0 < \theta < \pi/2)$. Then $x(w)$ and $z(w)$ are meaningful for any real number w, and the surface extends periodically as in the right-side figure of Fig. B.7.5.

The left-side (resp. right-side) figure in Fig. B.7.5 is the surface of revolution of constant Gaussian curvature obtained by the parallel surface of the unduloid (resp. the nodoid). The surface obtained from the unduloid has cone-like singular points, and singularities of the surface obtained from the nodoid are cuspidal edges (cf. Proposition B.9.4 in Appendix B.9).

B.8. Surfaces of revolution of negative constant Gaussian curvature

Constant Gaussian curvature surfaces of revolution are classified in Section 8 for the cases that the Gaussian curvature is non-negative. In this section we classify surfaces of revolution of negative constant Gaussian curvature. By applying similarity expansions and reductions (homotheties), it is sufficient to consider the case $K = -1$.

As seen on page 84 in Section 8, the surface of revolution obtained by rotating a curve $\gamma(s) = (x(s), z(s))$ $(x(s) > 0)$ in the xz-plane parametrized by arc-length s around the z-axis has constant Gaussian curvature -1 if and only if $x'' = x$ from (8.15). The general solutions of this differential equation are well-known to be

(B.8.1) $$x(u) = \alpha e^{u} + \beta e^{-u} \qquad (\alpha \text{ and } \beta \text{ are constants}).$$

We divide these into the following three cases by the sign of the product $\alpha\beta$:

(i) The case that $\alpha\beta = 0$: Since $x > 0$, at least one of α and β has a non-zero value. Replacing u by $-u$ if necessary, we may assume that $\alpha = 0$ without loss of generality. In this case, $x = \beta e^{-u}$, where $\beta > 0$. Replacing $u - \log\beta$ by u, the expression is turned into $x = e^{-u}$. By (8.13), we may set $z' = \sqrt{1-(x')^2}$. Then we have

(B.8.2) $$\begin{cases} x(u) = e^{-u}, \\ z(u) = \displaystyle\int_0^u \sqrt{1-e^{-2t}}\,dt \\ \qquad = u - \sqrt{1-e^{-2u}} + \log(1+\sqrt{1-e^{-2u}}), \end{cases}$$

where $u > 0$. The curve on the xz-plane given in (B.8.2) is called the *tractrix*, which is characterized by the length of the tangent line segment of the curve to the z-axis being constantly 1. The surface of revolution corresponding to this curve is called the *pseudosphere* (Fig. B.8.1, left).

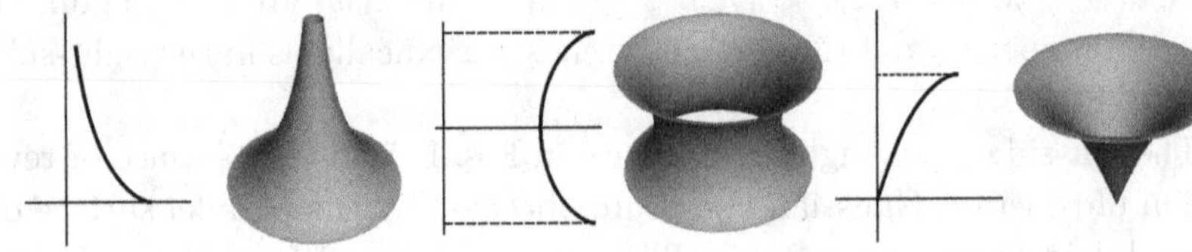

$\alpha\beta = 0$ (the pseudosphere) $\alpha\beta > 0$ $\alpha\beta < 0$

Fig. B.8.1 Surfaces of revolution of negative constant Gaussian curvature.

(ii) The case that $\alpha\beta > 0$: Since $x > 0$, both α and β must be positive. Then one can write

$$x(u) = \alpha e^u + \beta e^{-u} = \sqrt{\alpha\beta}\left(e^u\sqrt{\frac{\alpha}{\beta}} + e^{-u}\sqrt{\frac{\beta}{\alpha}}\right).$$

So, replacing $u + (1/2)\log(\alpha/\beta)$ by u, and setting $a = 2\sqrt{\alpha\beta}$, we may assume $x = a\cosh u$ without loss of generality. Similar to the case (i), we have

$$\text{(B.8.3)} \qquad x(u) = a\cosh u, \quad z(u) = \int_0^u \sqrt{1 - a^2\sinh^2 t}\,dt,$$

see Fig. B.8.1, middle.

(iii) The case that $\alpha\beta < 0$: Replacing u by $-u$ if necessary, we may assume $\alpha > 0$ and $\beta < 0$. Then one can write

$$x(u) = \alpha e^u - |\beta| e^{-u} = \sqrt{|\alpha\beta|}\left(e^u\sqrt{\left|\frac{\alpha}{\beta}\right|} - e^{-u}\sqrt{\left|\frac{\beta}{\alpha}\right|}\right).$$

So replacing $u + (1/2)\log(|\alpha/\beta|)$ by u, and setting $a = 2\sqrt{|\alpha\beta|}$, we may set $x = a\sinh u$ without loss of generality. Since $x > 0$, u must be positive, and since u is an arc-length parameter of γ, it holds that $|x'| = a\cosh u \leq 1$. Hence a must satisfy $0 < a < 1$, and we have

$$\text{(B.8.4)} \quad x(u) = a\sinh u, \quad z(u) = \int_0^u \sqrt{1 - a^2\cosh^2 t}\,dt \quad (0 < a < 1),$$

see Fig. B.8.1, right.

B.9. Criteria of typical singularities

In this section we introduce criteria for several types of singular points in curves and surfaces.

Singularities of plane curves. The plane curve

$$\gamma_0(t) := (t^2, t^3) \qquad (t \in \boldsymbol{R})$$

has a singular point at $t = 0$. This singular point is called the *standard cusp*. A curve $\gamma(t)$ $(a < t < b)$ has a *cusp* at $t = c$ if there exist a parameter change $t = t(u)$ and a coordinate change $x = x(\xi, \eta)$, $y = y(\xi, \eta)$ of $\boldsymbol{R}^2$ as on page 18 in Section 2 which maps a neighborhood of $\gamma(c)$ to one of the origin such that

$$\gamma(t(u)) = \Phi \circ \gamma_0(u),$$

where $\Phi(\xi, \eta) := (x(\xi, \eta), y(\xi, \eta))$. In particular, $\gamma_0(t)$ satisfies this condition for $t = 0$. That is, the standard cusp is a special case of cusps. Cusps appear most frequently in a parametrized plane curve.

To check that a singularity of a given plane curve is a cusp or not via direct use of a cusp's definition is not easy in general. However, the following convenient criterion is well-known (cf. [31]).

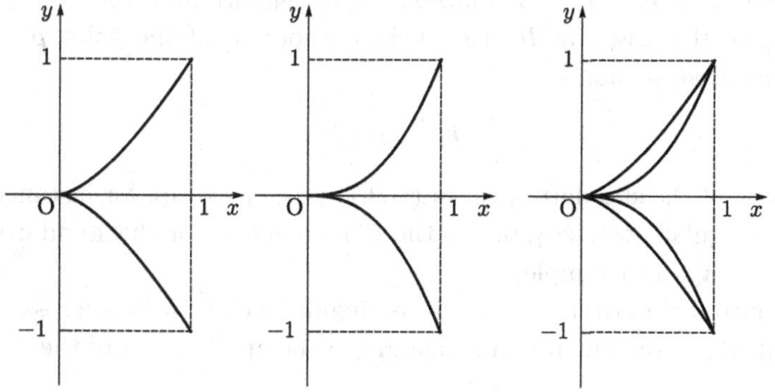

The cusp. The 5/2-cusp. The cusp and the 5/2-cusp.

Fig. B.9.1 The cusp and the 5/2-cusp.

Theorem B.9.1 (A criterion for cusps). *A curve $\gamma(t)$ has a cusp at $t = c$ if and only if $t = c$ is a singular point, that is, $\dot\gamma(c) = \mathbf{0}$, and $\ddot\gamma(c)$ and $\dddot\gamma(c)$ are linearly independent.*

One can easily check that the standard cusp $\gamma_0(t)$ satisfies this criterion at $t = 0$. In the parametrization (1.4) of the astroid in Section 1, the singular points on the interval $(-\pi, \pi]$ are $t = -\pi/2, 0, \pi/2, \pi$. It can be checked that these singular points are cusps, by differentiating the parametrization twice and then a third time.

Singular points different from cusps may appear in plane curves. For example, the curve $\sigma(t) = (t^2, t^5)$ has a singular point at $t = 0$, called a *5/2-cusp*, which does not satisfy the criterion in Theorem B.9.1. In fact, one can see that the shape of the 5/2-cusp is sharper than the cusp, as shown in Fig. B.9.1.

Singularities on surfaces. The following standard forms of four types of singularities on surfaces

$$\begin{aligned} &\boldsymbol{p}_0(u,v) := (u^2, uv, v) && \text{(cross cap)},\\ &\boldsymbol{p}_1(u,v) := (u, v^2, v^3) && \text{(cuspidal edge)},\\ &\boldsymbol{p}_2(u,v) := (3u^4 + u^2v, 4u^3 + 2uv, v) && \text{(swallowtail)},\\ &\boldsymbol{p}_3(u,v) := (u, v^2, uv^3) && \text{(cuspidal cross cap)} \end{aligned} \tag{B.9.1}$$

were introduced on page 65 of Section 6.

Assume the surface $\boldsymbol{p}(u,v)$ has a singular point at (u_0, v_0). If there exist two diffeomorphisms $\varphi\colon U_0 \to U$ and $\Phi\colon V_0 \to V$ from a neighborhood U_0 of the origin in $\boldsymbol{R}^2$ to a neighborhood U of (u_0, v_0), and from a neighborhood V_0 of the origin in $\boldsymbol{R}^3$ to a neighborhood V of the point $\boldsymbol{p}(u_0, v_0)$, respectively, such that

$$\Phi^{-1} \circ \boldsymbol{p} \circ \varphi$$

is just one of the standard form in (B.9.1) (the cross cap, for instance), we call the singular point (u_0, v_0) by the same name as for the standard form (the cross cap, for example).

We introduce criteria for a given singular point to be a cross cap, a cuspidal edge, swallowtail, and then give some applications of them.[1]

A criterion for cross caps. The following criterion for cross caps is well-known [43].

Proposition B.9.2 (Whitney's criterion). *Assume a surface $\boldsymbol{p}(u,v)$ has a singular point at $(u,v) = (u_0, v_0)$, and $\boldsymbol{p}_u(u_0, v_0) = \boldsymbol{0}$. Then (u_0, v_0) is a cross cap if and only if* $\det(\boldsymbol{p}_v, \boldsymbol{p}_{uu}, \boldsymbol{p}_{uv})$ *does not vanish at (u_0, v_0).*

[1] For a criterion of cuspidal cross cap, see [7].

Note that, for a cross cap singularity, one can choose a coordinate system such that $\boldsymbol{p}_u(u_0, v_0) = \mathbf{0}$, by a rotation of the uv-coordinate plane.

Example B.9.3. The standard cross cap $\boldsymbol{p}(u, v) = (u^2, uv, v)$ satisfies $\boldsymbol{p}_u(0, 0) = \mathbf{0}$, and

$$\boldsymbol{p}_v = (0, u, 1), \qquad \boldsymbol{p}_{uu} = (2, 0, 0), \qquad \boldsymbol{p}_{uv} = (0, 1, 0)$$

hold. In particular, these three vectors are linearly independent at $u = 0$. Then it satisfies the criterion in Proposition B.9.2.

Criteria for cuspidal edges and swallowtails. We first prepare some notations: For a function $f(u, v)$ of two variables, its gradient vector field is denoted by $\nabla f := (f_u, f_v)$. Moreover, we set

$$f_{\boldsymbol{w}} := \alpha f_u + \beta f_v (= \nabla f \cdot \boldsymbol{w}),$$

for each planar vector $\boldsymbol{w} = (\alpha, \beta)$, which is the *directional derivative*[2] interpreted as the partial derivative of f with respect to $\boldsymbol{w}$. Similarly, for a surface $\boldsymbol{p}(u, v)$ and a vector $\boldsymbol{w} = (\alpha, \beta)$, we denote

$$\boldsymbol{p}_{\boldsymbol{w}} := \alpha \boldsymbol{p}_u + \beta \boldsymbol{p}_v.$$

The following hold as common properties of cuspidal edges and swallowtails (see [33]):

(B.9.2) The *unit normal vector field* ν, that is the unit vector along the surface perpendicular to both $\boldsymbol{p}_u$ and $\boldsymbol{p}_v$, can be extended smoothly to the singular points.

(B.9.3) The function $\lambda := \det(\boldsymbol{p}_u, \boldsymbol{p}_v, \nu)$ vanishes on the singular points, but the gradient $\nabla\lambda$ does not vanish at the singular points.

We remark that the unit normal vector of a cross cap tends to various values as one approaches the singularity from various directions. That is, cross caps do not satisfy the first property (B.9.2).

Here, we assume the surface $\boldsymbol{p}(u, v)$ is defined on $U (\subset \boldsymbol{R}^2)$, and the singular point $(u_0, v_0) \in U$ satisfies the conditions (B.9.2) and (B.9.3). Using the function λ as in (B.9.3), the set of singular points (the singular set) of $\boldsymbol{p}$ on U is expressed as

$$\{(u, v) \in U \,|\, \lambda(u, v) = 0\},$$

[2]From the viewpoint of manifold theory, $\boldsymbol{p}_{\boldsymbol{w}}$ can be denoted as $d\boldsymbol{p}(\boldsymbol{w})$, which is the image of the tangent vector field $\alpha(\partial/\partial u) + \beta(\partial/\partial v)$ by the differential map $d\boldsymbol{p}$ of $\boldsymbol{p}$.

that is, it is expressed by the implicit function $\lambda(u,v)=0$. Since $\nabla\lambda\neq 0$ at (u_0,v_0), the implicit function theorem (Theorem A.1.6 in Appendix A.1, see also Section 1) yields that the singular set is a regular curve (cf. page 6) on a neighborhood of (u_0,v_0) in the uv-plane. We call this curve the *singular curve*. We parametrize the singular curve as $\gamma(t)=(u(t),v(t))$. Then at each point $\gamma(t)$ on the singular curve, $\boldsymbol{p}_u$ and $\boldsymbol{p}_v$ are linearly dependent, and do not vanish simultaneously, because of the condition (B.9.3).

In fact, by a differential formula for determinants (Proposition A.3.11 in Appendix A.3), we have

$$\lambda_u=\det(\boldsymbol{p}_u,\boldsymbol{p}_v,\nu)_u=\det(\boldsymbol{p}_{uu},\boldsymbol{p}_v,\nu)+\det(\boldsymbol{p}_u,\boldsymbol{p}_{vu},\nu)+\det(\boldsymbol{p}_u,\boldsymbol{p}_v,\nu_u),$$

where the right-hand side vanishes if $\boldsymbol{p}_u=\boldsymbol{p}_v=\boldsymbol{0}$. Similarly, $\lambda_v=0$ if $\boldsymbol{p}_u=\boldsymbol{p}_v=\boldsymbol{0}$. Hence, $\boldsymbol{p}_u=\boldsymbol{p}_v=\boldsymbol{0}$ implies $\lambda_u=\lambda_v=0$, contradicting (B.9.3). Hence $\boldsymbol{p}_u$ and $\boldsymbol{p}_v$ do not vanish at the same time.

Therefore, for each t, there exists a vector $\eta(t)=(a(t),b(t))(\neq(0,0))$ uniquely determined up to scalar multiplication, such that

$$\boldsymbol{p}_{\eta(t)}=a(t)\boldsymbol{p}_u(\gamma(t))+b(t)\boldsymbol{p}_v(\gamma(t))=\boldsymbol{0}. \tag{B.9.4}$$

We call $\eta(t)$ the *null vector* at the singular point $\gamma(t)$. Moreover, if $\eta(t)$ is chosen as a smooth (vector-valued) function in t, $\eta(t)$ is called the *null vector field*.

Using these notations, cuspidal edges and swallowtails satisfy

(B.9.5) The differential of the unit normal vector field ν with respect to the direction $\eta(t)$ does not vanish at the singularity $\gamma(t)$, that is, $\nu_{\eta(t)}\neq\boldsymbol{0}$.

In the situation above, we set

$$\mu(t):=\det(\dot\gamma(t),\eta(t))=\dot u(t)b(t)-\dot v(t)a(t), \tag{B.9.6}$$

where "$\cdot$" denotes the derivative with respect to t.

Proposition B.9.4 (Criteria for cuspidal edges and swallowtails [21]). *Assume that a surface* $\mathbf{p}(u,v)$ *has a singular point at* $(u,v)=(u_0,v_0)$, *and satisfies* (B.9.2), (B.9.3), (B.9.5) *in a neighborhood of the singular point. Then*

(1) (u_0,v_0) $(=\gamma(t_0))$ *corresponds to a cuspidal edge if and only if* $\mu(t_0)\neq 0$, *that is,* $\dot\gamma$ *and* η *are linearly independent at* $t=t_0$.
(2) (u_0,v_0) *corresponds to a swallowtail if and only if* $\mu(t_0)=0$ *and* $\dot\mu(t_0)\neq 0$.

It is easy to check that the standard cuspidal edge $\boldsymbol{p}_1$ and the standard swallowtail $\boldsymbol{p}_2$ as in (B.9.1) satisfy Proposition B.9.4 (Problems **3** and **4** in this section). We introduce some non-trivial examples here:

Example B.9.5 (Singularities of the pseudosphere). The generating curve (B.8.2) of the pseudosphere in Appendix B.8 is reparametrized as $(x(\tilde{u}), z(\tilde{u})) = (1/\cosh\tilde{u}, \tilde{u} - \tanh\tilde{u})$ by the parameter change $u = \log\cosh\tilde{u}$ ($\tilde{u} > 0$). These functions $x(\tilde{u})$, $z(\tilde{u})$ can be extended smoothly on $\tilde{u} \le 0$. Rotating this curve, the pseudosphere as in Fig. B.8.1 can be extended, as a C^∞-map to the surface that also includes its reflection across the xy-plane as

$$\text{(B.9.7)} \qquad \boldsymbol{p}(\tilde{u}, v) = \left(\frac{\cos v}{\cosh\tilde{u}}, \frac{\sin v}{\cosh\tilde{u}}, \tilde{u} - \tanh\tilde{u}\right).$$

This surface has singular points at the points where $\tilde{u} = 0$, and all of them are cuspidal edges (Fig. B.9.2, left). We show this fact in the following: Since

$$\boldsymbol{p}_{\tilde{u}} = \tanh\tilde{u}\left(-\frac{\cos v}{\cosh\tilde{u}}, -\frac{\sin v}{\cosh\tilde{u}}, \tanh\tilde{u}\right), \quad \boldsymbol{p}_v = \frac{1}{\cosh\tilde{u}}(-\sin v, \cos v, 0),$$

we have

$$\boldsymbol{p}_{\tilde{u}} \times \boldsymbol{p}_v = -\frac{\tanh\tilde{u}}{\cosh\tilde{u}}\left(\tanh\tilde{u}\cos v, \tanh\tilde{u}\sin v, \frac{1}{\cosh\tilde{u}}\right).$$

Then

$$\nu := \left(\tanh\tilde{u}\cos v, \tanh\tilde{u}\sin v, \frac{1}{\cosh\tilde{u}}\right)$$

is the unit normal vector, which is smoothly defined for all $(\tilde{u}, v)$, that is, (B.9.2) holds. Moreover, since

$$\lambda := \det(\boldsymbol{p}_{\tilde{u}}, \boldsymbol{p}_v, \nu) = -\frac{\tanh\tilde{u}}{\cosh\tilde{u}},$$

the singular set is $\tilde{u} = 0$, and (B.9.3) holds for each singular point. In addition, $\boldsymbol{p}_{\tilde{u}} = \boldsymbol{0}$ when $\tilde{u} = 0$, and the null vector field is expressed as $\eta = (1, 0)$ on the $\tilde{u}v$-plane, and $\nu_\eta = \nu_{\tilde{u}} \neq \boldsymbol{0}$ when $\tilde{u} = 0$, that is, (B.9.5) holds. Since the singular curve is parametrized as $\gamma(t) = (0, t)$, $\dot{\gamma}$ and η are linearly independent. Then by Proposition B.9.4, all singular points are cuspidal edges.

Two other classes of surfaces in Appendix B.8 can be extended as surfaces with singularities, as shown in Fig. B.9.2, center and right.

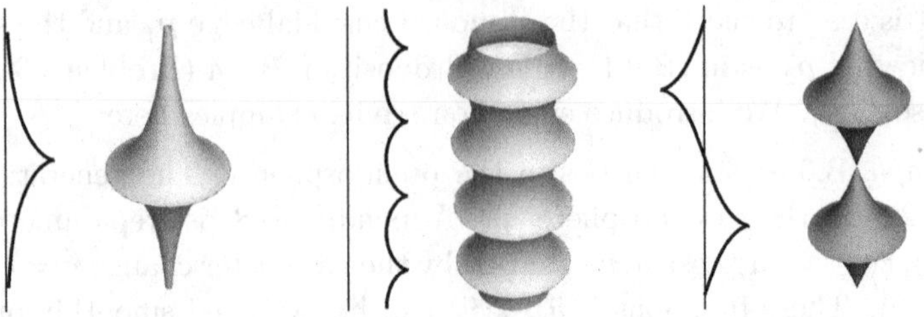

The pseudosphere. Bamboo-type. Japanese abacus-type.

Fig. B.9.2 Extensions of surface of revolution with $K = -1$.

Example B.9.6 (Kuen's surface). Define a surface $\boldsymbol{p} : \boldsymbol{R}^2 \to \boldsymbol{R}^3$ as

$$\boldsymbol{p}(u,v) = \left(\frac{2(\cos u + u \sin u)\cosh v}{u^2 + \cosh^2 v}, \frac{2(\sin u - u \cos u)\cosh v}{u^2 + \cosh^2 v}, v - \frac{2\sinh v \cosh v}{u^2 + \cosh^2 v} \right),$$

which is a famous surface called *Kuen's surface* [22] (see Fig. B.9.3). This surface has constant Gaussian curvature -1 at its regular points (cf. Problem **6** in this section). We verify this surface has two swallowtails and cuspidal edges.

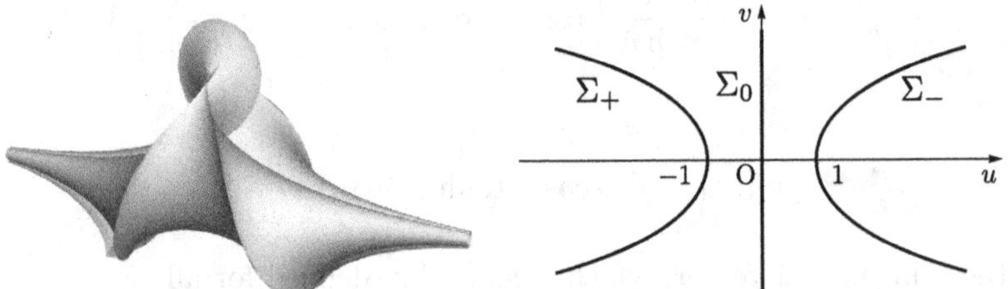

Fig. B.9.3 Kuen's surface and its singular set.

We set

$$\text{(B.9.8)} \qquad \alpha := u^2 + \cosh^2 v, \qquad \beta := u^2 - \cosh^2 v.$$

Then we have

$$\boldsymbol{p}_u = \frac{2u\cosh v}{\alpha^2}((\alpha - 2)\cos u - 2u \sin u, (\alpha - 2)\sin u + 2u\cos u, 2\sinh v),$$

$$\boldsymbol{p}_v = \frac{2\beta}{\alpha^2}(\sinh v(\cos u + u \sin u), \sinh v(\sin u - u\cos u), \beta + 2).$$

Thus,

$$\nu := \frac{1}{\alpha}(2u\cos u + \beta\sin u, 2u\sin u - \beta\cos u, -2u\sinh v) \tag{B.9.9}$$

satisfies $\nu\cdot\nu = 1$, $\nu\cdot\boldsymbol{p}_u = \nu\cdot\boldsymbol{p}_v = 0$, that is, ν is a unit normal vector field, and then (B.9.2) is satisfied. Moreover, since λ in (B.9.3) is computed as

$$\lambda = \det(\boldsymbol{p}_u, \boldsymbol{p}_v, \nu) = \frac{2u\beta\cosh v}{\alpha^2} = \frac{2u(u^2 - \cosh^2 v)\cosh v}{(u^2 + \cosh^2 v)^2},$$

the singular set is the union $\Sigma_0 \cup \Sigma_+ \cup \Sigma_-$ of three curves

$$\Sigma_0 := \{u = 0\}, \quad \Sigma_+ := \{u = \cosh v\}, \quad \Sigma_- := \{u = -\cosh v\}$$

on the uv-plane (Fig. B.9.3, bottom). In particular, $\lambda_u \neq 0$ holds on the singular set, and (B.9.3) is verified. Among these singular points, we show that two points $(u, v) = (\pm 1, 0)$ are swallowtails. We focus on the singular curve Σ_+. Since $u = \cosh v$ on Σ_+, $\beta = 0$ in (B.9.8), and we get $\boldsymbol{p}_v = \boldsymbol{0}$. Hence the null vector field is $\eta = (0, 1)$. Then

$$\begin{aligned}\nu_\eta = \nu_v = -\frac{4u\cosh v}{\alpha^2}(&(\cos u + u\sin u)\sinh v, \\ &(\sin u - u\cos u)\sinh v, 2 - \beta).\end{aligned}$$

Since $\cos u + u\sin u$ and $\sin u - u\cos u$ do not vanish at the same time, $\nu_\eta = \boldsymbol{0}$ happens only when $u = 0$. However, this is impossible, because $u = \cosh v \neq 0$ on Σ_+. Since $\gamma(t) = (\cosh t, t)^T$ gives a parametrization of Σ_+ and $\eta(t) = (0, 1)^T$ is a null vector for each t, we have

$$\mu = \det\begin{pmatrix}\sinh t & 0 \\ t & 1\end{pmatrix} = \sinh t.$$

In particular, $\mu(0) \neq 0$ for $t \neq 0$, which implies that Σ_+ consists of cuspidal edges except at $(1, 0)$. Since $\dot{\mu}(0) = 1 (\neq 0)$, the point $\gamma(0) = (1, 0)$ is a swallowtail (cf. (2) in Proposition B.9.4). Similarly, the point $(-1, 0)$ on Σ_- is also a swallowtail and the other points consists of cuspidal edges.

On the other hand, since Σ_0 is the v-axis and f_u vanishes on Σ_0, we can conclude that the null direction $\eta := \partial/\partial u$ is not proportional to the singular direction (i.e., tangential direction of the singular curve) $\partial/\partial v$ on Σ_0. Since

$$\nu_\eta|_{u=0} = \nu_u|_{u=0} = \frac{1}{2\cosh^2 v}(3 - \cosh 2v, 0, -4\sinh v)$$

does not vanish for all $v \in \boldsymbol{R}$, we can conclude that Σ_0 consists only of cuspidal edges.

Exercises B.9

1 Show that the singular points of the cycloid in Problem **4** in Section 1 are cusps.

2 A curve $\gamma(s) = (1-\sin s)(\cos s, \sin s)$ ($\sigma_{1/2}$ in Fig. 3.2 of Section 3, called the *cardioid*) has a singular point at $s = \pi/2$. Show that this singular point is a cusp.

3 Show that the standard cuspidal edge (p_1 in (B.9.1)) satisfies the criterion for cuspidal edges given in Proposition B.9.4.

4 Show that the singular set of the standard swallowtail ($\boldsymbol{p}_2$ in (B.9.1)) is a graph of $v = -6u^2$ on the uv-plane, which is a parabola. Moreover, show that $(0,0)$ satisfies the criterion for swallowtails given in Proposition B.9.4, and other singular points are all cuspidal edges, using Proposition B.9.4.

5 Consider the helix

$$\gamma(t) = (a\cos t, a\sin t, bt) \qquad (a, b \neq 0)$$

as in Example 5.1 in Section 5 and its tangent developable

$$\boldsymbol{p}(u,v) := \gamma(u) + v\dot\gamma(u)$$

(cf. Theorem B.6.3 in Appendix B.6, see right figure). Show that the singular points of $\boldsymbol{p}$ are all cuspidal edges.

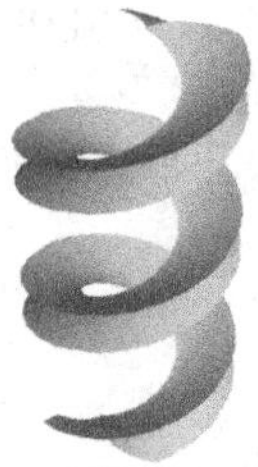

6 Show that the Gaussian curvature of Kuen's surface in Example B.9.6 is -1 on the set of regular points.

B.10. Proof of the fundamental theorem for surfaces

In this section, we give a proof of the fundamental theorem for surface theory (Theorem 17.2) in Section 17 of Chapter III.

Preliminaries. We prepare some notations: Let n be a positive integer. We denote the set of $n \times n$ real matrices by $\mathrm{M}_n(\boldsymbol{R})$. A map $\mathcal{F}\colon D \to \mathrm{M}_n(\boldsymbol{R})$ from a domain D in $\boldsymbol{R}^2$ to $\mathrm{M}_n(\boldsymbol{R})$ (a matrix-valued function) is said to be of class C^∞ if all n^2 components of $\mathcal{F}$ are C^∞-functions defined on D. By $\mathrm{GL}_n(\boldsymbol{R})$, we denote the set of $n \times n$ regular matrices, that is,

$$\mathrm{GL}_n(\boldsymbol{R}) := \{A \in \mathrm{M}_n(\boldsymbol{R}) \mid \det A \neq 0\}.$$

The set $\mathrm{GL}_n(\boldsymbol{R})$ can be considered as a group with respect to matrix multiplication, and called the *general linear group*.

For any two given matrix-valued C^∞-functions $\Omega(u,v)$ and $\Lambda(u,v)$ defined on a domain D in $\boldsymbol{R}^2$, consider the system of differential equation

$$\text{(B.10.1)} \qquad \begin{cases} \mathcal{F}_u = \mathcal{F}\Omega, \\ \mathcal{F}_v = \mathcal{F}\Lambda \end{cases}$$

for a matrix-valued unknown $\mathcal{F}\colon D \to \mathrm{M}_n(\boldsymbol{R})$. The equation (17.3) in Section 17 is a special case of (B.10.1) for $n = 3$. The following lemma holds.

Lemma B.10.1. *If* $\mathcal{F} : D \to \mathrm{GL}_n(\boldsymbol{R})$ *satisfies* (B.10.1), *then* Ω *and* Λ *satisfy*

$$\text{(B.10.2)} \qquad \Lambda\Omega + \Omega_v = \Omega\Lambda + \Lambda_u.$$

As we will see in Theorem B.10.4, the solution $\mathcal{F}$ of (B.10.1) does exist if (B.10.2) holds. Because of this fact, (B.10.2) is the *integrability condition*, also called the *compatibility condition*.

Proof. Differentiating the first equation of (B.10.1) with respect to v, we have

$$\mathcal{F}_{uv} = (\mathcal{F}_u)_v = (\mathcal{F}\Omega)_v = \mathcal{F}_v\Omega + \mathcal{F}\Omega_v = (\mathcal{F}\Lambda)\Omega + \mathcal{F}\Omega_v = \mathcal{F}(\Lambda\Omega + \Omega_v).$$

Similarly, differentiating the second equation with respect to u, we have

$$\mathcal{F}_{vu} = (\mathcal{F}_v)_u = (\mathcal{F}\Lambda)_u = \mathcal{F}_u\Lambda + \mathcal{F}\Lambda_u = \mathcal{F}(\Omega\Lambda + \Lambda_u).$$

Here we used the Leibniz formula for the matrix-valued functions (Proposition A.3.9 of Appendix A.3). Since $\mathcal{F}$ is of class C^∞, $\mathcal{F}_{uv} = \mathcal{F}_{vu}$ holds. Noticing that $\mathcal{F}$ is a regular matrix,

$$\Lambda\Omega + \Omega_v = \mathcal{F}^{-1}\mathcal{F}_{uv} = \mathcal{F}^{-1}\mathcal{F}_{vu} = \Omega\Lambda + \Lambda_u,$$

proving the lemma. □

By Lemma B.10.1, the integrability condition (B.10.2) is a necessary condition for the equation (B.10.1) to have solutions. Moreover, the converse assertion (Theorem B.10.4) holds. First, we show the following special case:

Proposition B.10.2. *Let* $D := \{(u,v) \in \boldsymbol{R}^2 \mid |u| < a,\ |v| < b\}$ $(a, b > 0)$ *be a rectangular domain in the uv-plane. Take a matrix* $R \in \mathrm{M}_n(\boldsymbol{R})$. *If two matrix-valued* C^∞*-functions* Ω, $\Lambda\colon D \to \mathrm{M}_n(\boldsymbol{R})$ *satisfy the integrability condition* (B.10.2), *there exists a unique matrix-valued function* $\mathcal{F}\colon D \to \mathrm{M}_n(\boldsymbol{R})$ *satisfying the differential equation* (B.10.1) *with initial condition*

$$\text{(B.10.3)} \qquad \mathcal{F}(0,0) = R.$$

Proof. First we prove the uniqueness: If there exists such an $\mathcal{F}$ as in the conclusion, the restriction $f(u) := \mathcal{F}(u,0)$ of $\mathcal{F}$ to the u-axis satisfies the initial value problem of the linear ordinary equation

$$\frac{df}{du}(u) = f(u)\Omega(u,0), \qquad f(0) = R. \tag{B.10.4}$$

By the uniqueness part of Theorem A.2.2 in Appendix A.2, $f(u)$ is determined uniquely by R. Next, we fix u and consider $\mathcal{F}(u,v)$ as a function of v. Then it satisfies the initial value problem of the linear ordinary differential equation

$$\frac{d\mathcal{F}}{dv}(u,v) = \mathcal{F}(u,v)\Lambda(u,v), \qquad \mathcal{F}(u,0) = f(u). \tag{B.10.5}$$

Then by the uniqueness part of Theorem A.2.2, $\mathcal{F}(u,v)$ is determined uniquely from $f(u)$. Thus, the uniqueness of $\mathcal{F}$ is proven.

Next, we show the existence of the solution $\mathcal{F}$: By the existence theorem of linear ordinary differential equations (Theorem A.2.2), there exists a function $f(u)$ ($|u| < a$) satisfying (B.10.4). Applying Theorem A.2.2 again, there exists $\mathcal{F}(u,v)$ ($|u| < a$, $|v| < b$) satisfying (B.10.5). Though the coefficient matrix Λ of the equation (B.10.5) contains the parameter u, the solution $\mathcal{F}(u,v)$ is C^∞ in (u,v), by the regularity of solutions of ordinary differential equations (the last statement in Theorem A.2.2). By (B.10.4) and (B.10.5), such an $\mathcal{F}$ satisfies the initial condition (B.10.3) and the second equation in (B.10.1). So it is sufficient to show that $\mathcal{F}$ satisfies the first equation of (B.10.1). Let

$$\mathcal{G} := \mathcal{F}_u - \mathcal{F}\Omega. \tag{B.10.6}$$

Since $\mathcal{F}$ satisfies the second equation in (B.10.1), it holds that

$$\begin{aligned}\mathcal{G}_v &= \mathcal{F}_{uv} - \mathcal{F}_v\Omega - \mathcal{F}\Omega_v = (\mathcal{F}_v)_u - \mathcal{F}\Lambda\Omega - \mathcal{F}\Omega_v \\ &= (\mathcal{F}\Lambda)_u - \mathcal{F}(\Lambda\Omega + \Omega_v).\end{aligned}$$

Using the integrability condition (B.10.2), we have

$$\mathcal{G}_v = \mathcal{F}_u\Lambda + \mathcal{F}\Lambda_u - \mathcal{F}(\Omega\Lambda + \Lambda_u) = \mathcal{F}_u\Lambda - \mathcal{F}\Omega\Lambda = \mathcal{G}\Lambda.$$

By (B.10.5) and (B.10.6), it holds that

$$\mathcal{G}(u,0) = \mathcal{F}_u(u,0) - \mathcal{F}(u,0)\Omega(u,0) = \frac{df(u)}{du} - f(u)\Omega(u,0) = O.$$

Then for each fixed u, $\mathcal{G}$ satisfies the initial value problem of the linear ordinary differential equation

$$\frac{d\mathcal{G}}{dv}(u,v) = \mathcal{G}(u,v)\Lambda(u,v), \qquad \mathcal{G}(u,0) = O. \tag{B.10.7}$$

On the other hand, (B.10.7) has a trivial solution which is identically O. Hence, by the uniqueness of the solution, we have $\mathcal{F}_u - \mathcal{F}\Omega = \mathcal{G} = O$, that is, the first equation of (B.10.1) is satisfied. □

Corollary B.10.3. *Assume Ω and Λ in Proposition B.10.2 are skew-symmetric matrices, and the initial matrix R is an orthogonal matrix. Then for each $(u,v) \in D$, the matrix $\mathcal{F}(u,v)$ is an orthogonal matrix.*

Proof. By (B.10.4) in the proof of Proposition B.10.2,

$$\begin{aligned}\frac{d}{du}(f(u)f(u)^T) &= \frac{df(u)}{du}f(u)^T + f(u)\frac{df(u)^T}{du} \\ &= f(u)\Omega(u,0)f(u)^T + f(u)\Omega(u,0)^T f(u)^T \\ &= f(u)\Omega(u,0)f(u)^T - f(u)\Omega(u,0)f(u)^T = O.\end{aligned}$$

In particular, $f(u)f(u)^T$ does not depend on u, that is,

$$f(u)f(u)^T = f(0)f(0)^T = RR^T = I,$$

where I is the $n \times n$ identity matrix. Hence $f(u)$ is an orthogonal matrix for each u.

Next, by (B.10.5),

$$\begin{aligned}(\mathcal{F}(u,v)\mathcal{F}(u,v)^T)_v &= \mathcal{F}_v(u,v)\mathcal{F}^T(u,v) + \mathcal{F}(u,v){\mathcal{F}_v}^T(u,v) \\ &= \mathcal{F}(u,v)\Lambda(u,v)\mathcal{F}(u,v)^T + \mathcal{F}(u,v)\Lambda(u,v)^T\mathcal{F}(u,v)^T \\ &= \mathcal{F}(u,v)\Lambda(u,v)\mathcal{F}(u,v)^T - \mathcal{F}(u,v)\Lambda(u,v)\mathcal{F}(u,v)^T \\ &= O,\end{aligned}$$

in particular $\mathcal{F}(u,v)\mathcal{F}(u,v)^T$ does not depend on v. So letting $v = 0$ and noticing that $f(u) = \mathcal{F}(u,0)$ is an orthogonal matrix, we have

$$\mathcal{F}(u,v)\mathcal{F}(u,v)^T = f(u)f(u)^T = I.$$

Hence $\mathcal{F}(u,v)$ is an orthogonal matrix. □

Proposition B.10.2 can be extended for general simply connected domains (cf. page 137), as follows:

Theorem B.10.4. *Let (u_0, v_0) be a fixed point in a simply connected domain $D \subset \boldsymbol{R}^2$ in the uv-plane, and take an $n \times n$ regular matrix R. If two matrix-valued C^∞-functions Ω, $\Lambda : D \to \mathrm{M}_n(\boldsymbol{R})$ satisfy* (B.10.2), *there exists a unique matrix-valued C^∞ function $\mathcal{F} : D \to \mathrm{M}_n(\boldsymbol{R})$ satisfying the initial condition $\mathcal{F}(u_0, v_0) = R$ and equation* (B.10.1).

Moreover, if Ω and Λ take values in the skew-symmetric matrices and R is orthogonal, then $\mathcal{F}$ takes values in the orthogonal matrices.

To prove this theorem, we prepare the following lemma.

Lemma B.10.5. *Let (u_0, v_0) be a fixed point in a simply connected domain $D \subset \boldsymbol{R}^2$ in the uv-plane. Then there exists a diffeomorphism $\Phi\colon D \to D_0$ mapping (u_0, v_0) to the origin $(0,0)$ of the st-plane, where*

$$D_0 := \{(s,t) \in \boldsymbol{R}^2 \mid s^2 + t^2 < 1\},$$

i.e., the unit disc on the st-plane.

Proof. We identify both the uv-plane and the st-plane with the complex plane $\boldsymbol{C}$, and write the unit disc as $D_0 = \{z \in \boldsymbol{C} \mid |z| < 1\}$. If D does not coincide with $\boldsymbol{R}^2$, Riemann's mapping theorem[1] implies that there exists a biholomorphic map Ψ which maps D to D_0, that is, the holomorphic map Ψ is a bijection and its inverse is also holomorphic. In particular, Ψ is a diffeomorphism from D to D_0. When $D = \boldsymbol{R}^2$, the map

$$\Psi(u,v) = \frac{1}{\sqrt{1+u^2+v^2}}(u,v)$$

gives a diffeomorphism $\Psi\colon \boldsymbol{R}^2 \to D_0$. Next, for a complex number $\alpha := \Psi(u_0, v_0) \in D_0$, the Möbius transformation (see Section 4)

$$T(z) = \frac{z-\alpha}{1-\bar{\alpha}z} \qquad (z \in D_0)$$

gives a diffeomorphism from D_0 to D_0 such that $T(\alpha) = 0$. Hence $\Phi := T \circ \Psi$ is the desired map. □

With the preparations above, we shall now prove Theorem B.10.4.

Proof of Theorem B.10.4. By Lemma B.10.5, there exists a diffeomorphism $\Phi_1 : D \to D_0$ such that $\Phi_1(u_0, v_0) = (0,0)$. Applying Lemma B.10.5 again, there exists a diffeomorphism $\Phi_2\colon D_1 \to D_0$ with $\Phi_2(0,0) = (0,0)$, where $D_1 := \{(\xi,\eta) \in \boldsymbol{R}^2 \mid |\xi| < 1, |\eta| < 1\}$ is a square domain. So we set

$$(u(\xi,\eta), v(\xi,\eta)) := \Phi_2^{-1} \circ \Phi_1(\xi,\eta). \tag{B.10.8}$$

By this coordinate change, the domain D in the uv-plane corresponds to the square domain D_1 in the $\xi\eta$-plane, and (u_0, v_0) corresponds to the origin on the $\xi\eta$-plane. Changing variables in (B.10.1) to (ξ,η), it is equivalent to

$$\mathcal{F}_\xi = \mathcal{F}\widetilde{\Omega}, \qquad \mathcal{F}_\eta = \mathcal{F}\widetilde{\Lambda}, \tag{B.10.9}$$

where

$$\widetilde{\Omega} := u_\xi \Omega + v_\xi \Lambda, \qquad \widetilde{\Lambda} := u_\eta \Omega + v_\eta \Lambda. \tag{B.10.10}$$

[1]This is one of the most important theorems in the complex function theory of one variable. For example, see Chapter 6 of [1].

In fact, if $\mathcal{F}$ satisfies (B.10.1), then

$$\mathcal{F}_\xi = u_\xi \mathcal{F}_u + v_\xi \mathcal{F}_v = \mathcal{F}(u_\xi \Omega + v_\xi \Lambda) = \mathcal{F}\widetilde{\Omega},$$
$$\mathcal{F}_\eta = u_\eta \mathcal{F}_u + v_\eta \mathcal{F}_v = \mathcal{F}(u_\eta \Omega + v_\eta \Lambda) = \mathcal{F}\widetilde{\Lambda},$$

which imply (B.10.9). Conversely, if (B.10.9) holds, then one can show that (B.10.1) holds, noticing that the Jacobi matrix of the coordinate change $(\xi, \eta) \mapsto (u, v)$ is a regular matrix (cf. Problem **1** in this section).

Here, for $\widetilde{\Omega}$ and $\widetilde{\Lambda}$ in (B.10.10), we have

$$\begin{aligned}\widetilde{\Omega}\widetilde{\Lambda} + \widetilde{\Lambda}_\xi &= (u_\xi \Omega + v_\xi \Lambda)(u_\eta \Omega + v_\eta \Lambda) + (u_\eta \Omega + v_\eta \Lambda)_\xi \\ &= u_\xi u_\eta \Omega^2 + v_\xi v_\eta \Lambda^2 + u_\xi v_\eta (\Omega\Lambda + \Lambda_u) + v_\xi u_\eta (\Lambda\Omega + \Omega_v) \\ &\quad + u_\xi u_\eta \Omega_u + v_\xi v_\eta \Lambda_v + u_{\xi\eta}\Omega + v_{\xi\eta}\Lambda,\end{aligned}$$

and similarly, we have

$$\begin{aligned}\widetilde{\Lambda}\widetilde{\Omega} + \widetilde{\Omega}_\eta &= u_\xi u_\eta \Omega^2 + v_\xi v_\eta \Lambda^2 + u_\xi v_\eta (\Lambda\Omega + \Omega_v) + v_\xi u_\eta (\Omega\Lambda + \Lambda_u) \\ &\quad + u_\xi u_\eta \Omega_u + v_\xi v_\eta \Lambda_v + u_{\xi\eta}\Omega + v_{\xi\eta}\Lambda.\end{aligned}$$

Since Ω and Λ satisfy the integrability condition (B.10.2) of (B.10.1), by assumption, the integrability condition for (B.10.9) also holds. Then by Proposition B.10.2, there exists a unique matrix-valued function $\mathcal{F}\colon D_1 \to \mathrm{M}_n(\boldsymbol{R})$ satisfying (B.10.9) and the initial condition $\mathcal{F}(0,0) = R$. Moreover, if Ω and Λ are skew-symmetric and the initial matrix R is an orthogonal matrix, then Corollary B.10.3 implies that $\mathcal{F}$ takes values in the orthogonal matrices. □

The following lemma can be obtained by a direct calculation (cf. Problem **2** in this section).

Lemma B.10.6. *Let $\zeta : D \to \mathrm{GL}_n(\boldsymbol{R})$ be regular matrix-valued C^∞-function defined on a domain $D \subset \boldsymbol{R}^2$. Then $\mathcal{F}\colon D \to \mathrm{M}_n(\boldsymbol{R})$ satisfies* (B.10.1) *if and only if $\hat{\mathcal{F}} := \mathcal{F}\zeta$ satisfies*

$$\text{(B.10.11)} \qquad \hat{\mathcal{F}}_u = \hat{\mathcal{F}}\hat{\Omega}, \qquad \hat{\mathcal{F}}_v = \hat{\mathcal{F}}\hat{\Lambda},$$

where

$$\text{(B.10.12)} \qquad \hat{\Omega} = \zeta^{-1}\Omega\zeta + \zeta^{-1}\zeta_u, \qquad \hat{\Lambda} = \zeta^{-1}\Lambda\zeta + \zeta^{-1}\zeta_v.$$

Moreover,

$$\text{(B.10.13)} \qquad \hat{\Lambda}\hat{\Omega} + \hat{\Omega}_v = \hat{\Omega}\hat{\Lambda} + \hat{\Lambda}_u$$

is equivalent to the integrability condition (B.10.2).

Proof of the fundamental theorem for surface theory. For given C^∞-functions E, F, G, L, M, N defined on a simply connected domain D in the uv-plane, set Ω and Λ as in the fundamental theorem (Theorem 17.2). We fix a point (u_0, v_0) in D.

Proof of the fundamental Theorem 17.2—uniqueness. Assume the coefficients of the first and second fundamental forms of two surfaces $\boldsymbol{p}$ and $\tilde{\boldsymbol{p}}$ are common, and denote them by E, F, G, L, M, N. It is sufficient to show that $\boldsymbol{p}$ and $\tilde{\boldsymbol{p}}$ are congruent. By a translation, we may assume $\boldsymbol{p}(u_0, v_0) = \tilde{\boldsymbol{p}}(u_0, v_0) = \boldsymbol{0}$. Moreover, by a rotation in $\boldsymbol{R}^3$, we may assume that both unit normal vectors ν and $\tilde{\nu}$ of $\boldsymbol{p}$ and $\tilde{\boldsymbol{p}}$, respectively, are $(0, 0, 1)$ at (u_0, v_0), and $\boldsymbol{p}_u(u_0, v_0)$ and $\tilde{\boldsymbol{p}}_u(u_0, v_0)$ are positive scalar multiples of $(1, 0, 0)$. Since $\boldsymbol{p}_v$ and $\tilde{\boldsymbol{p}}_v$ are perpendicular to ν and $\tilde{\nu}$, respectively, $\mathcal{F} = (\boldsymbol{p}_u, \boldsymbol{p}_v, \nu)$ and $\widetilde{\mathcal{F}} = (\tilde{\boldsymbol{p}}_u, \tilde{\boldsymbol{p}}_v, \tilde{\nu})$ satisfy

$$\text{(B.10.14)} \qquad \mathcal{F}(u_0, v_0) = \frac{1}{\sqrt{E_0}} \begin{pmatrix} E_0 & F_0 & 0 \\ 0 & \delta_0 & 0 \\ 0 & 0 & \sqrt{E_0} \end{pmatrix} = \widetilde{\mathcal{F}}(u_0, v_0)$$

by the definition of the first fundamental form (Problem **3** in this section). Here, E_0, F_0 and δ_0 are the values of E, F and $\delta := \sqrt{EG - F^2}$ at (u_0, v_0), respectively.

Hence the functions $\mathcal{F}$ and $\widetilde{\mathcal{F}}$ satisfy the same equation (B.10.1) and the same initial condition (B.10.14). Thus we have $\mathcal{F} = \widetilde{\mathcal{F}}$ by the uniqueness in Theorem B.10.4, and then $\boldsymbol{p}_u = \tilde{\boldsymbol{p}}_u$ and $\boldsymbol{p}_v = \tilde{\boldsymbol{p}}_v$. Since $\boldsymbol{p}(u_0, v_0) = \tilde{\boldsymbol{p}}(u_0, v_0) = \boldsymbol{0}$, we have $\boldsymbol{p} = \tilde{\boldsymbol{p}}$. □

Proof of the fundamental Theorem 17.2—existence. The existence of the solution $\mathcal{F} = (\varphi(u, v), \psi(u, v), \nu(u, v))$ of (B.10.1) with initial condition (B.10.14) itself is obvious. However, it is difficult to show the existence of $\boldsymbol{p}$ so that $\boldsymbol{p}_u = \varphi$ and $\boldsymbol{p}_v = \psi$, because showing that both φ and ψ are perpendicular to ν is not so easy. To remedy this, from now on, we use that the matrix-valued function ζ as in Lemma B.10.6 for the solution $\hat{\mathcal{F}} = \mathcal{F}\zeta$ is orthogonal matrix-valued, as follows.

First, since a_{ij} in (17.4) are given in (17.2),

$$\text{(B.10.15)} \qquad \begin{aligned} L &= Ea_{11} + Fa_{21}, \qquad N = Fa_{12} + Ga_{22}, \\ M &= Fa_{11} + Ga_{21} = Ea_{12} + Fa_{22} \end{aligned}$$

hold. We prepare the following lemma.

Lemma B.10.7. *The C^∞-functions Γ_{ij}^k determined from the functions E, F, G by* (10.6) *satisfy*

$$\begin{aligned} E_u &= 2(E\Gamma_{11}^1 + F\Gamma_{11}^2), \qquad E_v = 2(E\Gamma_{12}^1 + F\Gamma_{12}^2), \\ F_u &= E\Gamma_{12}^1 + F\Gamma_{12}^2 + F\Gamma_{11}^1 + G\Gamma_{11}^2, \\ F_v &= E\Gamma_{22}^1 + F\Gamma_{22}^2 + F\Gamma_{12}^1 + G\Gamma_{12}^2, \\ G_u &= 2(F\Gamma_{12}^1 + G\Gamma_{12}^2), \qquad G_v = 2(F\Gamma_{22}^1 + G\Gamma_{22}^2). \end{aligned}$$

Proof. The equation (10.6) is equivalent to

$$\begin{pmatrix} E & F \\ F & G \end{pmatrix} \begin{pmatrix} \Gamma_{11}^1 & \Gamma_{12}^1 & \Gamma_{22}^1 \\ \Gamma_{11}^2 & \Gamma_{12}^2 & \Gamma_{22}^2 \end{pmatrix} = \frac{1}{2} \begin{pmatrix} E_u & E_v & 2F_v - G_u \\ 2F_u - E_v & G_u & G_v \end{pmatrix},$$

as seen in (11.2) in Section 11. Then the conclusion is obtained. □

By the assumption of Theorem 17.2, $EG - F^2 > 0$ holds on D (cf. page 73). Then

$$\delta := \sqrt{EG - F^2} \tag{B.10.16}$$

is a C^∞-function defined on D. We set $\zeta\colon D \to \mathrm{M}_3(\boldsymbol{R})$ to be

$$\zeta := \frac{1}{\delta\sqrt{E}} \begin{pmatrix} \delta & -F & 0 \\ 0 & E & 0 \\ 0 & 0 & \delta\sqrt{E} \end{pmatrix}. \tag{B.10.17}$$

Since $\det \zeta = 1/\delta$ is positive, ζ takes values in the regular matrices.

Lemma B.10.8. *The function $\mathcal{F}\colon D \to \mathrm{M}_3(\boldsymbol{R})$ satisfies* (17.3) *if and only if $\hat{\mathcal{F}} := \mathcal{F}\zeta$ satisfies*

$$\hat{\mathcal{F}}_u = \hat{\mathcal{F}}\hat{\Omega}, \qquad \hat{\mathcal{F}}_v = \hat{\mathcal{F}}\hat{\Lambda}, \tag{B.10.18}$$

where

$$\begin{cases} \hat{\Omega} := \dfrac{1}{E} \begin{pmatrix} 0 & -\delta\Gamma_{11}^2 & -L\sqrt{E} \\ \delta\Gamma_{11}^2 & 0 & -\delta a_{21}\sqrt{E} \\ L\sqrt{E} & \delta a_{21}\sqrt{E} & 0 \end{pmatrix}, \\[2ex] \hat{\Lambda} := \dfrac{1}{E} \begin{pmatrix} 0 & -\delta\Gamma_{12}^2 & -M\sqrt{E} \\ \delta\Gamma_{12}^2 & 0 & -\delta a_{22}\sqrt{E} \\ M\sqrt{E} & \delta a_{22}\sqrt{E} & 0 \end{pmatrix}. \end{cases} \tag{B.10.19}$$

In particular, $\hat{\Omega}$ and $\hat{\Lambda}$ take values in the skew-symmetric matrices.

Proof. By Lemma B.10.6, setting the matrices $\hat{\Omega}$ and $\hat{\Lambda}$ as in (B.10.12) and given Ω, Λ in (17.4), the equations (17.3) and (B.10.18) are equivalent. So it is sufficient to show that $\hat{\Omega}$ and $\hat{\Lambda}$ can be written as (B.10.19). In fact, since

$$\zeta^{-1} = \frac{1}{\sqrt{E}} \begin{pmatrix} E & F & 0 \\ 0 & \delta & 0 \\ 0 & 0 & \sqrt{E} \end{pmatrix}, \tag{B.10.20}$$

(17.4) yields

$$\begin{aligned}
&\zeta^{-1}\Omega\zeta \\
&= \frac{1}{\delta E}\begin{pmatrix} \delta(E\Gamma_{11}^1 + F\Gamma_{11}^2) & -EF\Gamma_{11}^1 + E^2\Gamma_{12}^1 - F^2\Gamma_{11}^2 + EF\Gamma_{12}^2 & -\delta L\sqrt{E} \\ \delta^2\Gamma_{11}^2 & -\delta(F\Gamma_{11}^2 - E\Gamma_{12}^2) & -\delta^2 a_{21}\sqrt{E} \\ \delta L\sqrt{E} & \delta^2 a_{21}\sqrt{E} & 0 \end{pmatrix}, \\
&\zeta^{-1}\Lambda\zeta \\
&= \frac{1}{\delta E}\begin{pmatrix} \delta(E\Gamma_{12}^1 + F\Gamma_{12}^2) & -EF\Gamma_{12}^1 + E^2\Gamma_{22}^1 - F^2\Gamma_{12}^2 + EF\Gamma_{22}^2 & -\delta M\sqrt{E} \\ \delta^2\Gamma_{12}^2 & -\delta(F\Gamma_{12}^2 - E\Gamma_{22}^2) & -\delta a_{22}\sqrt{E} \\ \delta M\sqrt{E} & \delta^2 a_{22}\sqrt{E} & 0 \end{pmatrix}.
\end{aligned}$$

Here, (B.10.15) and (17.2) are applied to compute the $(1,3)$ and $(3,2)$-components, respectively. On the other hand, by $\delta = \sqrt{EG - F^2}$ and Lemma B.10.7,

$$\begin{aligned}
\zeta^{-1}\zeta_u &= \begin{pmatrix} -\dfrac{E_u}{2E} & \dfrac{-EF_u + FE_u}{E} & 0 \\ 0 & \dfrac{-F^2E_u - E^2G_u + 2EFF_u}{2E\delta^2} & 0 \\ 0 & 0 & 0 \end{pmatrix} \\
&= \frac{1}{\delta E}\begin{pmatrix} -\delta(E\Gamma_{11}^1 + F\Gamma_{11}^2) & EF\Gamma_{11}^1 - E^2\Gamma_{12}^1 + EF\Gamma_{12}^2 - (\delta^2 - F^2)\Gamma_{11}^2 & 0 \\ 0 & \delta(F\Gamma_{11}^2 - E\Gamma_{12}^2) & 0 \\ 0 & 0 & 0 \end{pmatrix}, \\
\zeta^{-1}\zeta_v &= \frac{1}{\delta E}\begin{pmatrix} -\delta(E\Gamma_{12}^1 + F\Gamma_{12}^2) & EF\Gamma_{12}^1 - E^2\Gamma_{22}^1 + EF\Gamma_{22}^2 - (\delta^2 - F^2)\Gamma_{12}^2 & 0 \\ 0 & \delta(F\Gamma_{12}^2 - E\Gamma_{22}^2) & 0 \\ 0 & 0 & 0 \end{pmatrix},
\end{aligned}$$

which are the conclusion. □

Using the above, we give a proof of the fundamental theorem (Theorem 17.2).

Proof of the fundamental Theorem 17.2—continued. Set $\hat{\Omega}$ and $\hat{\Lambda}$ be as in (B.10.19), with given functions E, F, G, L, M and N. Then by Theorem B.10.4, there exists a unique solution $\hat{\mathcal{F}}\colon D \to \mathrm{M}_3(\boldsymbol{R})$ of (B.10.18) with the initial condition $\hat{\mathcal{F}}(u_0, v_0) = I$, where I is the 3×3 identity matrix. Since $\hat{\Omega}$ and $\hat{\Lambda}$ are skew-symmetric matrices, $\hat{\mathcal{F}}$ takes values in the orthogonal matrices. Since the determinant of $\hat{\mathcal{F}}$ changes continuously in (u, v), and $\hat{\mathcal{F}}(u_0, v_0) = I$, $\det(\hat{\mathcal{F}}(u, v)) = 1$ for each (u, v). Set

(B.10.21) $$\mathcal{F} := \hat{\mathcal{F}}\zeta^{-1}.$$

Then by Lemma B.10.8, $\mathcal{F}$ satisfies (B.10.1) for $n = 3$. Decompose $\mathcal{F}(u, v)$ into column vectors as

(B.10.22) $$\mathcal{F}(u, v) = (\varphi(u, v), \psi(u, v), \nu(u, v)) = \begin{pmatrix} \varphi_1 & \psi_1 & \nu_1 \\ \varphi_2 & \psi_2 & \nu_2 \\ \varphi_3 & \psi_3 & \nu_3 \end{pmatrix},$$

where φ_j, ψ_j, ν_j $(j = 1, 2, 3)$ are real-valued functions defined on D. Looking at the second column of the first equation of (B.10.1) and the first column of the second equation of (B.10.1), we have

$$\psi_u = \Gamma_{21}^1\varphi + \Gamma_{21}^2\psi + M\nu, \qquad \varphi_v = \Gamma_{12}^1\varphi + \Gamma_{12}^2\psi + M\nu.$$

Since $\Gamma_{12}^k = \Gamma_{21}^k$ $(k = 1, 2)$ by the definition of Γ_{ij}^k, we have $\psi_u = \varphi_v$, that is,

$$(\psi_j)_u = (\varphi_j)_v \qquad (j = 1, 2, 3).$$

Since this is equivalent to $d\alpha_j = 0$ for $\alpha_j := \varphi_j\,du + \psi_j\,dv$ $(j = 1, 2, 3)$, the Poincaré Lemma (Theorem 12.3) implies the existence of a function $p_j\colon D \to \boldsymbol{R}$ $(j = 1, 2, 3)$ such that $dp_j = \alpha_j$. Set $\boldsymbol{p} := (p_1, p_2, p_3)$. Then by (B.10.22), we have

(B.10.23) $$\boldsymbol{p}_u = \varphi, \qquad \boldsymbol{p}_v = \psi.$$

Now, we shall prove that this $\boldsymbol{p}$ is the desired surface. Decompose $\hat{\mathcal{F}}$ into the column vectors as $\hat{\mathcal{F}} = (\hat{\varphi}, \hat{\psi}, \hat{\nu})$. Since $\hat{\mathcal{F}}$ is an orthogonal matrix with determinant 1, $\{\hat{\varphi}, \hat{\psi}, \hat{\nu}\}$ is an orthonormal frame field at each point in D. By the definition (B.10.21) of $\mathcal{F}$ and (B.10.20), we get

$$\boldsymbol{p}_u = \varphi = \sqrt{E}\hat{\varphi}, \qquad \boldsymbol{p}_v = \psi = \frac{F}{\sqrt{E}}\hat{\varphi} + \frac{\delta}{\sqrt{E}}\hat{\psi}, \qquad \nu = \hat{\nu}.$$

Since $\{\hat{\varphi}, \hat{\psi}, \hat{\nu}\}$ is an orthonormal frame field and $\delta = \sqrt{EG - F^2}$, we have

$$\begin{aligned} &\boldsymbol{p}_u \cdot \boldsymbol{p}_u = E(\hat{\varphi} \cdot \hat{\varphi}) = E, \qquad \boldsymbol{p}_u \cdot \boldsymbol{p}_v = \hat{\varphi} \cdot (F\hat{\varphi} + \delta\hat{\psi}) = F, \\ &\boldsymbol{p}_v \cdot \boldsymbol{p}_v = \frac{1}{E}(F\hat{\varphi} + \delta\hat{\psi}) \cdot (F\hat{\varphi} + \delta\hat{\psi}) = G, \\ &\boldsymbol{p}_u \cdot \nu = \boldsymbol{p}_v \cdot \nu = 0, \qquad \nu \cdot \nu = 1, \end{aligned}$$

and then the coefficients of the first fundamental form of $\boldsymbol{p}$ are E, F, G, and ν is a unit normal vector field. Moreover, by the first column of the first equation of (17.3) and (B.10.23), we have $\boldsymbol{p}_{uu} = \Gamma_{11}^1 \boldsymbol{p}_u + \Gamma_{11}^2 \boldsymbol{p}_v + L\nu$, and then $\boldsymbol{p}_{uu} \cdot \nu = L$. Similarly, one can show the coefficients of the second fundamental forms of $\boldsymbol{p}$ are L, M, N. □

Exercises B.10

1 In the proof of Theorem B.10.4 (page 273), derive (B.10.1) under the coordinate change $(\xi, \eta) \mapsto (u, v)$, using (B.10.9) and (B.10.10).

2 Prove Lemma B.10.6.

3 Show (B.10.14).

Answers to Exercises

Answers and/or hints to the exercises are given here. In the cases where the answers below are not complete, it is recommended to the readers that they work out complete answers for themselves.

Section 1 (page 9)

1 $\frac{1}{2}\log(2+\sqrt{5})$, by using the identity

$$\int \sqrt{1+4x^2}\,dx = \frac{1}{2}(x\sqrt{1+4x^2}) + \frac{1}{4}\log(2x+\sqrt{1+4x^2}).$$

2 Since the parameter change $t = t(u)$ is increasing, we have

$$\int_c^d \sqrt{\left(\frac{dx}{du}\right)^2 + \left(\frac{dy}{du}\right)^2}\,du = \int_c^d \sqrt{\left(\frac{dx}{dt}\right)^2 + \left(\frac{dy}{dt}\right)^2}\,\frac{dt}{du}\,du$$

by the chain-rule and $dt/du > 0$. Then the substitution rule for integrals yields the conclusion.

3 (2) Applying Taylor's formula (Theorem A.1.1 in Appendix A) to the function $f(x) = \sqrt{1-x}$, there exists θ such that

$$\sqrt{1-x} = 1 - \frac{1}{2}x - \frac{1}{8}x^2(1-\theta x)^{-3/2} \qquad (0<\theta<1).$$

Letting $x = \varepsilon^2\cos^2 t$, and noticing that $8/9 < \sqrt{1-\varepsilon^2} < \sqrt{1-\theta\varepsilon^2\cos^2 t}$ for $\varepsilon < 1/3$, we have

$$\pi\left(2 - \frac{1}{2}\varepsilon^2 - \frac{27}{256}\varepsilon^4\right) < \int_0^{2\pi} \sqrt{1-\varepsilon^2\cos^2 t}\,dt < \pi\left(2-\frac{1}{2}\varepsilon^2\right).$$

Substituting the values of a and b into this, the length l of the perimeter satisfies $40003.4 < l < 40003.6$. That is, it is approximately 40003.5 ± 0.1 km.

4 Notice that $|\dot{\gamma}(t)| = 2a \sin(t/2)$.

Section 2 (page 26)

1 Find the equations for the components of A using $A^T = A^{-1}$.

2 The unit vector perpendicular to $\boldsymbol{e} = \begin{pmatrix} \cos\theta \\ \sin\theta \end{pmatrix}$ is $\pm\begin{pmatrix} -\sin\theta \\ \cos\theta \end{pmatrix}$.

3 If one fixes the front wheels in place, then the front and back wheels draw concentric circles. The center of these circles lies on the intersection of the two lines that include the wheel axles, and the angle between these lines is θ. For example, the radius of the trajectory of the inner back wheel is $\Delta/\tan\theta - \varepsilon/2$, and the curvature is its reciprocal. On the other hand, the radius of the trajectory of the inner front wheel is $\Delta/\sin\theta - \varepsilon/2$. The difference between the two radii can be described as the *turning radius difference*.

4 We may assume that the straight line is the x-axis without loss of generality. Compare the curvatures of the two curves $\gamma(s) = (x(s), y(s))$ and $\tilde{\gamma}(s) = (x(s), -y(s))$.

5 (1) $\kappa(t) = -1/(4a\sin(t/2))$.
(2) $s(t) = 4a(1 - \cos(t/2))$, $t(s) = 2\arccos\{1 - (s/(4a))\}$ $(0 < s < 8a)$.
(3) Since

$$\gamma(s) = \left(2a\arccos\left(1 - \frac{s}{4a}\right) + \frac{(s-4a)\sqrt{s(8a-s)}}{8a}, \frac{s(8a-s)}{8a}\right),$$

we have $\kappa(s) = -1/\sqrt{s(8a-s)}$. Substituting $s(t)$ in (2) into this, $\kappa(t) = -1/(4a\sin(t/2))$, which coincides with the result in (1).

6 Let $\gamma_1(s)$ be a curve parametrized by the arc-length and set

$$\gamma_2(s) = A\gamma_1(s) + \boldsymbol{q} \quad \left(A = \begin{pmatrix} \cos\theta & -\sin\theta \\ \sin\theta & \cos\theta \end{pmatrix}, \ \boldsymbol{q} \text{ is a constant vector}\right).$$

Then s is the arc-length parameter of γ_2 and $\det(\gamma_2', \gamma_2'') = \det(\gamma_1', \gamma_1'')$.

7 Set $\theta(s) := \int_0^s \kappa(u)\,du$. Then

$$\gamma'(s) = (\cos\theta(s), \sin\theta(s)),$$
$$\gamma''(s) = \theta'(s)(-\sin\theta(s), \cos\theta(s)) = \kappa(s)(-\sin\theta(s), \cos\theta(s)u).$$

Hence $|\gamma'(s)| = 1$, and $\kappa(s)$ is the curvature.

8 Since s is the arc-length parameter, $\gamma'(0) = \boldsymbol{e}(0)$, $\gamma''(0) = \kappa(0)\boldsymbol{n}(0)$. Moreover, by differentiating the definition of the curvature, the Frenet formula yields that $\gamma'''(0) = -\kappa(0)^2\boldsymbol{e}(0) + \kappa'(0)\boldsymbol{n}(0)$. Then Taylor's formula (Theorem A.1.1 in Appendix A.1) yields the conclusion.

9 (1) Under the same notation as in Proposition 2.5, it holds that

$$\frac{d^3 f}{dx^3}(0) = \frac{\dddot{y}\dot{x}^2 - \dddot{x}\dot{x}\dot{y} - 3\dot{x}\ddot{x}\ddot{y} + 3\ddot{x}^2\dot{y}}{\dot{x}^5}.$$

That is, this can be expressed in terms of the derivatives of x and y up to third order.

(2) Apply the techniques used in the proof of Proposition 2.6.

(3) Take the xy-axis at $\gamma(s_0)$ as in Definition 2.3 and express the curve as a graph $y = f(x)$. Then we have $f(0) = \dot{f}(0) = 0$ ($\dot{} = d/dx$), and $\kappa(0) = \ddot{f}(0)$, $\dot{\kappa}(0) = \dddot{f}(0)$. Hence

$$f(x) = \frac{1}{2}\kappa(0)x^2 + \frac{1}{6}\dot{\kappa}(0)x^3 + o(x^3).$$

On the other hand, the osculating circle of the graph $y = f(x)$ at the origin can be written as the graph $y = g(x)$ where

$$g(x) = \frac{1}{2}\kappa(0)x^2 + o(x^3).$$

Then they have third order contact at the origin if and only if $\dot{\kappa}(0) = 0$. Since $\dot{\kappa} = d\kappa/dx = \kappa'/x'$, the conclusion follows.

Section 3 (page 38)

1 The total curvature of a curve parametrized by a general parameter t is $\int \kappa(s)\,ds = \int \kappa(t)\frac{ds}{dt}\,dt$.

2 A function $f(x)$ is of class C^1 if there exists a derivative $f'(x)$ when $x \neq c$, in a neighborhood of c, and there exists the limit $\lim_{x\to c} f'(x)$. The conclusion follows by an inductive application of this fact and L'Hôpital's rule (Theorem A.1.3 in Appendix A.1).
Alternatively, by the substitution $u = ts$, one can show

$$\varphi(t) = \varphi(t) - \varphi(0) = \int_0^t \dot{\varphi}(u)\,du = \int_0^1 \dot{\varphi}(ts)\,t\,ds = t\int_0^1 \dot{\varphi}(ts)\,ds.$$

The integral on the right-hand side is of class C^∞ in t, which is $\psi(t)$ in the conclusion.

3 (1) Apply Problem **2** to $\varphi(t) := f(t+s) - f(s)$ and make the change of variable $t \mapsto t - s$.

(2) Apply (1) to the vector-valued function $\gamma(t)$.

4 (1) Let $\boldsymbol{e} = (\xi, \eta)$ and set $\boldsymbol{n} = (-\eta, \xi)$. Since $\boldsymbol{e} \cdot \boldsymbol{e} = 1$, there exists a function f such that $\boldsymbol{e}' = f\boldsymbol{n}$. Here, define two orthogonal matrices as

$$A_1 := \begin{pmatrix} \cos\theta & -\sin\theta \\ \sin\theta & \cos\theta \end{pmatrix}, \qquad A_2 := (\boldsymbol{e}, \boldsymbol{n}) = \begin{pmatrix} \xi & -\eta \\ \eta & \xi \end{pmatrix}.$$

Then noticing $\theta' = f$, we have

$$\frac{d}{ds}(A_2 A_1{}^T) = A_2 \begin{pmatrix} 0 & -f \\ f & 0 \end{pmatrix} A_1{}^T + A_2 \begin{pmatrix} 0 & f \\ -f & 0 \end{pmatrix} A_1{}^T = O.$$

Hence $A_2 A_1{}^T$ is a constant matrix. Regarding the initial condition, it should be the identity matrix, and hence $A_2 = A_1$ holds. The first column of this equality implies the conclusion.

(2) Apply (1) to $\boldsymbol{e}(t) = \boldsymbol{w}(0, t)$ and $\alpha = 0$.

(3) Apply (2) to $\boldsymbol{e}(s) = \boldsymbol{w}(s, t)$ and $\alpha = \theta(t)$, where t is a fixed value.

Section 4 (page 49)

1 The area of the sector domain corresponding to the curve $r = r(\theta)$ ($\alpha \leq \theta \leq \beta$) represented in the polar coordinate system is given by $\frac{1}{2}\int_\alpha^\beta r^2(\theta)\,d\theta$.

2 Applying the similarity expansion to the spiral $r = a^\theta$ by a factor of k, one obtains the curve $r = ka^\theta$. By letting $b = \log_a k$, the curve is expressed as $r = ka^\theta = a^b a^\theta = a^{\theta+b}$. This curve is obtained by replacing the parameter θ of the original logarithmic spiral by $\theta + b$. On the other hand, for a given real number b, the similarity expansion to the spiral $r = a^\theta$ by a factor of $k := a^b$ with respect to the origin is congruent to the spiral $r = a^{\theta+b}$, proving the second assertion.

4 The curve $r = a\theta$ can be parametrized as $\gamma(\theta) = a\theta(\cos\theta, \sin\theta)$. Then by (2.7) in Section 2, the curvature is written as $(\theta^2+2)/(a(\theta^2+1)^{3/2})$. Similarly, the curvature of $r = a^\theta$ is $a^{-\theta}/\sqrt{1+(\log a)^2}$.

5 (1) The vertices are $(1/a, 0)$, $(-1/a, 0)$.

(2) Let $w = \xi + i\eta = 1/z = 1/(x+iy)$, that is, $x = \xi/(\xi^2+\eta^2)$, $y = -\eta/(\xi^2+\eta^2)$. Substituting these into the equation of the lemniscate in Example 1.1, we have $\xi^2 - \eta^2 = 1/a^2$.

(3) A vertex is a point where the curve turns from a positive spiral to a negative spiral. By Theorem 4.2, this is preserved by Möbius transformations. The mapping described in (2) describes a one-to-one correspondence between the points on the hyperbola and the lemniscate excluding the origin. Therefore, the points $(\pm a, 0)$ are the only vertices on the lemniscate excluding the origin. On the other hand, the origin on the lemniscate is the point corresponding to $t = \pi/2, 3\pi/2$ in the parametrization of Example 1.2, where the curvature changes sign, that is, the origin is not a vertex.

6 Consider any Möbius transformation $w = (az+b)/(cz+d)$ $(ad-bc \neq 0)$. If $c = 0$, it is written as $w = pz + q$ $(p \neq 0)$. Then setting $p = re^{i\theta}$ $(r > 0)$, one gets

$$w = re^{i\theta}\left(z + \frac{q}{p}\right),$$

then the transformation is a composition of a translation, a rotation and extension/shrinking. When $c \neq 0$, we may assume $ad - bc = 1$ without loss of generality. Then one can write

$$w = \frac{a}{c} + \frac{-1/c}{cz+d}.$$

7 A logarithmic spiral can be parametrized as $\gamma(\theta) = a^\theta(\cos\theta, \sin\theta)$. Then the evolute of it is $(\log a)a^\theta(-\sin\theta, \cos\theta)$. This curve is obtained by rotating the original spiral by 90° conterclockwise and applying the similarity transformation by a factor of $\log a$. Then it is congruent to the original spiral, by Problem **2**.

Section 5 (page 55)

1 Let $\gamma(s)$ be a curve parametrized by arc-length and $\tilde{\gamma}(s) = T\gamma(s) + \boldsymbol{a}$, where T is an orthogonal matrix with determinant 1 and $\boldsymbol{a} \in \boldsymbol{R}^3$ is a constant vector. Denote the unit tangent vector, the principal normal vector and the binormal vector of $\gamma(s)$ by $\boldsymbol{e}$, $\boldsymbol{n}$, $\boldsymbol{b}$, respectively. Then the unit tangent vector, the principal normal vector and the binormal vector of $\tilde{\gamma}(s)$, denoted by $\tilde{\boldsymbol{e}}$, $\tilde{\boldsymbol{n}}$ and $\tilde{\boldsymbol{b}}$, are given by $\tilde{\boldsymbol{e}} = T\boldsymbol{e}$, $\tilde{\boldsymbol{n}} = T\boldsymbol{n}$, $\tilde{\boldsymbol{b}} = T\boldsymbol{b}$. Note that the last equality is obtained by Corollary A.3.6 in Appendix A.3. These imply the conclusion.

2 The equality $\gamma' \times \gamma'' = (\dot{\gamma} \times \ddot{\gamma})/|\dot{\gamma}|^3$ is obtained from the first two of the following equalities:

$$\gamma' = \frac{\dot{\gamma}}{|\dot{\gamma}|}, \qquad \gamma'' = \left(\frac{1}{|\dot{\gamma}|}\right)'\dot{\gamma} + \frac{\ddot{\gamma}}{|\dot{\gamma}|^2},$$

$$\gamma''' = (\text{a linear combination of } \dot{\gamma} \text{ and } \ddot{\gamma}) + \frac{\dddot{\gamma}}{|\dot{\gamma}|^3}.$$

Then we have

$$\kappa = |\kappa\boldsymbol{b}| = |\gamma' \times (\kappa\boldsymbol{n})| = |\gamma' \times \gamma''| = \frac{|\dot{\gamma} \times \ddot{\gamma}|}{|\dot{\gamma}|^3}.$$

On the other hand, since $\boldsymbol{b}$ is proportional to $\gamma' \times \gamma''$, this is true of $\dot{\gamma} \times \ddot{\gamma}$ as well. Hence we have $\boldsymbol{b} = (\dot{\gamma} \times \ddot{\gamma})/|\dot{\gamma} \times \ddot{\gamma}|$.

Here, by the Frenet formula, it holds that $\tau\kappa^2 = \det(\gamma', \gamma'', \gamma''')$. Substituting the formula for κ^2 into this, and replacing γ', γ'', γ''' by $\dot{\gamma}$, $\ddot{\gamma}$, $\dddot{\gamma}$ using the above relation, the formula for τ is obtained.

4 Similar to the case of plane curves, (5.4) and Taylor's formula yield the conclusion.

5 Denote the curvature and the torsion by κ and τ, respectively. Then the helix in Example 5.1 with $a = \kappa/(\kappa^2+\tau^2)$, $b = \tau/(\kappa^2+\tau^2)$ has that given curvature and torsion. So by the fundamental theorem for space curves (Theorem 5.2), we have the conclusion.

6 Apply Problem **3** and the uniqueness property in the fundamental theorem (Theorem 5.2).

7 Let γ be a curve lying on the sphere of radius r centered at the origin. Differentiating $\gamma\cdot\gamma = r^2$ by the arc-length s, we have $\gamma'\cdot\gamma = 0$, $\gamma''\cdot\gamma = 1$. Then γ is perpendicular to $\boldsymbol{e} = \gamma'$, and $\gamma\cdot\boldsymbol{n} = 1/\kappa$, where $\boldsymbol{n}$ is the principal normal vector and κ is the curvature. Hence one can write $\gamma = (-1/\kappa)\boldsymbol{n} + a\boldsymbol{b}$ for some function a in s. Since κ is constant and $|\gamma| = r$, we have $a^2 = r^2 - \kappa^2$ is constant. Thus, we have $\boldsymbol{e} = \gamma' = (-1/\kappa)\boldsymbol{n}' + a\boldsymbol{b}'$. Applying the Frenet-Serret formula to this, we have that the torsion τ vanishes identically.

8 Let γ_2 be a curve obtained by reflecting γ_1 across a straight line d. Then γ_2 is obtained by 180° rotation of γ_1 about the axis d in $\boldsymbol{R}^3$.

Section 6 (page 67)

1 $\boldsymbol{p} = (\cos u(a+b\cos v), \sin u(a+b\cos v), b\sin v)$, where $0 \le u, v \le 2\pi$.

2 For example,

$$\boldsymbol{p} = (2\cos u, 2\sin u, 0) + v\left(\cos\frac{u}{2}(0,0,1) - \sin\frac{u}{2}(\cos u, \sin u, 0)\right),$$

where $-\pi \le u \le \pi$, $-1 \le v \le 1$.

3 The graph of $f(x,y)$ is parametrized as $\boldsymbol{p}(u,v) = (u, v, f(u,v))$.

4 By the chain rule,

$$\boldsymbol{p}_\xi \times \boldsymbol{p}_\eta = (u_\xi \boldsymbol{p}_u + v_\xi \boldsymbol{p}_v) \times (u_\eta \boldsymbol{p}_u + v_\eta \boldsymbol{p}_v) = (u_\xi v_\eta - v_\xi u_\eta)(\boldsymbol{p}_u \times \boldsymbol{p}_v).$$

5 By Problem **4**, $|\boldsymbol{p}_\xi \times \boldsymbol{p}_\eta| = |J|\,|\boldsymbol{p}_u \times \boldsymbol{p}_v|$ holds, where $J = \det\begin{pmatrix} u_\xi & u_\eta \\ v_\xi & v_\eta \end{pmatrix}$. By applying the substitution rule for double integrals, the conclusion follows.

6 Let $\tilde{\boldsymbol{p}} = T\boldsymbol{p} + \boldsymbol{b}$, where T is an orthogonal matrix. Since $\tilde{\boldsymbol{p}}_u = T\boldsymbol{p}_u$, $\tilde{\boldsymbol{p}}_v = T\boldsymbol{p}_v$, it holds that

$$|\tilde{\boldsymbol{p}}_u \times \tilde{\boldsymbol{p}}_v| = |(T\boldsymbol{p}_u) \times (T\boldsymbol{p}_v)| = |\det T(\boldsymbol{p}_u \times \boldsymbol{p}_v)| = |\boldsymbol{p}_u \times \boldsymbol{p}_v|.$$

7 It is obvious by definition that the map φ is a bijection. As seen in the proof of Theorem 8.7 on page 82, the surface can be expressed as a graph

$z = f(x, y)$ (changing the order of the coordinate axes, if necessary), and the coordinate change $\psi\colon (u,v) \mapsto (x(u,v), y(u,v))$ is a diffeomorphism. The map φ is expressed as $\varphi(\xi,\eta) = \psi^{-1}(x(\xi,\eta), y(\xi,\eta))$ by using the inverse ψ^{-1} of ψ. Though this representation is valid only on a neighborhood of P, one can show that φ is of class C^∞ on a neighborhood of each point of D. Exchanging roles of (u,v) and (ξ,η) one can see that φ^{-1} is of class C^∞, that is, φ is a diffeomorphism.

Section 7 (page 76)

1 In general, the eigenvalues of a real symmetric $n \times n$ matrix are reals, and the matrix can be diagonalized by an orthogonal matrix. Here, we give an alternative direct proof for the case $n = 2$:

(1) The discriminant of the characteristic polynomial $\lambda^2 - (a+c)\lambda + (ac - b^2)$ of the matrix A satisfies $(a+c)^2 - 4(ac - b^2) = (a-c)^2 + b^2 \geq 0$.

(2) $A\boldsymbol{e} = \lambda\boldsymbol{e}$ implies $ax + by = \lambda x$ and $bx + cy = \lambda y$ hold. Hence the first component of $A\boldsymbol{n}$ is computed as

$$a(-y) + bx = bx + cy - (a+c)y = (\lambda - (a+c))y = \mu(-y),$$

and similarly, the second component is μx.

(3) P is an orthogonal matrix with determinant 1, because of Problem **2** in Section 2. Moreover, $AP = (A\boldsymbol{e}, A\boldsymbol{n}) = (\lambda\boldsymbol{e}, \mu\boldsymbol{n})$, which implies the conclusion.

2 A quadratic equation $\lambda^2 - (a+c)\lambda + (ac - b^2) = 0$ has two positive real roots if and only if $a + c > 0$ and $ac - b^2 > 0$. The second inequality implies that a and c have the same sign.

3 We shall prove (7.4):

$$\begin{aligned} d(\varphi(f)) &= \frac{\partial}{\partial u}(\varphi(f(u,v)))\, du + \frac{\partial}{\partial v}(\varphi(f(u,v)))\, dv \\ &= \dot\varphi(f(u,v)) f_u\, du + \dot\varphi(f(u,v)) f_v\, dv = \dot\varphi(f)\, df. \end{aligned}$$

4 Let $\widetilde{E}$, $\widetilde{F}$, $\widetilde{G}$ be the coefficients of the first fundamental form of a parametrization $\boldsymbol{p}(\xi,\eta)$ of the surface obtained from $\boldsymbol{p}(u,v)$ with a parameter change $u = u(\xi,\eta)$, $v = v(\xi,\eta)$. Then by the chain-rule,

$$\begin{aligned} \widetilde{E} &= \boldsymbol{p}_\xi \cdot \boldsymbol{p}_\xi = (\boldsymbol{p}_u u_\xi + \boldsymbol{p}_v v_\xi) \cdot (\boldsymbol{p}_u u_\xi + \boldsymbol{p}_v v_\xi) \\ &= (u_\xi)^2 (\boldsymbol{p}_u \cdot \boldsymbol{p}_u) + 2(u_\xi v_\xi)(\boldsymbol{p}_u \cdot \boldsymbol{p}_v) + (v_\xi)^2 (\boldsymbol{p}_v \cdot \boldsymbol{p}_v) \\ &= u_\xi^2 E + 2u_\xi v_\xi F + v_\xi^2 G \end{aligned}$$

holds. The other components are computed in a similar way.

5 $E = 1 + f_x^2$, $F = f_x f_y$, $G = 1 + f_y^2$.

6 Under the representations as in pages 57 and 61:

(1) The elliptic paraboloid: $E = 1 + 4x^2/a^4$, $F = 4xy/(a^2b^2)$, $G = 1 + 4y^2/b^4$.
(2) The hyperbolic paraboloid: $E = 1 + 4x^2/a^4$, $F = -4xy/(a^2b^2)$, $G = 1 + 4y^2/b^4$.
(3) The ellipsoid: $E = a^2 \sin^2 u \cos^2 v + b^2 \sin^2 u \sin^2 v + c^2 \cos^2 u$, $F = (a^2 - b^2) \cos u \cos v \sin u \sin v$, $G = a^2 \cos^2 u \sin^2 v + b^2 \cos^2 u \cos^2 v$.
(4) The hyperboloid of one sheet: $E = a^2 \sinh^2 u \cos^2 v + b^2 \sinh^2 u \sin^2 v + c^2 \cosh^2 u$, $F = (b^2 - a^2) \cosh u \cos v \sinh u \sin v$, $G = a^2 \cosh^2 u \sin^2 v + b^2 \cosh^2 u \cos^2 v$.
(5) The hyperboloid of two sheets: $E = a^2 \cosh^2 u \cos^2 v + b^2 \cosh^2 u \sin^2 v + c^2 \sinh^2 u$, $F = (b^2 - a^2) \cosh u \cos v \sinh u \sin v$, $G = a^2 \sinh^2 u \sin^2 v + b^2 \sinh^2 u \cos^2 v$.

7 $E = \dot{x}^2 + \dot{z}^2$, $F = 0$, $G = x^2$.

Section 8 (page 86)

1 Since the matrix A is symmetric, it can be diagonalized by an orthogonal matrix (cf. Problem **1** in Section 7):

$$P^{-1}AP = P^TAP = \begin{pmatrix} \lambda & 0 \\ 0 & \mu \end{pmatrix} \qquad (\lambda, \mu > 0).$$

Let $C := P^{-1}(AB)P$ for P as above. Then the eigenvalues of C coincide with those of AB. Here, since $\widetilde{B} := P^TBP$ is symmetric, one can set $\widetilde{B} = \left(\begin{smallmatrix} a & b \\ b & c \end{smallmatrix}\right)$. Then

$$C = (P^TAP)\widetilde{B} = \begin{pmatrix} \lambda & 0 \\ 0 & \mu \end{pmatrix} \begin{pmatrix} a & b \\ b & c \end{pmatrix} = \begin{pmatrix} \lambda a & \lambda b \\ \mu b & \mu c \end{pmatrix},$$

and hence the discriminant of the characteristic polynomial of C is non-negative. Then the conclusion follows.

The proof above works only for 2×2-matrices. The following argument can be applied for general symmetric matrices of arbitrary size: Let $R = P\left(\begin{smallmatrix} \sqrt{\lambda} & 0 \\ 0 & \sqrt{\mu} \end{smallmatrix}\right)P^T$, where P is an orthogonal matrix which diagonalizes A. Then R is symmetric and $R^2 = A$. Here, $R^{-1}(AB)R = R^{-1}R^2BR = RBR$ is symmetric, because R^{-1} and B are symmetric. Then its eigenvalues are reals, and coincide with those of AB.

2 Positive on the convex rightward part, and negative on the convex leftward part.

3 $L = f_{xx}/\sqrt{1+f_x^2+f_y^2}$, $M = f_{xy}/\sqrt{1+f_x^2+f_y^2}$,
$N = f_{yy}/\sqrt{1+f_x^2+f_y^2}$,
$K = (f_{xx}f_{yy} - f_{xy}^2)/(1+f_x^2+f_y^2)^2$,
$H = \frac{1}{2}\big(f_{xx}(1+f_y^2) - 2f_{xy}f_xf_y + f_{yy}(1+f_x^2)\big)/(1+f_x^2+f_y^2)^{3/2}$.

4 Using the parametrizations on pages 57, 61:

(1) The elliptic paraboloid: $K = 4/(a^2b^2(1+4x^2/a^4+4y^2/b^4)^2)$,
$H = (a^2b^2(a^2+b^2)+4b^2x^2+4a^2y^2)/(a^4b^4(1+4x^2/a^4+4y^2/b^4)^{3/2})$.
(2) The hyperbolic paraboloid: $K = -4/(a^2b^2(1+4x^2/a^4+4y^2/b^4))$,
$H = (a^2b^2(b^2-a^2)-4b^2x^2+4a^2y^2)(a^4b^4(1+4x^2/a^4+4y^2/b^4)^{3/2})$.
(3) The ellipsoid: $K = a^2b^2c^2/\Delta^4$, $H = (abc/(2\Delta^3))\times$
$(a^2(\sin^2 v+\sin^2 u\cos^2 v)+b^2(\cos^2 v+\sin^2 u\sin^2 v)+c^2\cos^2 u))$, where
$\Delta = \sqrt{b^2c^2\cos^2 u\cos^2 v + c^2a^2\cos^2 u\sin^2 v + a^2b^2\sin^2 u}$.
In particular, when $a = b = c$ (i.e. the sphere), $K = 1/a^2$, $H = 1/a$.
(4) The hyperboloid of one sheet: $K = -a^2b^2c^2/\Delta^4$, $H = abc/(2\Delta^3)\times$
$\{a^2(\sinh^2 u\cos^2 v - \sin^2 v) + b^2(\sinh^2 u\sin^2 v - \cos^2 v) + c^2\cosh^2 u\}$,
where $\Delta = \sqrt{b^2c^2\cosh^2 u\cos^2 v + c^2a^2\cosh^2 u\sin^2 v + a^2b^2\sinh^2 u}$.
(5) The hyperboloid of two sheets: $K = a^2b^2c^2/\Delta^4$, $H = (abc/(2\Delta^3))\times$
$(a^2(\cosh^2 u\cos^2 v + \sin^2 v) + b^2(\cosh^2 u\sin^2 v + \cos^2 v) + c^2\sinh^2 u)$,
where $\Delta = \sqrt{b^2c^2\sinh^2 u\cos^2 v + c^2a^2\sinh^2 u\sin^2 v + a^2b^2\cosh^2 u}$.

5 Let $\tilde{\boldsymbol{p}}(u,v) = c\boldsymbol{p}(u,v)$. Then $\tilde{\boldsymbol{p}}_u = c\boldsymbol{p}_u$, $\tilde{\boldsymbol{p}}_v = c\boldsymbol{p}_v$ and thus, the coefficients of the first fundamental form of $\tilde{\boldsymbol{p}}$ are c^2 times of those of $\boldsymbol{p}$. On the other hand, $\boldsymbol{p}$ and $\tilde{\boldsymbol{p}}$ have a common unit normal vector. Then by $\tilde{\boldsymbol{p}}_{uu} = c\boldsymbol{p}_{uu}$ etc., the coefficients of the second fundamental form of $\tilde{\boldsymbol{p}}$ are c times those of $\boldsymbol{p}$.

6 Fix a point P on the surface, and represent the surface as a graph $z = f(x,y)$ as in the proof of Theorem 8.7, where P is the origin. Then the Weingarten matrix $A = \widehat{I}^{-1}\,\widehat{II}$ at the point P with respect to the parameter (x,y) is symmetric. Since the determinant and the trace of this matrix are 0, A is the zero matrix. Hence the coefficients of second fundamental form are all 0 under this coordinate system. By the coordinate change formula (8.3), the coefficients of second fundamental form are all 0 with respect to any coordinate system. Since the point P is arbitrary, L, M, N vanish identically. Then by the Weingarten formula (8.8), the unit normal vector ν is a constant vector. Hence the surface is (a part of) a plane.

7 By $(xx')' = 1$, we have $xx' = u + \alpha$ (α is a constant). Changing parameter if necessary, we may set $\alpha = 0$ without loss of generality. Since $2xx' = (x^2)'$, we have $x^2 = u^2 + c$ (c is a constant), that is, $x = \pm\sqrt{u^2+c}$. In particular, $u^2 + c$ must be positive. Here, $(z')^2 = 1 - (x')^2 = c/(u^2+c)$ must be positive, and then $c > 0$. So set $c = a^2$

(a is a real constant). Then

$$z(u) = \int \frac{a}{\sqrt{u^2 + a^2}}\, du = a \log(u + \sqrt{u^2 + a^2}) + b \qquad (b \text{ is a constant}).$$

One can set $b = 0$ by a suitable translation. Noticing that $u^2 = x^2 - a^2$, the equation of the curve is written as $z = a \log(x + \sqrt{x^2 - a^2})$. Solving this with respect to x, the conclusion follows.

8 The first and second fundamental forms are $ds^2 = du^2 + (1 + u^2)\, dv^2$, $I\!I = -(2/\sqrt{1 + u^2})\, du\, dv$, respectively.

9 The first and second fundamental forms are $ds^2 = (1 + u^2 + v^2)^2(du^2 + dv^2)$, $I\!I = -2(du^2 - dv^2)$, respectively.

10 By Problem **3**, the graph $z = f(x, y)$ is a minimal surface if and only if

$$(1 + f_y^2) f_{xx} - 2 f_x f_y f_{xy} + (1 + f_x^2) f_{yy} = 0.$$

By the property $f(x + \pi, y + \pi) = f(x + 2\pi, y) = f(x, y + 2\pi) = f(x, y)$, it is sufficient to check only on the domain $D := \{(x, y) \in \boldsymbol{R}^2 \mid |x| < \pi/2, |y| < \pi/2\}$. The left-side figure in page 88 shows the surface corresponding to this domain.

On the set D, $f(x, y) = \log(\cos y) - \log(\cos x)$ because $\cos x$, $\cos y > 0$. It is sufficient to show that this satisfies the equation above.

11 The translation $(x, y, z) \mapsto (x, y, z + 2n\pi)$ (n is an integer) preserves the surface. Moreover, the surface is symmetric with respect to the xy-plane, and the map $(x, y, z) \mapsto (x, y, \pi - z)$ preserves the surface. So it is sufficient to consider the part where $0 \leq z \leq \pi/2$. In this region, the surface is expressed as a graph of the function

$$f(x, y) = \arcsin((\sinh x)(\sinh y)).$$

Section 9 (page 98)

1 Assume that the curve $\gamma(s)$ is parametrized by the arclength s. Then by definition, $\gamma''(s) = \kappa(s)\boldsymbol{n}(s)$. On the other hand, denote by $\nu(s)$ the unit normal vector of the surface at $\gamma(s)$. Then $\kappa_n(s) = \gamma''(s) \cdot \nu(s) = \kappa(s)\boldsymbol{n}(s) \cdot \nu(s) = \kappa(s) \cos\theta$.

2 A surface of revolution can be parametrized as in Problem **7** in Section 7. Then the matrix $A = \widehat{I}^{-1}\, \widehat{I\!I}$ is diagonal, that is, the u-direction and the v-direction are principal directions. These are the directions of the generating curve and the rotation, respectively.

3 (1) Since $A = \lambda I$ (I is the identity matrix), the Weingarten formula (Proposition 8.5) yields the conclusion.

(2) Using (1) and $\nu_{uv} = \nu_{vu}$, we have $\lambda_u \boldsymbol{p}_v - \lambda_v \boldsymbol{p}_u = \boldsymbol{0}$. Since $\boldsymbol{p}_u$ and $\boldsymbol{p}_v$ are linearly independent, $\lambda_u = \lambda_v = 0$.

(3) Show that the derivatives $\boldsymbol{p} + (1/\lambda)\nu$ with respect to u and v are identically zero.

4 Let φ_1, φ_2 ($\varphi_1 < \varphi_2$, $-\pi/2 \leq \varphi_j \leq \pi/2$) be the angles between the two asymptotic directions and the positive direction of the x-axis, respectively. Then by Theorem 9.8, we have $\tan\varphi_1 = -\sqrt{|\lambda_1/\lambda_2|}$ and $\tan\varphi_2 = \sqrt{|\lambda_1/\lambda_2|}$. Hence $|\varphi_1| = |\varphi_2|$, that is, the x-axis bisects the two asymptotic directions. Letting μ be the angle between the two asymptotic directions, the conclusion follows, because $\varphi_2 = \mu/2$.

5 The principal curvatures of a minimal surface satisfy $\lambda_1 + \lambda_2 = 0$. Then the conclusion is obvious by Problem **4**.

Section 10 (page 115)

1 A line $\gamma(s)$ parametrized by the arclength s satisfies $\gamma''(s) = \mathbf{0}$.

2 Consider a cylinder parametrized as $\boldsymbol{p}(u,v) = (R\cos v, R\sin v, u)$ (R is a positive constant). Then its unit normal is $\nu = (-\cos v, -\sin v, 0)$. For a curve $\gamma(s) = \boldsymbol{p}(u(s), v(s))$ parametrized by the arclength s on the cylinder,

$$\gamma'(s) = u'(s)\boldsymbol{p}_u + v'(s)\boldsymbol{p}_v, \quad \gamma''(s) = u''(s)\boldsymbol{p}_u + v''(s)\boldsymbol{p}_v + R^2 v'(s)^2 \nu$$

hold. In particular, since s is arclength, $(u')^2 + R^2(v')^2 = 1$ holds. Hence γ is a geodesic if and only if $u'' = v'' = 0$, that is, $u = as + b$, $v = cs + d$, where a, b, c and d are constants. In particular, the curve is a circle perpendicular to the generating lines if $a = 0$, a generating line if $c = 0$, or a helix otherwise.

3 A curve $\gamma(s)$ parametized by arclength lying on the sphere of radius R centered at the origin is a geodesic if and only if γ'' is parallel to the normal of the sphere. Here, the normal vector of the sphere at $\gamma(s)$ is parallel to the position vector $\gamma(s)$. Then $\gamma'' \times \gamma = \mathbf{0}$ if γ is a geodesic. Moreover, by $(\gamma \times \gamma')' = \gamma \times \gamma''$, $\boldsymbol{v} := \gamma \times \gamma'$ is independent of s for a geodesic γ. Since $\boldsymbol{v}$ is perpendicular to γ, $\gamma(s)$ is contained in the plane passing through the origin (the center of the sphere) perpendicular to $\boldsymbol{v}$.

4 Set $\boldsymbol{p}(u,v) = (x(u)\cos v, x(u)\sin v, z(u))$, where $(x(s), z(s))$ is a curve on the xz-plane parametrized by the arclength s. Then a generating curve is a curve $\boldsymbol{p}(s, a)$ (a is constant). So it is sufficient to show that the acceleration vector of this curve is parallel to the normal vector of the surface.

5 By a rotation and a translation, one can assume that the surface is symmetric with respect to the xy-plane. Then the upward vector $(0,0,1)$ is tangent to the surface along the intersection of the surface and the xy-plane. Let $\gamma(s)$ be a parametrization of the intersection of the xy-plane and the surface, where s is arclength. Then any tangent vector of the surface at $\gamma(s)$ can be expressed as a linear combination of $\gamma'(s)$ and $(0,0,1)$. Moreover, $\gamma''(s)$ is also a vector parallel to the xy-plane and satisfies $\gamma'(s)\cdot\gamma''(s)=0$. Then $\gamma''(s)$ is parallel to the normal vector of the surface.

7 Apply the result of Problem **1** in Section 9 and equation (10.14).

8 Let s be the arclength of the curve $\gamma(t)$. Since $ds/dt = |\dot\gamma|$ ($\dot{} = d/dt$), we have $\gamma' = \dot\gamma/|\dot\gamma|$ and $\gamma'' = \frac{d}{ds}(1/|\dot\gamma|)\,\dot\gamma + \ddot\gamma/|\dot\gamma|$. Since the unit normal vector ν and the unit co-normal vector $\boldsymbol{n}_g$ do not depend on the parameter, (10.13) yields the conclusion.

9 Parametrize the curve as $\gamma(s)$ with arclength s, and let $\nu(s)$ be the unit normal vector of the surface at $\gamma(s)$. Then $\{\gamma'(s), \boldsymbol{n}_g(s), \nu(s)\}$ determines an orthonormal basis of $\boldsymbol{R}^3$. Since s is arclength parameter, $\gamma''(s)\cdot\gamma'(s)=0$ holds. On the other hand, $\kappa_g(s)=0$ implies $\gamma''(s)\cdot \boldsymbol{n}_g(s)=\boldsymbol{0}$. Hence $\gamma''(s)$ is parallel to $\nu(s)$, that is, $[\gamma''(s)]^{\mathrm{H}}=\boldsymbol{0}$.

10 By (10.16), (2.7),

$$\kappa_g = \frac{\det(\dot\gamma,\ddot\gamma,\nu)}{|\dot\gamma|^3} = \frac{1}{|\dot\gamma|^3}\det\begin{pmatrix}\dot x & \ddot x & 0\\ \dot y & \ddot y & 0\\ 0 & 0 & 1\end{pmatrix} = \frac{1}{|\dot\gamma|^3}\det\begin{pmatrix}\dot x & \ddot x\\ \dot y & \ddot y\end{pmatrix} = \kappa.$$

11 Since the outward unit normal vector of the sphere at $\gamma_\theta(t)$ is $\nu(t) = \gamma_\theta(t)$ and $\kappa_g = \tan\theta$ by (10.16).

13 For a spherical curve $\boldsymbol{e}(s)$, $\boldsymbol{e}'(s)=\kappa(s)\boldsymbol{n}(s)$ holds. Then the unit tangent vector of $\boldsymbol{e}(s)$ is the principal curvature vector $\boldsymbol{n}(s)$ of the curve $\gamma(s)$. On the other hand, the outward unit normal vector $\nu(s)$ of the sphere at $\boldsymbol{e}(s)$ coincides with $\boldsymbol{e}(s)$. Thus the geodesic curvature of $\boldsymbol{e}(s)$ is

$$\begin{aligned}\frac{\det(\boldsymbol{e}',\boldsymbol{e}'',\boldsymbol{e})}{|\boldsymbol{e}'|^3} &= \frac{\det(\kappa\boldsymbol{n},(\kappa\boldsymbol{n})',\boldsymbol{e})}{\kappa^3} = \frac{\det(\boldsymbol{n},\kappa'\boldsymbol{n}+\kappa\boldsymbol{n}',\boldsymbol{e})}{\kappa^2}\\ &= \frac{\det(\boldsymbol{n},\kappa(-\kappa\boldsymbol{e}+\tau\boldsymbol{b}),\boldsymbol{e})}{\kappa^2} = \frac{\tau}{\kappa}.\end{aligned}$$

14 Assume that γ is parametrized by the arclength s, and let $\boldsymbol{e}(s)$, $\boldsymbol{n}(s)$, $\boldsymbol{b}(s)$ be as in Section 5. Since γ is a spherical curve, $\gamma\cdot\gamma=1$ holds, implying that γ is a unit vector perpendicular to $\boldsymbol{e}=\gamma'$. Hence there exist two functions α, β ($\alpha^2+\beta^2=1$) satisfying $\gamma(s)=\alpha(s)\boldsymbol{n}(s)+\beta(s)\boldsymbol{b}(s)$. By differentiating $\gamma\cdot\gamma'=0$, it holds that

$$0=\gamma'\cdot\gamma'+\gamma\cdot\gamma''=1+(\alpha\boldsymbol{n}+\beta\boldsymbol{b})\cdot(\kappa\boldsymbol{n})=1+\alpha\kappa.$$

Hence $\alpha=-1/\kappa$, $\beta^2=1-1/\kappa^2$. On the other hand, since the unit normal vector $\nu(s)$ of the sphere at $\gamma(s)$ coincides with $\gamma(s)$, the unit co-normal vector is obtained by

$$\boldsymbol{n}_g = \nu \times \gamma' = \gamma \times \gamma' = (\alpha \boldsymbol{n} + \beta \boldsymbol{b}) \times \boldsymbol{e} = -\alpha \boldsymbol{b} + \beta \boldsymbol{n}.$$

Thus $\kappa_g = \gamma'' \cdot \boldsymbol{n}_g = \kappa \boldsymbol{n} \cdot (-\alpha \boldsymbol{b} + \beta \boldsymbol{n}) = \kappa\beta$. In particular, since $\kappa > 0$,

$$|\kappa_g| = \kappa\sqrt{1 - \frac{1}{\kappa^2}} = \sqrt{\kappa^2 - 1}.$$

Moreover, taking the inner product of the derivative of $\gamma = \alpha \boldsymbol{n} + \beta \boldsymbol{b}$ and $\boldsymbol{n}$, Frenet's formula implies that

$$\begin{aligned} 0 = \gamma' \cdot \boldsymbol{n} &= (\alpha \boldsymbol{n} + \beta \boldsymbol{b})' \cdot \boldsymbol{n} \\ &= \alpha'(\boldsymbol{n} \cdot \boldsymbol{n}) + \alpha(\boldsymbol{n}' \cdot \boldsymbol{n}) + \beta'(\boldsymbol{b} \cdot \boldsymbol{n}) + \beta(\boldsymbol{b}' \cdot \boldsymbol{n}) = \alpha' - \tau\beta. \end{aligned}$$

Since $\alpha = -1/\kappa$, we have $\kappa' = \tau\kappa^2\beta$. Hence

$$\kappa_g' = \pm(\sqrt{\kappa^2 - 1})' = \frac{\pm\kappa\kappa'}{\sqrt{\kappa^2 - 1}} = \frac{\pm\kappa'}{\sqrt{1 - 1/\kappa^2}} = \frac{\kappa'}{\beta} = \tau\kappa^2.$$

Thus, a point on a spherical curve satisfying $\kappa_g' = 0$ corresponds to a point where the torsion vanishes as a space curve. Identifying the sphere excluding one point with the plane via stereographic projection, a vertex (that is, a point where κ_g is extremal) corresponds to a vertex of the plane curve. Then by the four vertex theorem (Theorem 4.4 in Section 4), the torsion of a simple closed curve on the sphere changes sign at least four times.

Section 11 (page 129)

1 (1) Since $\boldsymbol{f}_s = (1 + t(\tau/\kappa)')\,\boldsymbol{e}$ and $\boldsymbol{f}_t = (\tau/\kappa)\boldsymbol{e} + \boldsymbol{b}$, it holds that $\boldsymbol{f}_s \times \boldsymbol{f}_t = -(1 + t(\tau/\kappa)')\,\boldsymbol{n}$. In particular, if $|t|$ is sufficiently small, $\boldsymbol{f}_s \times \boldsymbol{f}_t \neq \boldsymbol{0}$ holds, that is, $\boldsymbol{f}(s,t)$ gives a regular surface and $\nu(s,t) = -\boldsymbol{n}(s)$ is the unit normal vector field.

(2) $\gamma''(s) = \kappa(s)\boldsymbol{n}(s)$ is parallel to the unit normal of the surface.

Section 12 (page 137)

1 It is not hard to show (12.3) by (12.1) and (12.2). Here, it is necessary to show that $d\alpha$ satisfies the property $d\alpha(fX, Y) = f\alpha(X, Y)$. This follows from the property $[fX, Y] = f[X, Y] - df(Y)X$ of the bracket of vector fields.

2 For a 1-form $\omega = \alpha\, du + \beta\, dv$, $d\omega = (\beta_u - \alpha_v)\, du \wedge dv$.

5 Let ω be a 2-form which is written as $\omega = \lambda\, du \wedge dv$ with respect to a coordinate system (u, v). Then under another coordinate system, $\omega = \lambda\,(u_\xi v_\eta - u_\eta v_\xi)\, d\xi \wedge d\eta$ holds.

6 By (12.5), $\tilde{\omega}_1 \wedge \tilde{\omega}_2 = (\cos\theta\,\omega_1 + \sin\theta\,\omega_2) \wedge (-\sin\theta\,\omega_1 + \cos\theta\,\omega_2) = \omega_1 \wedge \omega_2$. Here we used $\omega_1 \wedge \omega_1 = \omega_2 \wedge \omega_2 = 0$, $\omega_1 \wedge \omega_2 = -\omega_2 \wedge \omega_1$.

7 Compute $\omega_1 \wedge \omega_2$ using the form as in Problem **4**.

8 (1) Take two curves $\gamma_1(t)$, $\gamma_2(t)$ $(0 \le t \le 1)$ joining $\mathrm{P_0}$ and P, and let γ be a curve joining γ_1 and the inverse curve of γ_2. If γ_1 does not intersect with γ_2, then γ is a simple closed curve. Since D is simply connected, γ bounds a domain Ω homeomorphic to the disc. Then by the Stokes theorem,

$$\int_{\gamma_1} \alpha - \int_{\gamma_2} \alpha = \int_{\gamma} \alpha = \int_{\Omega} d\alpha = 0,$$

holds, that is, the integral does not depend on the choice of path, but depends only on the terminal point P. When γ_1 and γ_2 intersect, take the third path γ_3 as shown in the right-side figure for this problem, and apply the Stokes theorem for γ_1 and γ_3, and γ_2 and γ_3, respectively.

(2) Let $\alpha = \alpha_1\,du + \alpha_2\,dv$, where (u,v) is a local coordinate system, and let the coordinates of $\mathrm{P_0}$ and P be (u_0, v_0) and (u,v), respectively. Then $f_u = \alpha_1$, because

$$f(u+\Delta u, v) - f(u,v) = \int_u^{u+\Delta u} \alpha_1(t,v)dt.$$

Similarly, $f_v = \alpha_2$. Hence $df = \alpha$.

Section 13 (page 147)

2 Differentiating $\hat{\mathbf{e}}_1 \cdot \hat{\mathbf{e}}_1 = \hat{\mathbf{e}}_2 \cdot \hat{\mathbf{e}}_2 = 1$ and $\hat{\mathbf{e}}_1 \cdot \hat{\mathbf{e}}_2 = 0$, one has $d\hat{\mathbf{e}}_1(X) \cdot \hat{\mathbf{e}}_1 = d\hat{\mathbf{e}}_2(X) \cdot \hat{\mathbf{e}}_2 = 0$ and $d\hat{\mathbf{e}}_2(X) \cdot \hat{\mathbf{e}}_1 = -d\hat{\mathbf{e}}_1(X) \cdot \hat{\mathbf{e}}_2$. Here, $d\boldsymbol{p}(X) = \omega_1(X)\hat{\mathbf{e}}_1 + \omega_2(X)\hat{\mathbf{e}}_2$, that is, $d\boldsymbol{p} = \omega_1\hat{\mathbf{e}}_1 + \omega_2\hat{\mathbf{e}}_2$, where $\{\omega_1, \omega_2\}$ is the dual basis of $\{\mathbf{e}_1, \mathbf{e}_2\}$. Then the exterior derivative of this yields

$$0 = d(d\boldsymbol{p}) = d\omega_1\hat{\mathbf{e}}_1 + d\omega_2\hat{\mathbf{e}}_2 + \omega_1 \wedge d\hat{\mathbf{e}}_1 + \omega_2 \wedge d\hat{\mathbf{e}}_2.$$

Taking inner products with $\hat{\mathbf{e}}_1$ and $\hat{\mathbf{e}}_2$, we have $d\omega_1 = \omega_2 \wedge \mu$ and $d\omega_2 = -\omega_1 \wedge \mu$ for μ as described in the problem.

3 Define $\nabla_{(\partial/\partial u)}(\partial/\partial u)$ etc. as in the problem, and define $\nabla\colon \mathfrak{X}^\infty(S) \times \mathfrak{X}^\infty(S) \to \mathfrak{X}^\infty(S)$ for general vector fields by having it satisfy the properties (13.6)–(13.9) of the covariant derivative. It is sufficient to show that ∇ is a covariant derivative satisfying (13.10), (13.11), because of Theorem 13.3. The properties (13.10), (13.11) follow from the property $\Gamma_{ij}^k = \Gamma_{ji}^k$ of the Christoffel symbol and relations $\Gamma_{11}^1 E + \Gamma_{11}^2 F = \frac{1}{2}E_u$, etc. as on page 118.

4 $e_1 = \partial/\partial r$ and $e_2 = h^{-1}\partial/\partial\theta$ form an orthonormal basis, and the dual basis is given by $\omega_1 = dr$ and $\omega_2 = h\,d\theta$.

Section 14 (page 152)

1 We define the orientation of the surface as follows: For a non-zero tangent vector $\boldsymbol{a}$ at a point P on the surface, $\boldsymbol{a}$ and $\nu \times \boldsymbol{a}$ form a positively oriented basis of the tangent plane. Then for a curve $\gamma(s)$ on the surface parametrized by the arclength s, the unit conormal vector $\boldsymbol{n}_g$ as in (10.15) is the left-ward unit normal vector of the curve. Since $\boldsymbol{n}_g$ is tangent to the surface,

$$\gamma'' \cdot \boldsymbol{n}_g = \left[\gamma''\right]^{\mathrm{H}} \cdot \boldsymbol{n}_g = \langle \nabla_{\gamma'}\gamma', \boldsymbol{n}_g\rangle = \langle \boldsymbol{\kappa}_g, \boldsymbol{n}_g\rangle$$

holds, by Theorem 13.5.

Section 15 (page 164)

1 (1) The origin is the only zero of the vector field X. Consider a loop $\gamma(t) = e^{it} = (\cos t, \sin t)$ $(0 \le t \le 2\pi)$ surrounding the origin. Since the value of X at $\gamma(t)$ is $X(\gamma(t)) = e^{ikt} = (\cos kt, \sin kt)$, the angle ψ between $\dot\gamma$ and X satisfies

$$\cos\psi(t) = (-\sin t, \cos t)\cdot(\cos kt, \sin kt) = \sin(k-1)t.$$

In particular, we can assume $\psi(0) = \pi/2$ without loss of generality. Then $\psi(t) = (k-1)t + \pi/2$, and hence $\psi(2\pi) - \psi(0) = 2(k-1)\pi$.

(2) Similarly, the value of the vector field Y at $\gamma(t)$ is $(\cos kt, -\sin kt)$. Then the angle between $\dot\gamma$ and Y is $-(k+1)t + \pi/2$.

Section 16 (page 171)

3 (1) Apply Problem **2**.

(2) Let $\gamma(t) = (u_0, t)$ $(a \le t \le b)$. The arclength parameter s with respect to the metric ds^2_{H} is given by $s = \int_a^t \frac{1}{\tau}\,d\tau = \log\frac{t}{a}$. Using this parameter, compute $\nabla_{\gamma'}\gamma'$.

(6) Take an isometry which maps both z and w to the imaginary axis.

4 (1) For the cycloid, the parameter t as in Problem **4** in Section 1 is proportional to the arclength.

Section 17 (page 180)

1 Use an isothermal coordinate system.

2 The vector field $\boldsymbol{H} := H\nu$ is a normal vector field defined on the entire surface which has no zeros. For a local coordinate system (u, v) of S, $\boldsymbol{p}_u \times \boldsymbol{p}_v = \lambda\boldsymbol{H}$, where λ is a non-vanishing function on the coordinate neighborhood. If λ is negative, then by replacing (u, v) with (v, u), S can be covered by coordinate neighborhoods such that $\boldsymbol{p}_u \times \boldsymbol{p}_v$ points in the same direction as $\boldsymbol{H}$. For such an atlas, the Jacobian of each coordinate change is positive. This means that S is orientable.

3 The first fundamental form, the unit normal vector field, and the second fundamental form are

$$\begin{aligned} ds^2 &= (1 + (u^2 + v^2)^2)^2 (du^2 + dv^2), \\ \nu &= \frac{1}{1 + (u^2 + v^2)^2} (2(u^2 - v^2), 4uv, (u^2 + v^2)^2 - 1), \\ II &= 4(-u\, du^2 + 2v\, du\, dv + u\, dv^2), \end{aligned}$$

respectively, and then the Hopf differential is $-2(u + iv)\, dz^2 = -2z\, dz^2$. Then by (17.17), $\boldsymbol{p}$ has an umbilic of index $-1/2$ at the origin.

4 Since the Christoffel symbols are computed as

$$\Gamma^1_{11} = \frac{\theta_u}{\tan\theta}, \; \Gamma^2_{11} = -\frac{\theta_u}{\sin\theta}, \; \Gamma^1_{12} = \Gamma^2_{12} = 0, \; \Gamma^1_{22} = -\frac{\theta_v}{\sin\theta}, \; \Gamma^2_{22} = \frac{\theta_v}{\tan\theta},$$

the necessary and sufficient condition for (17.5) is $\theta_{uv} = \sin\theta$. Moreover, the Gaussian curvature is $\det \widehat{II} / \det \widehat{I} = -1$.

5 Since the Christoffel symbols are computed as

$$\begin{aligned} \Gamma^1_{11} &= \frac{\theta_u}{2} \tanh\frac{\theta}{2}, \quad \Gamma^1_{12} = \frac{\theta_v}{2} \tanh\frac{\theta}{2}, \quad \Gamma^1_{22} = -\frac{\theta_u}{2} \tanh\frac{\theta}{2}, \\ \Gamma^2_{11} &= -\frac{\theta_v}{2} \coth\frac{\theta}{2}, \quad \Gamma^2_{12} = \frac{\theta_u}{2} \coth\frac{\theta}{2}, \quad \Gamma^2_{22} = \frac{\theta_v}{2} \coth\frac{\theta}{2}, \end{aligned}$$

where $\coth\frac{\theta}{2} := 1/\tanh\frac{\theta}{2}$, the equivalency of (17.5) and the equation of the problem follows. The Gaussian curvature is $\det \widehat{II} / \det \widehat{I} = 1$.

Appendix B.1 (page 217)

1 Let $(u, 0)$, $(0, v)$ be the endpoints of the segment on the x-axis and y-axis, respectively, where $u, v \geq 0$. Since the length of the segment is 1, $u^2 + v^2 = 1$ holds, that is, one can write $u = \cos t$, $v = \sin t$ $(0 \leq t \leq \pi/2)$. The equation of the line segment is written as

$$F(x, y, t) = x \sin t + y \cos t - 2 \sin t \cos t.$$

Then the equation $F(x, y, t) = F_t(x, y, t) = 0$ is turns out to be

$$\begin{aligned} x \sin t + y \cos t &= \sin t \cos t, \\ x \cos t - y \sin t &= \cos^2 t - \sin^2 t \end{aligned} \qquad \left(0 \leq t \leq \frac{\pi}{2}\right).$$

Solving these with respect to x and y, we have the parametrization of the astroid $(x, y) = (\cos^3 t, \sin^3 t)$ as the case $a = 1$ of Example 1.3 in Section 1.

2 Let $\boldsymbol{n}(s)$ be the left-ward unit normal vector of $\sigma(s)$. Noticing that the leftward normal vector of γ is $(-\operatorname{sgn}\kappa)\sigma'$, we have $\gamma'(s) = (\delta - s)\kappa(s)\boldsymbol{n}(s)$ and $\gamma''(s) = \{(\delta - s)\kappa(s)\}'\boldsymbol{n}(s) - (\delta - s)\kappa(s)^2\sigma'(s)$ hold. Hence the curvature of $\gamma(s)$ is obtained by

$$\frac{\det(\gamma', \gamma'')}{|\gamma'(s)|^3} = \frac{\kappa^3(\delta - s)^2}{|\kappa^3(\delta - s)|^3} = \frac{\operatorname{sign}(\kappa)}{\delta - s} \qquad \left(\operatorname{sign}(\kappa) = \frac{\kappa}{|\kappa|}\right).$$

3 The involute of the cycloid $\gamma(t) = (a(t - \sin t), a(1 - \cos t))$ is

$$\begin{aligned}\sigma(t) &= (a(t + \sin t), a(\cos t - 1)) \\ &= (a(\pi + t - \sin(\pi + t)), a(1 - \cos(\pi + t))) - (\pi a, 2a).\end{aligned}$$

4 (1) Let $\gamma(w(t)) = (u(w(t)), v(w(t)))$ be the position at time t of the object which starts at $\gamma(b)$ at $t = 0$, and has speed $V(t)$. By the conservation law of mechanical energy, $mV^2/2 + mgv$ must be constant. In particular, the constant must be 0 because $V = 0$ and $v = 0$ at $t = 0$. Hence $V = \sqrt{2g|v|}$. On the other hand, since

$$V = \sqrt{\left(\frac{du}{dt}\right)^2 + \left(\frac{dv}{dt}\right)^2} = \sqrt{\left(\frac{du}{dw}\right)^2 + \left(\frac{dv}{dw}\right)^2}\,\frac{dw}{dt},$$

it holds that

$$\frac{dt}{dw} = \frac{1}{\sqrt{2g|v|}}\sqrt{\left(\frac{du}{dw}\right)^2 + \left(\frac{dv}{dw}\right)^2}.$$

Integrating this from $w = b$ to $w = c$, the conclusion follows.

(2) Parametrizing the cycloid as shown in the figure of the problem as $\gamma(s) = (u(s), v(s))$, where s is the arclength, then $v(s) = \{b(8a - b) - s(8a - s)\}/8a$ holds, as seen in Problem **5** in Section 2. In particular, the object travels from $s = b$ to $s = 4a$, and the integral in (1) gives the result. The period of the cycloid pendulum is four times this value: $2\pi\sqrt{l/g}$.

(3) Consider the cycloid for $2\pi a = 4.03 \times 10^5$m. The time until the object reaches the bottom is $\pi\sqrt{a/g}$, and the desired time is two times this value. Assuming the gravitational acceleration to be $g = 9.8\text{m/s}^2$, the answer is about 8 and a half minutes.

Appendix B.3 (page 230)

1 Let $\gamma(t) = (x(t), y(t))$ $(a \le t \le b)$ be a smooth curve and $\tilde{\gamma}(u) = (\tilde{x}(u), \tilde{y}(u))$ $(\alpha \le u \le \beta)$ a parameter change, that is, $\tilde{\gamma}(u) = \gamma(t(u))$, where $t = t(u)$ $(\alpha \le u \le \beta)$ is an increasing smooth function such that $t(\alpha) = a$ and $t(\beta) = b$. Then integrating, using substitution, we have

$$\int_a^b f(x(t), y(t)) \frac{dx}{dt}(t)\, dt = \int_\alpha^\beta f(x(t(u)), y(t(u))) \frac{dx}{dt}(t(u)) \frac{dt}{du}(u)\, du$$
$$= \int_\alpha^\beta f(\tilde{x}(u), \tilde{y}(u)) \frac{d\tilde{x}}{du}(u)\, du,$$
$$\int_a^b g(x(t), y(t)) \frac{dx}{dt}(t)\, dt = \int_\alpha^\beta g(\tilde{x}(u), \tilde{y}(u)) \frac{d\tilde{y}}{du}(u)\, du,$$

and the conclusion holds.

2 Since the segments C_1, C_2 and C_3 are parametrized as $\gamma_1(t) = (t, t)$, $\gamma_2(t) = (0, t)$ and $\gamma_3(t) = (t, 1)$ $(0 \le t \le 1)$, respectively,

$$\int_{C_1} x\, dy = \int_0^1 t\, dt = \frac{1}{2},$$
$$\int_{C_2 + C_3} x\, dy = \int_{C_2} x\, dy + \int_{C_3} x\, dy = \int_0^1 0\, dt + \int_0^1 t \times 0\, dt = 0.$$

3 The convex hull $\widetilde{C}$ of the simple closed curve C is obtained as follows: Take any tangent line L to C which touches the curve at more than one point $\mathrm{P}_1, \ldots, \mathrm{P}_n$ (ordered by their appearance along the line), and replace the arc of C between P_1 and P_n by the line segment $\mathrm{P}_1\mathrm{P}_n$. Repeating this procedure for all tangent lines, we have a piecewise smooth convex curve $\widetilde{C}$ bounding a convex domain $\widetilde{D}$, which is the convex hull of the domain D bounded by C. Since $D \subset \widetilde{D}$, the area $\widetilde{A}$ of $\widetilde{D}$ is not less than the area A of D. On the other hand, since $\widetilde{C}$ is obtained by replacing arcs in C by line segments, the length $\tilde{l}$ of $\widetilde{C}$ is not greater than the length l of C. Here, if the isoperimetric inequality holds for convex curves, it holds that $4\pi\widetilde{A} \le \tilde{l}^2$. Then

$$4\pi A \le 4\pi \widetilde{A} \le \tilde{l}^2 \le l^2$$

holds, which is the isoperimetric inequality for the curve C.

4 Let $T \in \mathcal{T}$ be a triangle with sides of lengths a, b and c. Since $a + b + c = 1$, the area S of T satisfy

$$16S^2 = f(a, b) := (1 - 2a)(1 - 2b)(2a + 2b - 1)$$

by Heron's formula. By the triangle inequality, the parameters a, b runs over the set

$$D := \left\{(a,b) \,\middle|\, a>0,\ b>0,\ \frac{1}{2}<a+b<1\right\} \subset \boldsymbol{R}^2.$$

One can find easily that $f(a,b)$ attains its maximum at $a=b=1/3$.

Appendix B.4 (page 236)

1 Derive $\cos^2 v = 1/\cosh^2\eta$ from $e^\eta = \cos v/(1-\sin v)$.

2 For a parametrization $\boldsymbol{p}(u,v)$ as in the answer of Problem **1** in Section 6, we have $E=b^2$, $F=0$ and $G=(b\cos u+a)^2$. Let $\xi = b\sin u + au$, $\eta = v$. Then by the inverse function theorem, the map $(u,v)\mapsto(\xi,\eta)$ locally has an inverse, that is, this gives a coordinate change. Then the first fundamental form is expressed by

$$ds^2 = \frac{1}{(b\cos u+a)^2}\,d\xi^2 + (b\cos u+a)^2\,d\eta^2.$$

Hence the parameter (ξ,η) gives an equi-area projection. We remark that the choice of equi-area projection is not unique.

3 Using the polar coordinate system (r,θ) of the xy-plane, let $\xi = \sqrt{2r}\cos(\theta/\sqrt{2})$, $\eta=\sqrt{2r}\sin(\theta/\sqrt{2})$. Then the coordinate system (ξ,η) is the desired one.

4 Let Π be the tangent plane at the south pole. A great circle is the intersection of the sphere and a plane Γ passing through the center of the sphere. Hence the image of the great circle by the central projection is the intersection of Γ and Π, a line.

Appendix B.5 (page 244)

1 Suppose the surface is parametrized as $\boldsymbol{p}(u,v)$ with principal curvature line coordinates (u,v). It is sufficient to show that the coordinate system (u,v) is a principal curvature line coordinate system of $\tilde{\boldsymbol{p}}(u,v) := T\circ \boldsymbol{p}(u,v)$. For an arbitrary vector $\boldsymbol{a}\in\boldsymbol{R}^3\setminus\{\boldsymbol{0}\}$, the map $U_{\boldsymbol{a}}\colon \boldsymbol{R}^3\to\boldsymbol{R}^3$ given by

$$U_{\boldsymbol{a}}(\boldsymbol{x}) := \boldsymbol{x} - \frac{2(\boldsymbol{a}\cdot\boldsymbol{x})}{|\boldsymbol{a}|^2}\boldsymbol{a} \qquad (\boldsymbol{x}\in\boldsymbol{R}^3)$$

is a linear map preserving the inner product. Moreover, $U_{\boldsymbol{a}}(\boldsymbol{a}) = -\boldsymbol{a}$, and $U_{\boldsymbol{a}}(\boldsymbol{x}) = \boldsymbol{x}$ holds for each vector $\boldsymbol{x}$ perpendicular to $\boldsymbol{a}$, that is, $U_{\boldsymbol{a}}$ is a reflection with respect to the plane passing through the origin and perpendicular to $\boldsymbol{a}$. In particular, there exists an orthogonal matrix $A_{\boldsymbol{a}}$ with determinant -1 such that $U_{\boldsymbol{a}}(\boldsymbol{x}) = A_{\boldsymbol{a}}\boldsymbol{x}$.

Differentiating $\tilde{\boldsymbol{p}} = \boldsymbol{p}/|\boldsymbol{p}|^2$, $\tilde{\boldsymbol{p}}_u = A_{\boldsymbol{p}}\boldsymbol{p}_u/|\boldsymbol{p}|^2$ and $\tilde{\boldsymbol{p}}_v = A_{\boldsymbol{p}}\boldsymbol{p}_v/|\boldsymbol{p}|^2$ hold. Since (u,v) is a curvature line coordinate system of $\boldsymbol{p}$, Theorem B.5.1 implies

$$\tilde{\boldsymbol{p}}_u \cdot \tilde{\boldsymbol{p}}_v = \frac{1}{|\boldsymbol{p}|^4}(A_{\boldsymbol{p}}\boldsymbol{p}_u \cdot A_{\boldsymbol{p}}\boldsymbol{p}_v) = \frac{1}{|\boldsymbol{p}|^4}(\boldsymbol{p}_u \cdot \boldsymbol{p}_v) = 0.$$

On the other hand,

$$\tilde{\boldsymbol{p}}_u \times \tilde{\boldsymbol{p}}_v = \frac{1}{|\boldsymbol{p}|^4}(A_{\boldsymbol{p}}\boldsymbol{p}_u \times A_{\boldsymbol{p}}\boldsymbol{p}_v) = -\frac{1}{|\boldsymbol{p}|^4}A_{\boldsymbol{p}}(\boldsymbol{p}_u \times \boldsymbol{p}_v)$$

holds, because of Corollary A.3.6 in Appendix A.3. Thus, the unit normal vector $\tilde{\nu}$ of $\tilde{\boldsymbol{p}}$ is expressed as

$$\tilde{\nu} = -A_{\boldsymbol{p}}\nu = -\nu + \frac{2(\nu \cdot \boldsymbol{p})}{|\boldsymbol{p}|^2},$$

where ν is the unit normal vector of $\boldsymbol{p}$. Hence it holds that

$$\tilde{\nu}_v = -\left(\nu - \frac{2(\nu \cdot \boldsymbol{p})}{|\boldsymbol{p}|^2}\right)_v = -A_{\boldsymbol{p}}\nu_v + \frac{2(\nu \cdot \boldsymbol{p})}{|\boldsymbol{p}|^2}F_{\boldsymbol{p}}(\boldsymbol{p}_v).$$

Here, noticing that (u,v) is a principal curvature line coordinate system, that is, $\boldsymbol{p}_u \cdot \boldsymbol{p}_v = 0$ and $\boldsymbol{p}_u \cdot \nu_v = 0$ hold, we have $\tilde{\boldsymbol{p}}_u \cdot \tilde{\nu}_v = 0$. By Theorem B.5.1, (u,v) is a principal curvature line coordinate system of $\tilde{\boldsymbol{p}}$.

Appendix B.6 (page 251)

1 Since $\dot{\gamma}$ and ξ are linearly independent, $\boldsymbol{n} := \dot{\gamma} \times \xi \neq \boldsymbol{0}$. Here, it holds that $\det(\dot{\gamma}, \xi, \dot{\xi}) = 0$, because $\boldsymbol{q}$ is developable (cf. (B.6.3)). Hence $\dot{\xi}$ can be expressed as a linear combination of $\dot{\gamma}$ and ξ, and then $\dot{\xi}$ is perpendicular to $\boldsymbol{n}$. Thus, both $\boldsymbol{q}_u = \dot{\gamma} + v\dot{\xi}$ and $\boldsymbol{q}_v = \xi$ are perpendicular to $\boldsymbol{n}$. Hence the unit normal vector is $\pm\boldsymbol{n}/|\boldsymbol{n}|$.

Appendix B.9 (page 268)

1 The singular points are $t = 2n\pi$ (n runs over the integers). Apply Theorem B.9.1 at these points.

3 For $\boldsymbol{p}_1$ in (B.9.1), $(\boldsymbol{p}_1)_u = (1,0,0)$, $(\boldsymbol{p}_1)_v = v(0,2,3v)$, and then $(\boldsymbol{p}_1)_u \times (\boldsymbol{p}_1)_v = v(0,-3v,2)$. Hence $\nu := (0,-3v,2)/\sqrt{4+9v^2}$ gives the unit normal vector field, that is, the condition (B.9.2) holds. Then λ in (B.9.3) is computed by $\det((\boldsymbol{p}_1)_u, (\boldsymbol{p}_1)_v, \nu) = v\sqrt{4+9v^2}$. Thus the set of singular points is $\{(u,v) \in \boldsymbol{R}^2 | v = 0\}$, i.e., the u-axis. On the u-axis, $\lambda_v = 2 \neq 0$ holds, and then the condition (B.9.3) holds. Moreover, $(\boldsymbol{p}_1)_v = \boldsymbol{0}$ on the singular curve (the u-axis). Then $\eta = (0,1)$ is a null vector field as in (B.9.4). Therefore, $\nu_\eta = \nu_v = (0, -3/2, 0) \neq \mathbf{0}$ holds on the singular set $\{v = 0\}$, that is, the condition (B.9.5) holds. Parametrizing the singular curve as $\gamma(t) = (t, 0)$, $\dot\gamma(t) = (1, 0)$. Then μ in (B.9.6) is identically 1, because $\eta = (0, 1)$. Therefore, each singular point is a cuspidal edge, because of Proposition B.9.4.

4 For $\boldsymbol{p}_2$ in (B.9.1), $(\boldsymbol{p}_2)_u = 2(6u^2 + v)(u, 1, 0)$, $(\boldsymbol{p}_2)_v = v(u^2, 2u, 1)$ and $(\boldsymbol{p}_2)_u \times (\boldsymbol{p}_2)_v = -2(6u^2 + v)(1, -u, u^2)$ hold. Then the unit normal vector $\nu := (1, -u, u^2)/\sqrt{1 + u^2 + u^4}$ can be defined as a smooth vector field, proving (B.9.2). Then λ in (B.9.3) is computed as $2(6u^2 + v)(1 + u^2 + u^4)^{1/2}$, that is, the singular points are on a parabola $v = -6u^2$. Here, $\lambda_v \neq 0$ on the parabola, proving (B.9.3). On the other hand, $(\boldsymbol{p}_2)_u = \mathbf{0}$ on the singular curve, and then $\eta = (1, 0)$ is the null vector field. Then

$$\nu_\eta = \nu_u = \frac{1}{\sqrt{1 + u^2 + u^4}^3}\left(-u(1 + 2u^2), u^4 - 1, u(2 + u^2)\right) \neq \mathbf{0},$$

that is, the condition (B.9.5) holds. Parametrizing the singular curve as $\gamma(t) = (t, -6t^2)$, $\dot\gamma(t) = (1, -12t)$. Combined with $\eta = (1, 0)$, μ in (B.9.6) is $\mu(t) = 12t$. Then a singular point is a swallowtail at $t = 0$ (i.e., $(0, 0)$), and otherwise a cuspidal edge.

5 Under the notation of Proposition B.9.4,

$$\nu = \frac{1}{\sqrt{a^2 + b^2}}(-b \sin u, b \cos u, -a), \qquad \lambda = a\sqrt{a^2 + b^2}v.$$

Then the singular set is $\{(u, v) \in \boldsymbol{R}^2 | v = 0\}$. In particular, if $v = 0$, then $\lambda_v \neq 0$ holds. Moreover, $\eta = (1, -1)$ is a null vector, because $\boldsymbol{p}_u - \boldsymbol{p}_v = \mathbf{0}$ on $\{v = 0\}$. Then $\mu(t) = -1$ on the singular curve $\gamma(t) = (t, 0)$. Hence the singular points are cuspidal edges.

6 Take the unit normal vector as in (B.9.9). Then the first and second fundamental forms are

$$ds^2 = \frac{1}{(u^2 + \cosh^2 v)^2}(4u^2 \cosh^2 v\, du^2 + (u^2 - \cosh^2 v)^2\, dv^2),$$
$$II = \frac{2u(u^2 - \cosh^2 v) \cosh v}{(u^2 + \cosh^2 v)^2}(-du^2 + dv^2).$$

Appendix B.10 (page 278)

1 The Jacobian matrix of the inverse map $(u, v) \mapsto (\xi, \eta)$ of the coordinate change $(\xi, \eta) \mapsto (u, v)$ satisfies

$$\begin{pmatrix} \xi_u & \xi_v \\ \eta_u & \eta_v \end{pmatrix} = \begin{pmatrix} u_\xi & u_\eta \\ v_\xi & v_\eta \end{pmatrix}^{-1}.$$

Then $\Omega = \xi_u \widetilde{\Omega} + \eta_u \widetilde{\Lambda}$, $\Lambda = \xi_v \widetilde{\Omega} + \eta_v \widetilde{\Lambda}$ hold. So exchanging the roles of (u, v) and (ξ, η), similarly to the way we derived (B.10.9) from (B.10.1), the conclusion follows.

2 If $\mathcal{F}$ satisfies (B.10.1),

$$\begin{aligned}\hat{\mathcal{F}}_u &= (\mathcal{F}\zeta)_u = \mathcal{F}_u\zeta + \mathcal{F}\zeta_u = \mathcal{F}\Omega\zeta + \mathcal{F}\zeta_u = \hat{\mathcal{F}}\zeta^{-1}\Omega\zeta + \hat{\mathcal{F}}\zeta^{-1}\zeta_u \\ &= \hat{\mathcal{F}}(\zeta^{-1}\Omega\zeta + \zeta^{-1}\zeta_u) = \hat{\mathcal{F}}\hat{\Omega}\end{aligned}$$

and $\hat{\mathcal{F}}_v = \hat{\mathcal{F}}\hat{\Lambda}$, proving (B.10.11). Conversely, assume that $\hat{\mathcal{F}}$ satisfies (B.10.11). Differentiating $I = \zeta^{-1}\zeta$ (I is the identity matrix) by u, we have $O = (\zeta^{-1})_u\zeta + \zeta^{-1}\zeta_u$, that is, $(\zeta^{-1})_u = -\zeta^{-1}\zeta_u\zeta^{-1}$. Then differentiating $\mathcal{F} = \hat{\mathcal{F}}\zeta^{-1}$,

$$\begin{aligned}\mathcal{F}_u &= (\hat{\mathcal{F}})_u\zeta^{-1} - \hat{\mathcal{F}}(\zeta^{-1}\zeta_u\zeta^{-1}) = \hat{\mathcal{F}}\hat{\Omega}\zeta^{-1} - \hat{\mathcal{F}}\zeta^{-1}\zeta_u\zeta^{-1} \\ &= \mathcal{F}(\zeta\hat{\Omega}\zeta^{-1} - \zeta_u\zeta^{-1}) = \mathcal{F}(\zeta(\zeta^{-1}\Omega\zeta + \zeta^{-1}\zeta_u)\zeta^{-1} - \zeta_u\zeta^{-1}) = \mathcal{F}\Omega\end{aligned}$$

holds, and similarly $\mathcal{F}_v = \mathcal{F}\Lambda$, proving (B.10.1).

The equivalence of (17.5) and (B.10.13) can be proved by a direct calculation.

3 Considering $\boldsymbol{a} := \boldsymbol{p}_u(u_0, v_0)$, $\boldsymbol{b} := \boldsymbol{p}_v(u_0, v_0)$, $\boldsymbol{c} := \nu(u_0, v_0)$ as column vectors, we write $\mathcal{F}(u_0, v_0) = (\boldsymbol{a}, \boldsymbol{b}, \boldsymbol{c})$. Since we set $\boldsymbol{c} = \nu(u_0, v_0) = (0, 0, 1)^T$, (B.10.14) implies the result for the third column. Since $\boldsymbol{p}_u(u_0, v_0)$ is a positive real multiple of $(1, 0, 0)^T$ and $E_0 = E(u_0, u_0) = |\boldsymbol{a}|^2$, the first column of (B.10.14) follows. Moreover, $\boldsymbol{b}$ is perpendicular to ν, it is in the form $\boldsymbol{b} = (\alpha, \beta, 0)^T$. Here, $\boldsymbol{a} \cdot \boldsymbol{b} = F_0$ implies $\alpha = F_0/\sqrt{E_0}$, and then $\beta^2 = (E_0G_0 - F_0^2)/E_0$, because $\boldsymbol{b} \cdot \boldsymbol{b} = G_0$. In particular, since $\boldsymbol{c} = (\boldsymbol{a} \times \boldsymbol{b})/|\boldsymbol{a} \times \boldsymbol{b}|$ and the determinant of $\mathcal{F}(u_0, v_0)$ is positive, $\beta = \delta_0/\sqrt{E_0}$.

Bibliography

[1] Ahlfors, L. V. (1978). *Complex Analysis*, 3rd edn. (McGraw-Hill Book Co., New York), ISBN 0-07-000657-1.

[2] Alexandrov, A. D. (1965). The method of normal map in uniqueness problems and estimations for elliptic equations, in *Seminari 1962/63 Anal. Alg. Geom. e Topol., Vol. 2, Ist. Naz. Alta Mat* (Ediz. Cremonese, Rome), pp. 744–786.

[3] Cheeger, J. and Ebin, D. G. (2008). *Comparison Theorems in Riemannian Geometry* (AMS Chelsea Publishing, Providence, RI), ISBN 978-0-8218-4417-5, revised reprint of the 1975 original.

[4] Courant, R. and Robbins, H. (1996). *What Is Mathematics? An Elementary Approach to Ideas and Methods*, 2nd edn. (Oxford University Press), ISBN 0195105192.

[5] do Carmo, M. P. (1992). *Riemannian Geometry*, Mathematics: Theory & Applications (Birkhäuser Boston, Inc., Boston, MA), ISBN 0-8176-3490-8, doi:10.1007/978-1-4757-2201-7, translated from the second Portuguese edition by Francis Flaherty.

[6] Euclid (1956). *The Thirteen Books of Euclid's Elements Translated from the Text of Heiberg. Vol. I: Introduction and Books I, II. Vol. II: Books III–IX. Vol. III: Books X–XIII and Appendix* (Dover Publications, Inc., New York), translated with introduction and commentary by Thomas L. Heath, 2nd ed.

[7] Fujimori, S., Saji, K., Umehara, M. and Yamada, K. (2008). Singularities of maximal surfaces, *Math. Z.* **259**, 4, pp. 827–848, doi:10.1007/s00209-007-0250-0.

[8] Greenberg, M. J. and Harper, J. R. (1981). *Algebraic Topology*, Mathematics Lecture Note Series, Vol. 58 (Benjamin/Cummings Publishing Co., Inc., Advanced Book Program, Reading, Mass.), ISBN 0-8053-3558-7; 0-8053-3557-9, a first course.

[9] Hartman, P. (1964). *Ordinary differential equations* (John Wiley & Sons, Inc., New York-London-Sydney).

[10] Hartshorne, R. (2000). *Geometry: Euclid and Beyond*, Undergraduate Texts in Mathematics (Springer-Verlag), ISBN 0-387-98650-2, doi:10.1007/978-0-387-22676-7.

[11] Hopf, H. (1935). Über die Drehung der Tangenten und Sehnen ebener Kurven, *Compositio Math.* **2**, pp. 50–62.
[12] Hopf, H. (1951). Über Flächen mit einer Relation zwischen den Hauptkrümmungen, *Math. Nachr.* **4**, pp. 232–249.
[13] Hopf, H. (1989). *Differential Geometry in the Large*, Lecture Notes in Mathematics, Vol. 1000, 2nd edn. (Springer-Verlag), ISBN 3-540-51497-X, doi: 10.1007/3-540-39482-6, notes taken by Peter Lax and John W. Gray, With a preface by S. S. Chern, With a preface by K. Voss.
[14] Jaglom, I. M. and Boltjanskiĭ, V. G. (1960). *Convex Figures*, Translated by Paul J. Kelly and Lewis F. Walton (Holt, Rinehart and Winston, New York).
[15] Kapouleas, N. (1991). Compact constant mean curvature surfaces in Euclidean three-space, *J. Differential Geom.* **33**, 3, pp. 683–715.
[16] Kenmotsu, K. (2003). *Surfaces with Constant Mean Curvature*, Translations of Mathematical Monographs, Vol. 221 (American Mathematical Society, Providence, RI), ISBN 0-8218-3479-7, translated from the 2000 Japanese original by Katsuhiro Moriya and revised by the author.
[17] Klingenberg, W. (1978). *A Course in Differential Geometry* (Springer-Verlag), ISBN 0-387-90255-4, translated from the German by David Hoffman, Graduate Texts in Mathematics, Vol. 51.
[18] Kneser, A. (1912). Bemerkungen über die Anzahl der Extrema des Krümmung auf geschlossenen kurven und uber verwandte Fragen in einer nichteuklidischen Geometrie, in *Festschrift Heinrich Weber zu seinem siebzigsten Geburtstag an 5. März 1912* (Druck und Verlag von B. G. Taubner), pp. 170–180.
[19] Kobayashi, O. (1990). *Yamabe no mondai ni tsuite*, Seminar on Mathematical Sciences, Vol. 16 (Keio University, Department of Mathematics, Yokohama), in Japanese.
[20] Kobayashi, O. and Umehara, M. (1996). Geometry of scrolls, *Osaka J. Math.* **33**, 2, pp. 441–473.
[21] Kokubu, M., Rossman, W., Saji, K., Umehara, M. and Yamada, K. (2005). Singularities of flat fronts in hyperbolic space, *Pacific J. Math.* **221**, 2, pp. 303–351, doi:10.2140/pjm.2005.221.303.
[22] Kuen, T. (1884). Flächen von constantem krümmungsmaass, *Sitsungsbar. d. königl. Beyer. Akad. Wiss. Math-Phys. KlassePacific J. Math.* **Heft II**, pp. 193–206.
[23] Lawson, H. B., Jr. (1980). *Lectures on Minimal Submanifolds. Vol. I*, Mathematics Lecture Series, Vol. 9, 2nd edn. (Publish or Perish, Inc., Wilmington, Del.), ISBN 0-914098-18-7.
[24] Marsden, J. and Tromba, A. (2003). *Vector Calculus* (W. H. Freeman), ISBN 9780716749929.
[25] Massey, W. S. (1991). *A Basic Course in Algebraic Topology*, Graduate Texts in Mathematics, Vol. 127 (Springer-Verlag), ISBN 0-387-97430-X.
[26] Milnor, J. W. (1950). On the total curvature of knots, *Ann. of Math. (2)* **52**, pp. 248–257.

[27] Mukhopadhyaya, S. (1909). New methods in the geometry of a plane arc.–I cyclic and sextactic points, *Bull. Cal. Math. Soc.* **1**, 1, pp. 31–37.

[28] Munkres, J. R. (1966). *Elementary Differential Topology*, Lectures given at Massachusetts Institute of Technology, Fall, Vol. 1961 (Princeton University Press, Princeton, N.J.).

[29] Osserman, R. (1986). *A Survey of Minimal Surfaces*, 2nd edn. (Dover Publications, Inc., New York), ISBN 0-486-64998-9.

[30] Pinkall, U. (1987). On the four-vertex theorem, *Aequationes Math.* **34**, 2–3, pp. 221–230, doi:10.1007/BF01830673.

[31] Porteous, I. R. (2001). *Geometric Differentiation*, 2nd edn. (Cambridge University Press, Cambridge), ISBN 0-521-00264-8, for the intelligence of curves and surfaces.

[32] Riemann, B. (2004). *Collected Papers* (Kendrick Press, Heber City, UT), ISBN 0-9740427-2-2; 0-9740427-3-0, translated from the 1892 German edition by Roger Baker, Charles Christenson and Henry Orde.

[33] Saji, K., Umehara, M. and Yamada, K. (2009). The geometry of fronts, *Ann. of Math. (2)* **169**, 2, pp. 491–529, doi:10.4007/annals.2009.169.491.

[34] Schmidt, E. (1939). Über das isoperimetrische Problem im Raum von n Dimensionen, *Math. Z.* **44**, 1, pp. 689–788, doi:10.1007/BF01210681.

[35] Singer, I. M. and Thorpe, J. A. (1976). *Lecture Notes on Elementary Topology and Geometry* (Springer-Verlag), reprint of the 1967 edition, Undergraduate Texts in Mathematics.

[36] Spivak, M. (1965). *Calculus on manifolds. A modern approach to classical theorems of advanced calculus* (W. A. Benjamin, Inc., New York-Amsterdam).

[37] Spivak, M. (2006). *Calculus* (Cambridge Univ. Press), ISBN 9780521867443.

[38] Thorbergsson, G. and Umehara, M. (1999). A unified approach to the four vertex theorems. II, in *Differential and symplectic topology of knots and curves*, American Mathematical Society Translation Series 2, Vol. 190 (American Mathematical Society, Providence, RI), pp. 229–252.

[39] Umehara, M. (1999). A unified approach to the four vertex theorems. I, in *Differential and Symplectic Topology of Knots and Curves*, American Mathematical Society Translation Series 2, Vol. 190 (American Mathematical Society, Providence, RI), pp. 185–228.

[40] Warner, F. W. (1983). *Foundations of Differentiable Manifolds and Lie Groups*, Graduate Texts in Mathematics, Vol. 94 (Springer-Verlag), ISBN 0-387-90894-3, corrected reprint of the 1971 edition.

[41] Wente, H. C. (1986). Counterexample to a conjecture of H. Hopf, *Pacific J. Math.* **121**, 1, pp. 193–243.

[42] Whitney, H. (1937). On regular closed curves in the plane, *Compositio Math.* **4**, pp. 276–284.

[43] Whitney, H. (1944). The singularities of a smooth n-manifold in $(2n-1)$-space, *Ann. of Math. (2)* **45**, pp. 247–293.

List of Symbols

$\approx$; approximately equal, 10, 64, 216
$[*, *]$; the Lie bracket, 132
$:=$; definition, 4
$* \cdot *$; the inner product, 12, 70, 200
$\dot{*}$; derivative, 5
$\hat{X}$, 145
∇; the covariant derivative, 140
$*'$; derivative in arc-length, 11, 50
$\star$; the Hodge star operator, 165
$\times$; the vector product, 51, 205
$\wedge$; the exterior product, 132

A; the Weingarten matrix, 78
$\mathcal{A}^k(s)$; the set of k-forms, 132
$\angle\mathrm{A}$; interior angle, 106

$\boldsymbol{b}$; the binormal vector, 51

$\chi(S)$; the Euler number, 108
C^∞, 2, 193
$C^\infty(S)$; the set of smooth functions on S, 132
C^n, 193
cosh; the hyperbolic cosine, 195
sinh; the hyperbolic sine, 195
tanh; the hyperbolic tangent, 195

d; the exterior derivative, 132
dA; the area element, 64, 73, 134
$d\hat{A}$; the oriented area element, 134
Δ_{ds^2}; the Laplacian, 165

δ_{ij}; the Kronecker delta, 202
d_{E}; the Euclidean distance, 112
d_{H}; the hyperbolic distance, 114
ds^2_{H}; the hyperbolic metric, 113
ds^2; the first fundamental form, 70

$\boldsymbol{e}$; the unit tangent vector, 12, 50
E, F, G; the coefficients of the first fundamental form, 70

Γ_{ij}^k; the Christoffel symbols, 105

H; the mean curvature, 79
$[*]^{\mathrm{H}}$; the tangential part, 99

I; the identity matrix, 22
$\widehat{I}$; the first fundamental matrix, 70
$\mathrm{ind}_{\mathrm{P}}\, X$; the index of a vector field, 153
$*^{-1}$; the inverse matrix, 201

K; the Gaussian curvature, 79
κ; the curvature, 12, 51
κ_g; the geodesic curvature, 111, 148
κ_n; the normal curvature, 90
$\boldsymbol{\kappa}_g$; the geodesic curvature vector, 89, 148
$\boldsymbol{\kappa}_n$; the normal curvature vector, 89

$\mathcal{L}$; the length of curves, 7, 74
λ_1, λ_2; the principal curvatures, 79

Index

Printed in the United States
By Bookmasters

Printed in the United States
By Bookmasters